Study Guide and Problems Workbook

Featuring Complete Answers and Solutions to all Text Questions and Problems

H. JAMES PRICE
Texas A&M University

to accompany

PRINCIPLES OF GENETICS

D. PETER SNUSTAD
University of Minnesota

MICHAEL J. SIMMONS
University of Minnesota

JOHN B. JENKINS
Swarthmore College

John Wiley & Sons, Inc.

New York • Chichester • Weinheim • Brisbane • Singapore • Toronto

ISBN 0-471-15284-6

Printed in the United States of America

10 9 8 7 6 5 4 3

Printed and bound by Bradford & Bigelow, Inc.

Table of Contents

PART I

PART II

To The Student

This companion to *Principles of Genetics* , 1st Edition, by Peter Snustad, Michael Simmons, and John Jenkins is written to be a hands-on workbook to help you learn and reinforce terminology and concepts, develop problem solving skills, and challenge your mind. Learning introductory genetics requires both communication of knowledge and solution of problems. To communicate your newly acquired knowledge involves learning new terminology. Solving genetics problems requires the ability to analyze and interpret data. The science of genetics becomes fascinating and interesting as your confidence in understanding and using your newly acquired knowledge and skills grows.

How to Use This Problems Workbook and Study Guide

You should use this workbook after you have read your textbook and attended the lectures on each topic. The manual is divided into two parts. **Part I** contains a series of chapters designed to help review concepts and develop problem solving skills. **Part II** contains the complete answers and solutions to the text questions and answers.

In part I, each chapter begins with an outline of the corresponding chapter in Snustad et al. A summary of **Important Concepts** follows to help you organize your studies.

Next are lists of **Important Terms** and **Important Names**. In the space allotted, you are asked to define concisely what each term means or to identify the contribution(s) made by an individual or group. A few short and well-worded sentences are better than rambling answers. More importantly, try to understand the real importance of the terms, and the contributions of the scientists.

The **Testing Your Knowledge** section is designed to allow you to determine your understanding of genetic concepts and terminology and to help develop problem-solving skills. As you progress through this section, the questions become more complex and involve more problem working. The more problems you work, the better you will become at problem solving. Space is allotted in the workbook for your calculations.

A **Thought Challenging Exercise** is included in each chapter. It is intended to stimulate additional thought and discussion. I encourage you to discuss these with your fellow students.

A **Key Figures** section follows. It contains important figures from the textbook for easy reference during your studies.

A **Summary of Key Points** section lists the Key Points presented in the text, thereby providing another summary of important concepts to help you study.

Answers to Questions and Problems are given. The **Approaches to Problem Solving** section is designed to help you learn genetics problem-solving skills. In many cases, a problem is much easier to work if it is broken down into simpler components. This approach is emphasized throughout.

Now that you have the resources to study, i.e., your professor, *Principles of Genetics*, 1st Edition, this workbook, and your inquisitive mind, it is time to embark on your fascinating study of genetics. Study hard, devote time to study each day, and watch your interest and knowledge grow. I encourage you to study in small groups after you have used this book and have attempted to work all the problems. Explaining concepts and discussing solutions to problems helps the learning process.

Acknowledgments

I thank Robert E. Hanson and Michael S. Zwick for proofreading a draft of the text and checking problems. I also thank Bonnie Cabot and Jennifer Yee of John Wiley & Sons, Inc., for their highly professional editorial assistance. I appreciate most Patricia, Susan, Ginger, and Rick, for their understanding and patience during the many evenings and weekends I spent writing this book.

PART I

1

The Science of Genetics

IMPORTANT CONCEPTS

A. Two fundamental laws that form the foundation of the modern science of genetics were discovered and reported by Gregor Johann Mendel in 1865.
  1. Mendel proposed that cells had pairs of factors (genes) that determine a specific trait.
     a. Members of each gene pair segregate from each other during the process of sex cell formation, so that each gamete contains one member of each gene pair.
     b. The segregation of each pair of factors was independent of the segregation of other pairs of factors.
  2. Mendel's observations and interpretations were not recognized for 35 years.

B. The spectacular unfolding of modern genetic concepts has occurred during the 20th century.
  1. Genetics has grown from Mendel's obscure units of segregation and independent assortment, that affected the appearance of the organism, to an understanding of the organization, composition and expression of genes.
  2. Genetics is a vital and dynamic science that touches all facets of our being.
  3. Genetics has given us powerful molecular tools for the study of genes.
     a. One application of genetics, *DNA fingerprinting*, provides a powerful forensic science tool.
        (1) A molecular genetic analysis of a small sample of tissue can be used to positively identify or exclude a person as a crime suspect.

C. Modern genetics has had a profound impact on medicine.
  1. Geneticists now understand the metabolic basis for several hundred inherited diseases.
  2. Mutant genes have been isolated that cause inherited disorders such as cystic fibrosis, Duchene muscular dystrophy, and Huntington's disease.
  3. Gene therapy, made possible only after a gene has been isolated, provides a new approach to treating some genetic disorders.
     a. It has been successfully used to treat a devastating immune system disorder called combined immunodeficiency disease.
     b. Gene therapy may soon prove successful in the treatment of other genetic diseases such as cystic fibrosis, various types of cancer, hemophilia and AIDS.
  4. Cancer is a genetic disease and mutations that cause cancer are being identified and studied intensely.

a. The isolation of the breast cancer genes BRCA1 and BRCA2 and their mutant forms is creating important social and ethical issues.
D. To date, the greatest impact of genetics has been on modern agriculture.
   1. The development of hybrid corn was the first great achievement in the application of genetics principles to agriculture.
      a. During the period from 1940 to 1980, the average corn yield increased over 250%.
   2. The use of genetic principles in plant breeding has resulted in dramatic increases in yield of nearly all important food crops.
      a. Selective breeding has altered the growth form of plants such as the tomato to make them agronomically more desirable.
      b. Genes that confer resistance to pests or pathogens such as insects, nematodes, and fungi have been bred into modern plant varieties.
   3. In the 1950s through the 1970s, Norman Borlaug's group used classical genetic principles in developing Mexican wheat strains that perform well under stressed conditions.
      a. Borlaug launched what is called the "green revolution." He was awarded a Nobel Prize in 1970.
   4. The use of genetic principles in selective breeding programs has produced improvements in domesticated food animals.
      a. Modern chickens are meatier, grow faster, are disease resistant, and lay more eggs.
      b. Cattle and pigs grow faster, are more efficient in converting feed to meat, and are better adapted to regional environments.
      c. Selective breeding has resulted in dramatic increases in milk production per cow.
   5. The genetic engineering of crop plants has become a reality and has enormous potential.
      a. Inserting genes conferring resistance to insects and pathogens is becoming a major weapon in fighting devastating pests.
      b. Genetic technology is also being used to improve nutritional quality of plants, and to help plants synthesize their own nitrogen.
      c. The *flavr savr* tomato is an example of a genetically manipulated plant.
E. Genetic discoveries often have a direct impact on society.
   1. They may hold promise for cures of fatal diseases, or may help create new food products.
   2. Sometimes, genetic discoveries create complex, moral dilemmas.
      a. Do insurance companies have the right to deny health insurance to families at risk of developing certain inherited diseases?
      b. What rights or obligations do insurance companies have in the prevention of genetic birth defects?
      c. What responsibilities do parents who are at risk have in the prevention of genetic disorders?
F. Genetics has great potential for good, but also may be misused.
   1. Eugenics movements, based upon prejudice and misuse and misunderstanding of genetics, gained strength in the United States in the early part of the twentieth century.
   2. The eugenics movements took its most perverted form in Nazi Germany where the Hitler regime attempted to exterminate individuals of "inferior" genetic material.
   3. While the application of genetic principles resulted in dramatic increases in agricultural production in the United States in the period from 1937 to 1964, agricultural production in the Soviet Union was stagnant.
      a. The stagnant agricultural program in the Soviet Union resulted from its control by one individual, T. D. Lysenko, who rejected the principles of genetics, and based his plant improvement program on an erroneous belief in the inheritance of environmentally - induced characters.
G. Three major questions at the core of modern genetics are addressed in the text.
   1. What is the chemical nature of genetic material?
   2. How is genetic material transmitted?
   3. What does the genetic material do?

IMPORTANT TERMS

In the space allotted, concisely define each term.

classical genetics:

molecular genetics:

polymerase chain reaction:

human genome project:

eugenics:

green revolution:

IMPORTANT NAMES

In the space allotted, concisely state the major contribution made by the following individual.

Charles Darwin:

Gregor Johann Mendel:

Kary Mullis:

Sir Archibald Garrod:

Nancy Wexler:

Norman Borlaug:

Francis Galton:

T. D. Lysenko:

TESTING YOUR KNOWLEDGE

In this section, fill in the blank or answer the question in the space allotted.

1. The two fundamental genetics laws were discovered by _____.

2. The study of the relationship between the units of inheritance and the physical appearance of an organism falls into the discipline of _____.

3. The study of the biochemical nature of genes and how genes express their encoded information is called _____.

4. The international effort with the goal of mapping and sequencing all human genes is called the _____.

5. The first great success from applying genetic principles to plant breeding was _____.

6. The first commercially produced fruit which included genetic engineering in the breeding program was the _____.

7. Why weren't Mendel's observations and interpretations recognized as such for 35 years?

THOUGHT CHALLENGING EXERCISE

Before you continue further into your study of genetics, prepare a list of the ways in which you think genetics has directly and indirectly influenced your life. Save this list. At the end of your genetics course, prepare another list of how genetics has affected your life. Prepare the last list without referring to the first one. Then compare the two lists.

ANSWERS TO QUESTIONS

1) Gregor Mendel **2)** classical genetics **3)** molecular genetics **4)** human genome project **5)** hybrid corn **6)** *flavr savr* tomato **7)** A major reason that Mendel's work was essentially unrecognized for 35 years was due to the state of biological knowledge at the time in which he lived. In the mid-1800s, little was known about chromosomes and the process by which cells divide. However, mitosis and meiosis had been characterized by 1900. When Mendel's laws were rediscovered at the beginning of the 20th century, it was soon hypothesized that the segregation and independent assortment of chromosomes were the physical bases of segregation and independent assortment of genes located on the chromosomes. Another reason why Mendel's laws were not appreciated was that Mendel did not aggressively bring them to the attention of the scientific world. Also, other scientists simply did not understand his numerical analysis of data. Mendel's primary duties were as a priest and an abbot. Although he was well-educated in science, science was his hobby.

2
Reproduction as the Basis of Heredity

IMPORTANT CONCEPTS

A. The two basic types of cells are prokaryotic and eukaryotic.
1. Prokaryotic cells are the simpler of the two types.
   a. They have a cell wall constructed of peptidoglycan surrounding a cytoplasmic membrane that encloses the cytoplasm.
   b. There is no membrane-bound nucleus. Their DNA is concentrated in a region of the cytoplasm called the nucleoid.
   c. An example of a prokaryotic organism is bacteria.
2. Eukaryotic cells are more complex.
   a. They contain a membrane-bound true nucleus which houses the chromosomes, that is comprised of DNA, RNA, and various proteins.
   b. In the nucleoplasm are found one or more nucleoli which function in the production of a specific class of RNAs called ribosomal RNAs.
   c. The components of eukaryotic cells are encapsulated by a phospholipid plasma membrane which has a variety of glycoproteins embedded in it. The plasma membrane functions as a barrier between the extracellular and intracellular matrices, and in regulating the flow of molecules into and out of the cell.
   d. Membrane-bound organelles such as mitochondria, lysosomes, Golgi complex, peroxisomes, and vacuoles occur in the cytoplasm. The endoplasmic reticulum is a cytoplasmic system of membranes that functions in protein synthesis.
   e. All eukaryotic cells have a network of protein filaments, called a cytoskeleton, which gives the cell its shape, its ability to move, and its ability to organize its organelles

within the cytoplasm. The two most important filaments are microfilaments and microtubules.

    f.   Some eukaryotic cells such as those of plants have a wall surrounding the exterior to the plasma membrane that is comprised of cellulose and other constituents, but never peptidoglycan.

B. Chromosomes, which contain the genetic material, function in the transmission of genetic information and the ordered release of this information to control cellular function and development.

    1.   As viewed using light microscopy, chromosomes at prophase and metaphase are made up of two sister chromatids held together by a centromere.

    2.   The kinetochore is a protein structure at the centromere that functions in chromosome movement during the cell cycle.

    3.   The ends of the chromosome are called telomeres.

C. The eukaryotic cell cycle is characterized by duplication of the DNA and other chromosomal material (the S phase), followed by a G2 phase during which the nucleus prepares for division. The G2 phase is followed by the division phase (M), and the G1 phase during which the cell grows preceding the next S phase.

D. There are two types of eukaryotic cell division, mitosis and meiosis.

    1.   The key feature of mitosis is that the two daughter cells are identical to each other and to the parent cell.

        a.   In interphase of mitosis, the DNA of the chromosomes replicates and the synthesis of a variety of proteins necessary for mitosis occurs.

        b.   In prophase, the chromosomes become progressively more condensed and each chromosome appears as two rod-shaped identical sister chromatids that are held together at the centromere. In late prophase, the nuclear membrane and nucleolus disappear and the chromosomes move toward the equator of the cell. Microtubules invade the nuclear region and become attached to each chromatid at the kinetochore.

        c.   The fully contracted chromosomes line up on the equatorial plate at metaphase.

        d.   At anaphase, the centromere divides and the sister chromatids (now chromosomes) move to opposite poles.

        e.   Telophase begins when the chromosomes reach the poles. During telophase the chromosomes decondense, the microtubules disappear, and the nuclear membrane reforms. Cytokinesis follows in which the cytoplasm divides to produce two identical daughter cells. Plant cell cytokinesis involves the formation of a cell plate between the daughter cells, on which the cellulose walls are deposited.

    2.   Meiosis is a process in which diploid cells divide to produce genetically different haploid cells. One round of DNA replication is followed by two rounds of cell division.

        a.   During prophase of meiosis I, replicated homologous chromosomes synapse (pair), condense, and crossover.

        b.   The homologous pairs (bivalents) line up on the equatorial plane during metaphase.

        c.   During anaphase I the members of each bivalent segregate and go to opposite poles, completing their migration at telophase I.

        d.   The chromatids of each chromosome do not separate from each other during meiosis I.

        e.   During meiosis II, the chromosomes line up on the equatorial plane at metaphase II.

        f.   At anaphase II, the centromere divides and the sister chromatids migrate to opposite poles.

        g.   After completing two meiotic divisions, a single diploid nucleus has produced four haploid nuclei, each containing one member of each chromosome pair.

        h.   Meiosis recombines the maternal and paternal genetic material into haploid gametes or spores. Recombination results from both independent assortment of chromosome pairs and from crossing over between non-sister homologous chromatids.

        I.   Meiosis and fertilization have the potential to produce an almost infinite variety of new genetic combinations upon which evolutionary forces can act.

3. Occasionally, chromosomes or chromatids may fail to separate from each other during mitosis or meiosis. This phenomenon, called nondisjuction, leads to nuclei with abnormal numbers of chromosomes.
   a. Nuclei with abnormal chromosomal constitutions generally result in abnormal phenotypes, such as Down syndrome in humans who have three instead of the normal two chromosomes number 21.
4. In animals, meiosis accompanies gamete formation; oogenesis in females and spermatogenesis in males.
   a. In females, primary oocytes undergo meiosis I to form a secondary oocyte and a primary polar body, and meiosis II to produce the egg and another polar body.
   b. In spermatogenesis, primary spermatocytes undergo meiosis I to produce two secondary spermatocytes and meiosis II to produce four spermatids. The spermatids differentiate into spermatozoa.
5. In plants, meiosis produces haploid spores that in turn undergo mitotic divisions and differentiate into gametophytes.
   a. The haploid gametophytes produce the haploid gametes.
   b. Fertilization produces a diploid zygote that divides and differentiates into the sporophyte.
6. The life cycles of species such as *Neurospora*, corn, *Drosophila*, and humans represent strategies that have evolved for the reshuffling of the genetic material as it is transmitted from one generation to the next.

IMPORTANT TERMS

In the space allotted, concisely define each term.

prokaryotic cell:

eukaryotic cell:

cytoplasm:

peptidoglycan:

nucleoid:

nucleus:

chromosome:

nucleolus:

ribosomes:

nucleoplasm:

endoplasmic reticulum (ER):

Golgi complex:

lysosomes:

peroxisomes:

vacuoles:

mitochondria:

chloroplasts:

cytoskeleton:

chromosome:

chromatid:

centromere:

kinetochore:

telomere:

cell cycle:

mitosis:

cytokinesis:

centrosome:

meiosis:

homolog:

reduction division:

equatorial division:

synapsis:

crossing over:

chiasmata:

tetrad:

bivalent:

nondisjunction:

oogenesis:

oogonium:

primary oocyte:

secondary oocyte:

polar body:

ovum (egg):

zygote:

Spermatogenesis:

spermatogonium:

primary spermatocyte:

secondary spermatocyte:

spermatids:

spermatozoa:

spores:

gametophyte:

sporophyte:

alternation of generations :

conidia:

ascus:

ascospores:

microspore mother cell:

megaspore mother cell:

endosperm:

embryo:

Drosophila melanogaster:

pupa:

imaginal disks:

placenta:

fetus:

TESTING YOUR KNOWLEDGE

In this section, answer the questions, fill in the blanks, or solve the the problems in the space allotted. Problems noted with an * are solved in the Approaches to Problem Solving section at the end of the chapter.

1. A cross-shaped structure formed between nonsister homologous chromatids by crossing over that is visible during the diplonema stage of meiosis is called a _____.

2. The stage of meiosis at which synapsis occurs is called _____.

*3. Four gametes result from a cell that has undergone meiosis. If the nucleus of the cell at pachynema had 5 picograms of DNA, the total amount of DNA in each gamete must be _____ picograms.

4. Segregation of homologous chromosomes occurs during _____ of meiosis.

*5. A sunflower has a diploid chromosome number of 2N = 34. How many chromatids are present in a secondary meiocyte at prophase II?

6. A pair of synapsed chromosomes in the first meiotic division is called a _____.

7. The failure of a homologous chromosome pair to segregate during meiosis, therefore resulting in meiotic products with either an extra or a missing chromosome is called _____.

8. The process of pairing of homologous chromosomes during meiosis is called _____.

*9. A human has 46 chromosomes in somatic (body) cells. How many chromatids are present in a secondary polar body nucleus?

10. Metaphase I of meiosis and mitosis differ in regard to the presence of _____.

11. For a human, the life cycle has the sequence
    (a) 1n = 23 -> meiosis -> 2n = 46 -> fertilization -> 1n = 23
    (b) 2n = 46 -> meiosis -> 1n = 23 -> fertilization -> 2n = 46
    (c) 1n = 23 -> mitosis -> 2n = 46 -> fertilization -> 1n = 23
    (d) 2n = 46 -> mitosis -> 1n = 23 -> fertilization -> 2n = 46

*12. A geneticist cytologically examined a corn plant and found that it had one member of chromosome # 10 possessing a very large terminal knob and the other member lacking this knob. A knob is a darkly staining enlarged region on a chromosome. When the geneticist looked at the much larger chromosome number 1, it was observed that only one chromosome of this pair possessed a terminal knob.

   (a) In regard to these two heteromorphic chromosome pairs, draw all possible types of meiotic products that can be produced by this plant.

   (b) Can all of these types of gametes be produced by one meiosis? Why or why not?

*13. A particular animal species has four chromosomes (2n = 4). For each of the following, indicate whether the nucleus or cell is in mitosis or meiosis, the particular stage, and the substage if in meiosis.

(a)  _____

(b)  _____

(c)  _____

(d) 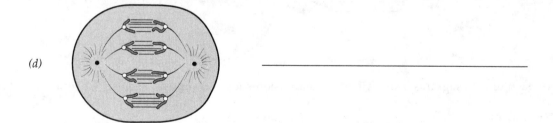 _____

(e)  _____

*14. Cells from a diploid organism (2n = 4) are shown undergoing division in the diagrams, (I), (II), (III), and (IV), shown below:

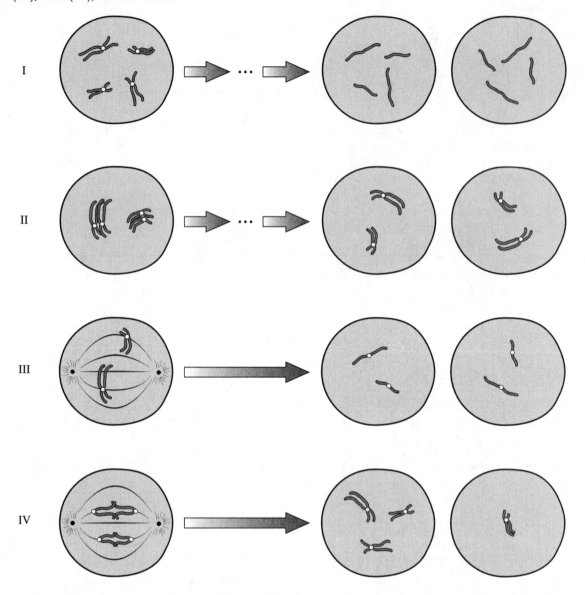

(a) For each of the diagrams, state whether the division shown is mitosis, meiosis I, or meiosis II?

Division # I: _____   Division # III: _____

Division # II: _____   Division # IV: _____

(b) Which diagrams show synapsed chromosomes?

(c) Which of the above diagrams starts with a haploid cell?

(d) In which diagram did segregation of both chromosome pairs occur?

(e) What is the term that best describes the phenomenon that occurred in the division of diagram # IV?

15. For each of the following stated phenomena, indicate whether it occurs in mitosis and/or meiosis and the phase or subphase at which it occurs.

(a) Chromosomes line up with their centromeres on the equatorial plate.

(b) DNA replication occurs.

(c) Homologous chromosomes undergo synapsis.

(d) Centromeres divide and sister chromosomes move to opposite poles.

(e) The cell is haploid and the chromosomes consist of one chromatid.

(f) Homologous chromosomes segregate to opposite poles.

(g) Nonsister homologous chromosomes undergo crossing over.

(h) The first appearance of haploid cells, with each chromosome consisting of two chromatids.

(i) Chromosomes are maximally condensed.

(j) Bivalents line up with their centromeres on either side of the equatorial plate.

(k) Chromatids are first visually recognized through a light microscope.

(l) Two nuclei are formed, each identical to that which underwent division.

(m) Independent assortment of chromosomes occurs.

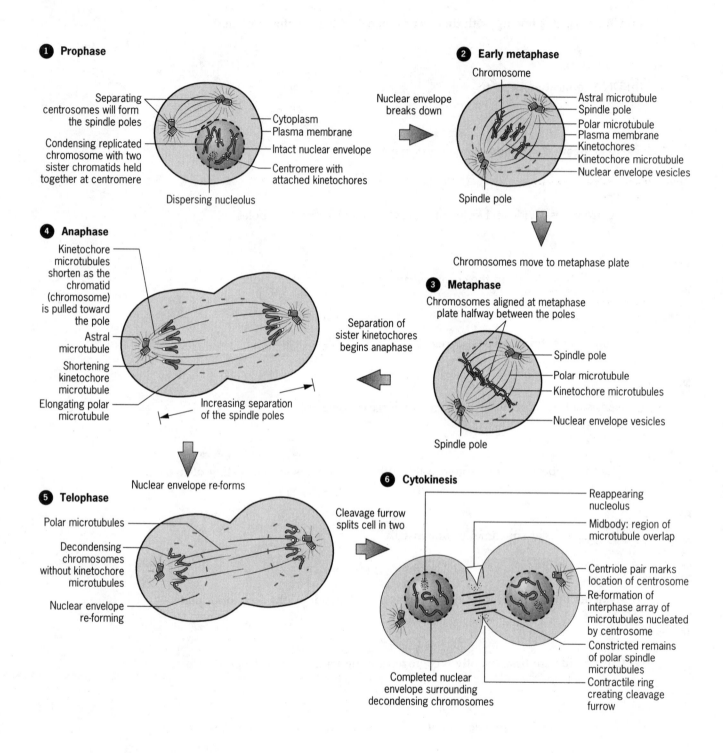

Figure 2.7   Mitosis in detail.

Early prophase I

Replicated chromosomes become visible.

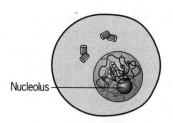

Nucleolus

Middle prophase I

Homologous chromosomes shorten and thicken. The chromosomes synapse and crossing over occurs.

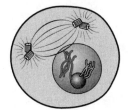

Late prophase I

Results of crossing over become visible as chiasmata. Nuclear membrane begins to disappear. Spindle apparatus begins to form. Nucleolus disperses.

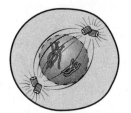

Metaphase I

Assembly of spindle is completed. Each chromosome pair aligns across the metaphase plate of the spindle.

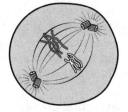

Anaphase I

Homologous chromosome pairs separate and migrate toward opposite poles.

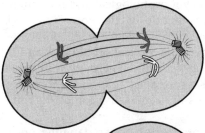

Telophase I

Chromosomes (each with two sister chromatids) complete migration to the poles, and new nuclear membranes may form.

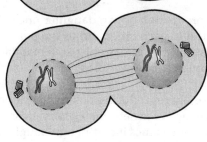

Cytokinesis

In most species, cytokinesis produces two daughter cells. Chromosomes do not replicate before meiosis II.

Prophase II

Chromosomes condense and move to metaphase plate.

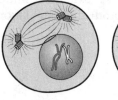 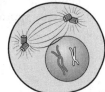

Metaphase II

Kinetochores attach to spindle fibers. Chromosomes line up on metaphase plate.

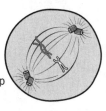 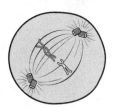

Anaphase II

Sister chromatids separate and move to opposite poles as separate chromosomes.

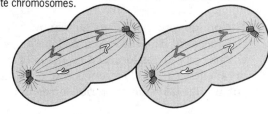

Telophase II

Nuclear membrane forms around chromosomes and chromosomes uncoil. Nucleolus reforms.

Reforming nucleolus

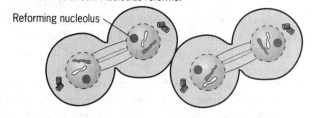

Four haploid cells form after cytokinesis.

Nucleolus

Figure 2.10   Meiosis in detail.

THOUGHT CHALLENGING EXERCISE

It is generally agreed that meiosis and fertilization evolved because they produce an enormous variety of new genetic combinations upon which evolutionary forces can act. Although most higher organisms have meiosis and fertilization as part of their reproductive cycle, some plants and animals skip meiosis and fertilization and reproduce parthenogenically by the division and differentiation of a cell produced asexually by mitosis. Discuss possible reasons for the evolution of parthenogenesis from sexually reproducing populations.

SUMMARY OF KEY POINTS

The prokaryotic cell has no membrane-bound nucleus. The prokaryotic equivalent of the true nucleus is the nucleoid. The eukaryotic cell has a membrane-bound nucleus that houses the chromosomes. The eukaryotic cell has localized many of its metabolic processes into membrane-bound compartments or organelles, such as lysosomes, Golgi apparatus, mitochondria, and chloroplasts. Protein synthesis in eukaryotes takes place on ribosomes in the cytoplasm, in mitochondria, and in chloroplasts.

The eukaryotic chromosome is a thread-like structure consisting mainly of a complex of DNA (the genetic material) and proteins. Chromosomes have key morphological and molecular features such as centromere, telomere, chromatid, and DNA sequences that designate replication origin.

The cell cycle is a sequence of events involving the periodic replication of DNA and the segregation of the replicated DNA with cellular constituents to daughter cells. There are four phases of the cell cycle: G1, S, G2, and M or division. The two G phases are called "gaps"; the S phase is the period of DNA replication; and the division or M phase signals the actual division of the cell.

Mitosis, a mode of nuclear division, provides for the production of two daughter nuclei that contain identical chromosome sets and are genetically identical to each other and to the parent cell from which they arose. The division is divided into five phases: interphase, prophase, metaphase, anaphase, and telophase. Cytokinesis, the division of the cytoplasm, follows telophase.

Meiosis, a fundamental process of cell division in sexually reproducing eukaryotes, involves three main events: pairing of homologous chromosomes; the exchange of genetic material by crossing over; and the segregation of the members of a homologous pair of chromosomes into different daughter nuclei. Meiosis involves one round of DNA replication followed by two separate divisions. The result of meiosis in a diploid cell is four haploid cells. Meiotic nondisjunction is the failure of sister chromatids or homologous chromosomes to disjoin.

The most important consequence of meiosis is that it produces an enormous amount of genetic variation through the production of different combinations of chromosomes.

Gametogenesis is the formation of male and female gametes or sex cells. Spermatogenesis and oogenesis are the formation of male and female gametes, respectively, by the process of mitosis and meiosis. Gametes unite in the process of fertilization. In plants, meiosis in the sporophyte produces spores that develop into haploid gametophytes. Gametophytes produce haploid gametes that fuse to produce a diploid zygote, which divides mitotically to produce a diploid sporophyte.

The life cycle is the sequence of events from the individual's origin as a zygote to its ultimate death. It can also be described as the stages through which an organism passes between the production of gametes by one generation and the next. In many plants and some animals, the life cycle involves a regular alteration of generations between haploid and diploid phases.

ANSWERS TO QUESTIONS AND PROBLEMS

1) chiasma **2)** zygonema **3)** 1.25 **4)** anaphase I **5)** 34 **6)** bivalent **7)** nondisjunction **8)** synapsis **9)** 23 **10)** paired chromosomes (bivalents) at MI of meiosis **11)** b **12)** see answer in next section **13) a.** mitosis, metaphase **b.** meiosis, prophase I, diplonema; **c.** meiosis, prophase II **d.** mitosis, anaphase **e.** meiosis, anaphase I **14) a.** (I). mitosis; (II) meiosis I; (III) meiosis II; (IV). meiosis I **b.** II, IV; **c.** III; **d.** II; **e.** nondisjunction. **15) a.** mitosis, metaphase; meiosis, metaphase II **b.** interphase before mitosis and meiosis **c.** meiosis, prophase I, zygonema **d.** mitosis, anaphase; meiosis, anaphase II **e.** meiosis, telophase II **f.** meiosis, anaphase I **g.** meiosis, prophase I, pachynema **h.** meiosis, telophase I **i.** metaphase of mitosis; metaphase I and metaphase II of meiosis **j.** meiosis, metaphase I **k.** meiosis, leptonema; mitosis, prophase **l.** mitosis, telophase **m.** meiosis, anaphase I

Comments on the thought challenging exercise:

In plants, parthenogenesis generally occurs following hybridization of divergent populations. The parthenogenetic individuals often have a broader range of adaptation than either parent and occupy different ecological niches. If the heterozygosity of the hybrids results in highly adaptive individuals, then ways to restrict recombination and pass the adaptive genotypes to progeny would have a selective advantage. Parthenogenesis is a way of eliminating meiosis and thereby maintaining the genotype of the mother. Many other mechanisms have evolved that limit the recombination of chromosomes while still utilizing meiosis and fertilization. Also, organisms that reproduce by parthenogenesis generally retain a low frequency of sexual reproduction.

APPROACHES TO PROBLEM SOLVING

3. The cell at pachynema is diploid and has 5 pg of DNA. Each chromosome consists of two chromatids. At anaphase I the bivalents segregate to opposite poles into secondary meiocytes with a complete set of chromosomes, each consisting of two chromatids This reduces the DNA amount to 2.5 pg or 1/2 that of the cell at pachynema. The anaphase of meiosis II involves a division of the centromeres and the separation of sister chromatids to opposite poles. This reduces the DNA amount by 1/2 again and, thus, produces gametes with 1.25 pg DNA.

5. The first meiotic division reduces the chromosome number from 2n = 34 to 1n = 17. Since prophase II chromosome still have two chromatids each, there is a total of 34 chromatids.

9. The first meiotic division reduces the chromosome number from 2n = 46 to 1n = 23. A secondary polar body is a product of a second meiotic division during oogenesis. Therefore, a secondary polar body is haploid, 1n = 23, with each chromosome having just one chromatid. There are 23 chromatids present.

12. The cell that will undergo meiosis has the following chromosome makeup in regard to chromosomes 1 and 10.

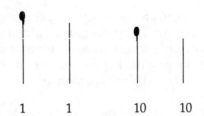

1   1   10   10

(a) The independent assortment of the two heteromorphic chromosome pairs results in the following four kinds of gametes produced in equal frequencies.

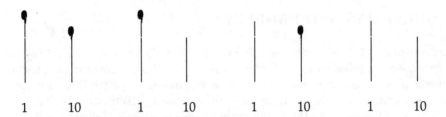

1   10   1   10   1   10   1   10

(b) The four types of gametes could be produced from one meiosis only if crossing over simultaneously occurred between the knob and the centromere in each of the heteromorphic chromosome pairs. If so, then both chromosome pairs at prophase II would consist of one knobbed and one unknobbed chromatid as drawn below.

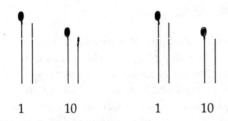

1   10   1   10

The random disjunction of the chromatids at anaphase II could result in all four gametic types being produced.

13. (a) This figure shows four chromosomes lined up with their centromeres on the equatorial plate. Since the diploid number of this cell is 2n = 4, it must be a mitotic metaphase.

(b) This figure shows four unpaired chromosomes, each chromosome consisting of two chromatids. The nuclear membrane in still intact and spindle fibers have not formed. Therefore, this cell is prophase of mitosis.

(c) Two different chromosomes are apparent in the nucleus. Since this is the haploid number, the cell must have at least completed the first meiotic division. Each chromosome still consists of two chromatids. Therefore, the nucleus must be át prophase II of meiosis.

(d) Four chromosomes have migrated toward, but have not reached, each pole. Therefore, the nucleus is at anaphase of mitosis.

(e) This drawing depicts two chromosomes migrating to each pole. Since the diploid chromosome number is 2n = 4, this must be a meiotic anaphase. Each chromosome consists of two chromatids indicating that this must be anaphase I.

14. (a) I. The first cell has the diploid number (2n = 4) of chromosomes. Each chromosome consists of two chromatids, and the chromosomes are not paired. The initial cell must be at mitotic prophase. The two chromatids of each chromosome have migrated to separate cells during the division depicted as would be expected of a mitotic division.

   II. This figure starts with paired chromosomes at pachynema and ends with the haploid cells of prophase II. This depicts meiosis I.

   III. The initial cell shown has the haploid number of chromosomes (n = 2). Each chromosome consists of two chromatids and is lined up on the equatorial plate. This cell is in metaphase II, this diagram depicts a cell in meiosis II.

   IV. The initial cell has paired chromosomes (at metaphase I) and therefore is in the first meiotic division.

   (b) The chromosomes in II and IV are paired or synapsed.

   (c) The division shown in III has two chromosomes which is the haploid number. Therefore, the division shown starts with a haploid cell.

   (d) Since the members of each chromosome pair have gone to different poles in diagram II, it depicts the segregation of both chromosome pairs.

   (e) In diagram IV, both members of the larger chromosome pair have ended up in the same cell after the first meiotic division. This illustrates the phenomenon of nondisjunction.

3

Mendelism: The Basic Principles of Inheritance

IMPORTANT CONCEPTS

A. Through carefully designed hybridization experiments with the garden pea and numerical analyses of progeny, Gregor Mendel deduced fundamental laws of transmission genetics.
  1. The principle of segregation states that alleles of a gene pair segregate from each other during gamete formation so that gametes have only one member of a gene pair.
  2. The principle of independent assortment states that alleles of different genes segregate, or assort, independently of each other, and recombine at random at fertilization. In other words, the segregation of one gene pair does not interfere with or influence the segregation of other gene pairs.
  3. Mendel also discovered another principle involving gene function, referred to as dominance.
B. Mendel's discoveries went essentially unnoticed until 35 years later.
  1. In 1900, three scientists (Correns, deVries, and von Tschermak-Seysenegg) independently rediscovered Mendel's laws.
  2. The discipline of genetics was born in 1900 when William Bateson championed the cause of Mendelian genetics and introduced it to the English-speaking world.
C. The results of genetic crosses can be predicted by the Punnett square, forked line, or probability methods.
D. Since deviations from expected genetic ratios result from elements of chance, statistical methods such as the Chi-square analysis are used to determine the "goodness of fit" of data.
E. The probabilities of various combinations of events in groups of a specific size can be calculated by the binomial expansion $(p + q)^n$.
  1. If the probabilities of p and q are known, then $(p + q)^n$ represents the probabilities of all combinations of events p and q in a sample size of n.
  2. In F_2 and testcross progeny, the genotypes resulting from segregation and independent assortment of heterozygous gene pairs are binomially distributed.
F. Genetics in humans has traditionally been aided by pedigree analysis.

1. Pedigree are diagrams that show relationships among members of a family and the occurrence of the trait(s) of interest among these family members.

IMPORTANT TERMS

In the space alloted, concisely define each term.

genetics:

true-breeding:

cross-fertilized:

dominant:

recessive:

monohybrid cross:

gene:

allele:

homozygous:

heterozygous:

genotype:

phenotype:

segregate:

parental:

F_1:

F_2:

principle of dominance:

principle of segregation:

dihybrid cross:

principle of independent assortment:

Punnett square method:

backcross:

testcross:

forked-line method:

probability:

Chi-square (X^2):

degrees of freedom:

pedigree:

multiplicative rule:

additive rule:

binomial distribution:

IMPORTANT NAMES

In the space allotted, concisely state the major contribution made by the individual.

Gregor Johann Mendel:

Hugo de Vries:

Carl Correns:

Eric von Tschermak-Seysenegg:

William Bateson:

W. Johannsen:

TESTING YOUR KNOWLEDGE

In this section, answer the questions, fill in the blanks, and solve the problems in the space allotted. Problems noted with an * are solved in the Approaches to Problem Solving section at the end of the chapter.

1. Alternative forms of genes are called _____.

2. The expression of one allele to the exclusion of the other allele of a heterozygous gene pair is called _____.

3. Name the three rediscoverers of Mendel's laws.

4. The person who gave genetics its name was _____.

5. Which one of the following is <u>true</u> when genotype *Cc Dd* undergoes independent assortment?

    (a) the parental combinations will be most frequent among the gametes.
    (b) four different gametic genotypes will occur in equal frequency.

(c) the gametes will be either *CD* or *cd*.
(d) the segregation of one gene pair will determine how the other gene pair segregates.

*6. In maize, a color gene and a height gene control the following phenotypes: *CC* and *Cc*, purple; *cc*, white; *TT*, tall; *Tt* intermediate; tt short. If a dihybrid is self-fertilized, what is the resulting phenotypic ratio?

   (a) 1:2:1:2:4:2:1:2:1   (b) 2:3:6:3:2   (c) 3:3:3:3:3:1   (d) 3:6:3:1:2:1   (e) 9:3:3:1

*7. From a cross, *Aa Bb cc Dd Ee ff* x *AA Bb Cc Dd ee Ff*, what is the probability of obtaining the genotype *Aa bb Cc DD Ee ff* in the progeny?

*8. What is the probability of rolling a combination totaling eight on two dice thrown simultaneously?

*9. If both parents were known to be carriers (*Aa*) for the recessive albino allele, what is the chance of three albino and two normal children if they are to have five children?

10. In a family of six children, what is the probability that 4 will be girls and 2 will be boys?

11. A man and wife are both heterozygous for the recessive albino allele. What is the probability, if they have six children, that three will be albino and three will be normal?

*12. A phenotypically normal couple has one albino child. What is the probability that their other phenotypically normal child is a heterozygote?

   (a) 1/4   (b) 1/3   (c) 1/2   (d) 3/4   (e) 2/3

13. A mouse of the *CcDd* genotype is testcrossed. What is the probability that the first three offspring are dihybrid females? What is the probability that the first four offspring will be homozygous?

*14. Two strains of sorghum (one homozygous for all dominant alleles and the other homozygous for all recessive alleles at the same six genes) are crossed. The six gene pairs at which they differ segregate independently. The F_1 hybrid is selfed to produce F_2s.

(a) How many kinds of gametes can be produced by the F_1 plants?

(b) What is the number of possible genotypes among the F_2s?

(c) What proportion of the F_2 progeny are expected to be homozygous at all six gene pairs?

(d) What proportion of the F_2 progeny are expected to show all the dominant phenotypes?

(e) What proportion of the F_2 progeny are expected to be homozygous for all dominant alleles?

*15. In squash, fruit color is determined by one gene, and fruit shape by another independently segregating gene. The white fruit allele (*W*) is dominant to yellow fruit allele (*w*) and disk-shaped fruit allele (*D*) is dominant to round-shaped fruit allele (*d*). From progenies listed in the following table, provide probable genotypes for the parents of each cross.

Parents	Phenotypes of Progeny			
	white disk	white round	yellow disk	yellow round
(a) white, round x white, round	0	31	0	9
(b) white, disk x yellow, round	52	48	0	0
(c white, disk x white, round	31	29	9	10
(d) white, disk x white, disk	88	30	27	10
(e) white, disk x yellow, round	16	14	13	15

*16. The following is a legend for maize.

v^+_ = nonvariegated leaves b^+_ = yellow seed
$v\ v$ = variegated leaves $b\ b$ = bronze seed

g^+_ = nonglossy leaves D _ = dwarf
$g\ g$ = glossy leaves d^+d^+ = tall

l^+_ = liguled $w+$_ = starchy endosperm
$l\ l$ = liguleless $w\ w$ = waxy endosperm

From a cross:

$vv\ g^+g\ l\ l\ b^+b\ Dd^+\ w^+w$ (female) X $v^+v\ gg\ l\ l\ b^+b\ Dd^+\ w^+w$ (male)

(a) How many kinds of gametes can be formed by the male parent of the above cross?

(b) How many different genotypes are possible in the progeny of the above cross?

(c) How many different phenotypes are possible in the progeny of the above cross?

(d) What is the probability of obtaining the following genotype from the above cross?

$vv\ gg\ l\ l\ b^+b\ Dd^+\ ww$

(e) What is the probability of obtaining the following phenotype in the progeny of the above cross?

nonvariegated, glossy, liguleless, bronze, dwarf, waxy

*17. The following is a legend for chickens.

$B\,B$ = black	$C\,C$ = embryonic lethal	$P\_$ = pea comb
$B\,b$ = blue	$C\,c$ = creeper	$p\,p$ = single comb
$b\,b$ = white	$c\,c$ = normal	

$M\_$ = bearded	$S\_$ = silky feathers	$W\_$ = white skin
$m\,m$ = beardless	$s\,s$ = normal feathers	$w\,w$ = yellow skin

From a cross:

$Bb\ Cc\ pp\ Mm\ ss\ Ww$ (female) x $Bb\ Cc\ Pp\ Mm\ Ss\ Ww$ (male)

(a) How many kinds of gametes can be formed by the female parent of the above cross?

(b How many genotypes are possible in the **living** progeny of the above cross?

(c) How many phenotypes are possible in the **living** progeny of the above cross?

(d) What is the probability of obtaining the following genotype in the **living** progeny from the above cross?

$Bb\ Cc\ pp\ MM\ Ss\ ww$

(e) What is the probability of obtaining the following phenotype in the **living** progeny of the above cross?

black, non-creeper, pea comb, bearded, silky, yellow skin

*18. To this point we have been considering the basic F_2 phenotypic ratios of 3:1, 1:2:1, and 9:3:3:1. In later chapters you will learn that modifications of these ratios may occur due to various types of gene interaction. In a plant species, red , yellow and white seed colors are found. A plant from a line breeding true for red seeds crossed to a plant from a line breeding true for white seeds produces F_1 progeny that are all red. The 820 plants of the F_2 generation consisted of 450 red seeded, 160 yellow seeded, and 210 white seeded individuals. Using the Chi-square procedure and the following Chi-square table, test whether these data better fit a 2 red: 1 yellow: 1 white or a 9: red: 3 yellow: 4 white ratio.

Table of Chi-square (X^2)[a]

Degrees of Freedom	Probability (P)						
	0.99	0.95	0.80	0.50	0.20	0.05	0.01
1	0.0002	0.004	0.064	0.455	1.642	3.841	6.635
2	0.020	0.103	0.446	1.386	3.219	5.991	9.210
3	0.115	0.352	1.005	2.366	4.642	7.815	11.345
4	0.297	0.711	1.649	3.357	5.989	9.488	13.277
5	0.554	1.145	2.343	4.351	7.289	11.070	15.086
6	0.872	1.635	3.070	5.348	8.558	12.592	16.812

[a]Selected data from R. A. Fisher and F. Yates, *Statistical Tables for Biological, Agriculture and Medical Research.* Oliver & Boyd, London, 1943.

*19. The following is a pedigree for a rare trait:

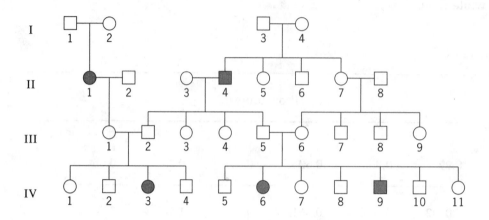

(a) What is the most probable mode of inheritance for the trait indicated above?

(b) If individuals III-3 and III-8 marry, what is the probability of them having a child with the trait?

(c) If individuals IV-1 and IV-10 marry, what is the probability of them having a child who expresses the trait?

*20. The following is a pedigree for a rare trait:

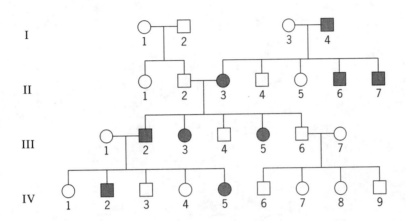

(a) What is the most probable mode of inheritance for the trait indicated above?

(b) If individuals IV-4 and IV-6 marry, what is the probability of them having a child homozygous for the allele determining the trait?

(c) If individuals IV-5 and IV-6 marry, what is the probability of them having a child with the trait?

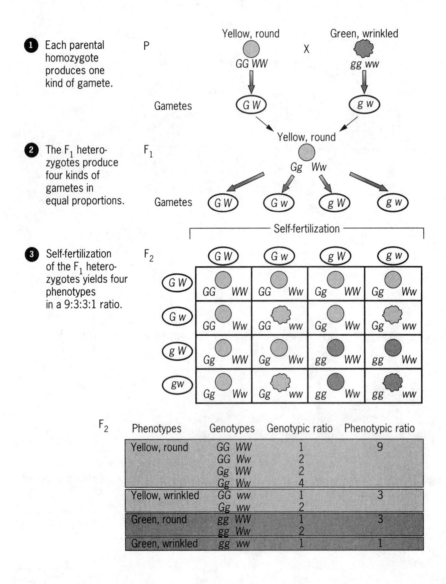

Figure 3.4 Symbolic representation of the results of a cross between a variety of peas with yellow, round seeds and a variety with green, wrinkled seeds.

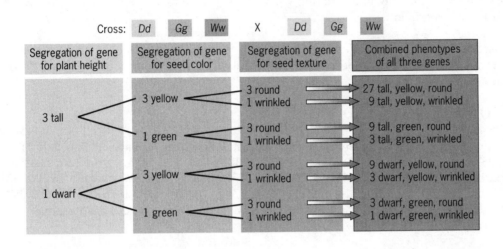

Figure 3.7   The forked-line method for predicting the outcome of an intercross involving three independently assorting gene pairs.

THOUGHT CHALLENGING EXERCISE

A beginning genetics student gathered data from many phenotypically normal couples to whom one or more albinos had been born. The pooled data involving their children showed 61 normal and 39 albino children. These data were subjected to Chi-square analysis to test a "goodness of fit" to a 3:1 ratio.

Much to the student's surprise, the ratio was not consistent with the data ($X^2 = 10.41$; df = 1; $P < 0.01$). What factor(s) could have contributed to this highly significant variation from a 3:1 ratio of normal to albino in children of heterozygous parents?

SUMMARY OF KEY POINTS

Mendel's experiments established three basic genetic principles: (1) Some alleles are dominant, others recessive. (2) During gamete formation, different alleles segregate from each other. (3) Different genes assort independently.

The outcome of a cross can be predicted by the systematic enumeration of the genotypes in a Punnett Square. However, when more than two genes are involved, the forked-line or probability methods are used to predict the outcome of a cross.

The chi-square test is a simple way of evaluating whether the predictions of a genetic hypothesis agree with the data from an experiment.

Pedigrees are used to identify dominant and recessive traits in human families. The analysis of pedigrees allows genetic counselors to determine the probability that an individual will inherit a particular trait.

ANSWERS TO QUESTIONS AND PROBLEMS

1) alleles 2) dominance 3) Correns, deVries, von Tschermak-Seyenegg 4) Bateson 5) b 6) d 7) 1/256; 8) 5/36 9) 45/512 10) 15/64 11) 135/1024 12) e 13) 1/512, 1/256 14) a, 64; b. 729; c. 1/64; d. 729/4096; e. 1/4096 15) a. *Ww dd* x *Ww dd*; b. *WW Dd* x *ww dd*; c. *Ww Dd* x *Ww dd*; d. *Ww Dd* x *Ww Dd*; e. *Ww Dd* x *ww dd* 16) a. 16; b. 108; c. 32; d. 1/64; e. 3/256 17) a 16; b. 216; c. 96; d. 1/192; e. 1/256 18) data fit a 9:3:4 ratio; 19) a. autosomal recessive; b. 1/8; c. 1/9; 20) a. autosomal dominant; b. 0; c. 1/2.

Some ideas concerning thought challenging exercise:

The first thought may be that one genotype shows less viability than the others, e.g., *aa*. But if albinos had a higher death rate than normal, the ratio would be skewed in excess to 3:1. A more careful investigation indicates a bias toward albinos due to the method of identifying heterozygous couples; the couples were identified by having produced one or more albino children. Therefore, the heterozygous couples producing only normal children went unidentified. If the albino children that led the investigator to the family were excluded from the data, a 3:1 normal:albino ratio should be obtained.

APPROACHES TO PROBLEM SOLVING

6. This problem can be worked by the probability method by analyzing the inheritance of one gene pair at a time and then multiplying their probabilities. From the *Cc Tt* selfed plant a progeny will be either purple, *C_* (p = 3/4), or white, *cc* (p = 1/4). A progeny will be either Tall, *TT* (p = 1/4), intermediate, *Tt* (p = 2/4), or short, *tt* (p = 1/4). Since the genes segregate independently of each other, the expected phenotypic ratio of the progeny can be calculated by assembling all combinations of phenotypes and multiplying their individual probabilities.

> purple (3/4) and tall (1/4) = 3/16
> purple (3/4) and intermediate (2/4) = 6/16
> purple (3/4) and short (1/4) = 3/16
> white (1/4) and tall (1/4) = 1/16
> white (1/4) and intermediate (2/4) = 2/16
> white (1/4) and short (1/4) = 1/16

> The phenotypic ratio is 3:6:3:1:2:1

7. Assume that the six gene pairs segregate independently. To work the problem, it is best to break it down into a series of monohybrid crosses and then multiply the probabilities of obtaining the individual genotypes, e.g., the probability of obtaining Aa from $Aa \times AA$ is 1/2; bb from $Bb \times Bb$ is 1/4, etc.

$$
\begin{array}{cccccc}
A\,a & bb & Cc & DD & Ee & ff \\
1/2 & 1/4 & 1/2 & 1/4 & 1/2 & 1/2 = 1/256 \\
\end{array}
$$
$1/2 \ \times \ 1/4 \ \times \ 1/2 \ \times \ 1/4 \ \times \ 1/2 \ \times \ 1/2 = 1/256$

8. There are five random combinations on two dice that equal eight; 2 and 6, 6 and 2, 5 and 3, 3 and 5, and 4 and 4. Each combination has a $1/6 \times 1/6$ or 1/36 probability of occurring. Since each combination is a mutually exclusive event, the individual probabilities are summed. The answer is 5/36.

9. This problem may be solved by use of the binomial expansion $(B + b)^5$. The terms of the expanded binomial represent all combinations of events B (normal phenotype, P = 3/4) and b (albino, P =1/4) in a sample size of 5. The term utilized is $10\,B^2b^3$. The probability is $10(3/4)^2(1/4)^3 = 45/512$.

A way to expand a binomial and assign the coefficient to each term without the aid of Pascal's triangle is shown using the example of $(B + b)^n$, where n = 5. The first term is the first letter of the binomial to the power of n, or B^5 in this example. For each subsequent term, the exponent of the first letter is decreased by a power of 1, and the exponent of the second letter increases by one. The six terms of the expansion of $(B + b)^5$ are:

$$ B^5 \ + \ B^4b^1 \ + \ B^3b^2 \ + \ B^2b^3 \ + \ B^1b^4 \ + \ b^5 $$

The coefficients are symmetrically distributed. The coefficient of the first and last terms is 1. The coefficient of the second term is "**n**" or $5B^4b^1$. The proceeding coefficient is determined by multiplying the exponent of the first letter in the current term by the coefficient of the term, and dividing by the term number. In our example, the coefficient of the term following $5B^4b^1$ is $\dfrac{5 \times 4}{2} = 10$, and the term is $10\,B^3b^2$. The coefficients of the six terms are 1, 5, 10, 10, 5, 1.

A way to represent one term of an expanded binomial involves the use of the factorial method and the formula

$$ P = \frac{N!}{x!\,(N-x)!}\,p^x q^y $$

The best way to start is to define the terms of the factorial. From the example in this problem

P = probability
N = sample size (5)
x = number of event one (2 normal)

y = number of event two (3 albino)
p = probability of event one (3/4)
q = probability of event two (1/4)

Entering these numbers into the terms of the equation results in

$$ P = \frac{5!}{2!\,(5-2)!}\,(3/4)^2(1/4)^3 = \frac{5 \times 4 \times 3 \times 2 \times 1}{2 \times 1 \times 3 \times 2 \times 1}\,(9/16)(1/64) = 45/512. $$

12. The genotypes and their probabilities produced by $Aa \times Aa$ mating is 1/4 AA: 2/4 Aa: 1/4 aa. Since the child in question is normal, it is not aa. Therefore, it has twice the probability of being heterozygous Aa than homozygous AA. The probability of being heterozygous, if normal, is 2/3.

14. (a) The F_1 is heterozygous at six gene pairs, *Aa Bb Cc Dd Ee Ff*. Assume that each gene pair assorts independently. The segregation of each heterozygous gene pair produces 2 kinds of gametes. The total number of gametes produced by the F1 plant is therefore 2 x 2 x 2 x 2 x 2 x 2 = 64. A formula that denotes the number of gametes formed is $(2)^n$, where "n" is the number of heterozygous gene pairs.

(b) The F_1 is heterozygous at six gene pairs. Selfing of the F1 results in three genotypes in the F_2 for each gene pair, e.g., *AA*, *Aa*, *aa*. The formula $(3)^n$ can be used to determine the number of genotypes possible in the F_2 generation, i.e., $3^6 = 729$.

(c) Half of the F_2 progeny are expected to be homozygous for each gene pair, e.g, **1/4 *AA***, 2/4 *Aa*, **1/4 *aa***. Therefore, the proportion of F_2 progeny expected to be homozygous at all six gene pairs is $(1/2)^6$ or 1/64.

(d) Three-quarters of the F_2 progeny are expected to show the dominant phenotype at each gene pair, e.g., **1/4 *AA***, **2/4 *Aa***, 1/4 *aa*. The proportion of the F_2 progeny expected to show all the dominant phenotypes is $(3/4)^6$ or 729/4096.

(e) The probability that the F_2 progeny will be homozygous for any dominant allele, e.g., *AA*, is 1/4. The proportion of progeny expected to be homozygous for all six dominant alleles is $(1/4)^6$ or 1/4096.

15. (a) The progeny segregate white to yellow in a 3:1 ratio. Round and disk do not segregate. The parents must be *Ww dd* x *Ww dd*.

(b) Disk and round segregate in the progeny, while white and yellow do not segregate. The parents must be *WW Dd* x *ww dd*.

(c) White and yellow segregate in a 3:1 ratio and disk and round segregate in a 1:1 ratio. The parents must be *Ww Dd* x *Ww dd*.

(d) Both gene pairs are segregating to give a 9:3:3:1 ratio. The parents are dihybrids of the genotype *Ww Dd* x *Ww Dd*.

(e) Both gene pairs are segregating in a 1:1 ratio. The parents must be *Ww Dd* x *ww dd*.

16. (a) The male parent is heterozygous at 4 gene pairs. Therefore $(2)^4$ or 16 kinds of gametes can be produced.

(b) To determine the number of genotypes and phenotypes possible in the progeny, we analyze this multiple hybrid as a series of individual monohybrid crosses and multiply the probabilities.

vv x v^+v -> $v^+ v$ and vv = 2 genotypes and 2 phenotypes
g^+g x gg -> g^+g and gg = 2 genotypes and 2 phenotypes
$l\,l$ x $l\,l$ -> $l\,l$ = 1 genotype and 1 phenotype
b^+b x b^+b -> b^+b^+, b^+b, and bb = 3 genotypes and 2 phenotypes
Dd^+ x Dd^+ -> DD, Dd^+, and d^+d^+ = 3 genotypes and 2 phenotypes
w^+w x w^+w -> w^+w^+, w^+w, and ww = 3 genotypes and 2 phenotypes

The number of genotypes possible in the progeny is 2 x 2 x 1 x 3 x 3 x 3 = 108.

(c) The number of different phenotypes possible in the progeny is 2 x 2 x 1 x 2 x 2 x 2 = 32.

(d) To calculate the probability of obtaining the $vv\ gg\ l\ l\ b^+b\ Dd^+\ ww$ genotype, determine the probability of obtaining from the parents the genotype at each gene pair, and multiply these. This is $1/2 \times 1/2 \times 1 \times 1/2 \times 1/2 \times 1/4 = 1/64$

(e) Multiply the probabilities of obtaining the six respective individual phenotypes, i.e., $1/2 \times 1/2 \times 1 \times 1/4 \times 3/4 \times 1/4 = 3/256$

17. This problem is worked by the same procedure as problem 16, except that there is a lethal segregating in the cross. The fact that the CC genotype and phenotype don't appear in the living progeny must be taken into account when calculating genotypic and phenotypic ratios.

(a) $2^4 = 16$

(b) $3 \times 2 \times 2 \times 3 \times 2 \times 3 = 216$

(c) $3 \times 2 \times 2 \times 2 \times 2 \times 2 = 96$

(d) $1/2 \times \mathbf{2/3} \times 1/2 \times 1/4 \times 1/2 \times 1/4 = 1/192$

(e) $1/4 \times \mathbf{1/3} \times 1/2 \times 3/4 \times 1/2 \times 1/4 = 1/256$

18. Chi-square analysis for 2:1:1 ratio:

	observed (O)	expected (E)	(O - E)	(O - E)^2	(O - E)^2/E
red	450	410	40	1,600	3.9
yellow	160	205	- 45	2,025	9.9
white	210	205	5	25	0.1
	820	820			$X^2 = 13.9$

Obtain the probability from the chi-square table with d.f. = 2.   P = < 0.001
This means that the probability of obtaining deviations from the expected ratio, due to chance alone, as large as were observed is less than 1/1000. Therefore, a 2:1:1 ratio is not consistent with the data.

Chi-square analysis for 9:3:4 ratio:

	observed (O)	expected (E)	(O - E)	(O - E)^2	(O - E)^2/E
red	450	461	- 11	121	0.262
yellow	160	154	6	36	0.234
white	210	205	5	25	0.122
	820	820			$X^2 = 0.618$

Obtain the probability from the chi-square table with d.f. = 2. 0.70 > P > .50
This means that the probability of obtaining deviations from the expected ratio as large as observed is between 50% and 70%. Therefore, a 9:3:4 ratio is compatible with the data.

19. (a) The inheritance of the trait appearing in the pedigree is most probably due to a recessive allele. This assessment is based upon several features of the pedigree, i.e., the trait appears to occur equally in both males and females, individuals with the trait are not present ever generation, and the trait appears among the progeny of some parents who are both phenotypically normal.

(b) Individual III-3 is heterozygous (*Aa*) for the recessive allele. She inherited the recessive allele from her father who displayed the recessive phenotype and thus was homozygous (*aa*). III-6 is heterozygous (*Aa*) because she had two homozygous recessive offspring. It follows that she inherited the recessive allele from II-7 (*Aa*). III-8 is a son of II-7 and there is a 1/2 chance that he is heterozygous (*Aa*). The probability of having a child with the trait is the probability of III-3 being heterozygous (P = 1) times the probability of III-8 being heterozygous (P = 1/2) times 1/4 = 1/8.

(c) For IV-1 and IV-10 to have a child that shows the trait (homozygous recessive, aa) requires that both be heterozygous (*Aa*). Since IV-1 has a homozygous recessive sibling (*aa*) and phenotypically normal parents (*A_*), both parents must be heterozygous (*Aa*). It follows that IV-1 has a 2/3 chance of being heterozygous (see problem 12 solution above). For the same reasons, IV-10 has a 2/3 chance of being heterozygous for the recessive allele. The probability of IV-1 and IV-10 having a child with the trait (*aa*) is 2/3 x 2/3 x 1/4 = 1/9.

20. (a) The mode of inheritance of the trait shown in the pedigree is most probably due to an autosomal dominant allele. This was concluded because the trait (progressing upward through the pedigree from affected individuals) is not skipping generations and it occurs equally in both males and females.

(b) Since individuals IV-4 and IV-6 do not express the trait and, therefore, do not have the dominant allele, they cannot have a child homozygous for the allele determining the dominant trait.

(c) Individual IV-5 shows the dominant trait. Her mother was unaffected (*aa*) and her father showed the dominant phenotype and was heterozygous (*Aa*). IV-5 must be heterozygous (*Aa*) for the dominant allele. IV-6 is phenotypically normal and does not carry a dominant allele. If IV-5 (*Aa*) and IV-6 (*aa*) marry, the probability of producing a child (*Aa*) with the trait is 1/2.

4

Extensions of Mendelism

IMPORTANT CONCEPTS

    A. Mendel's principle of dominance has been modified by the discovery of incompletely dominant and codominant alleles.
        1. Incomplete or partial dominance is when a heterozygous allelic pair results in a phenotype intermediate to that of the respective homozygous genotypes.
        2. Codominance is when both members of a heterozygous allelic pair are expressed.
    B. Alternative forms of genes are called alleles.
        1. Some genes exist in multiple allelic states.
        2. Most mutant alleles are recessive to the wild-type alleles because they result in a loss of gene function; however, some mutant alleles are dominant to the wild-type allele because they supersede or interfere with its function.
        3. Recessive mutations can be tested for allelism by combining them in the same individual. If a mutant phenotype occurs in the hybrid, the mutants are alleles of the same gene. If a dominant phenotype occurs in the hybrid, then the mutations are generally nonallelic.
        4. Visual mutations affect some aspect of the phenotype.
        5. Sterile mutations limit reproductive ability.
        6. Lethal mutations kill their carriers.
    C. The function of most genes is to produce a polypeptide which may be an enzyme.
        1. Enzymes catalyze specific steps in biochemical pathways.
    D. Gene action can be influenced by environmental factors.
    E. Different genes may interact to determine a phenotype.
        1. Epistatic interactions, in which a gene has an overriding effect on the phenotype, may indicate that the genes involved control different steps in a pathway.
            a. Epistasis can result in many different F_2 dihybrid ratios, depending upon the type of dominance displayed. Some common F_2 dihybrid ratios involving full dominance are 9:3:4 (recessive epistasis), 12:3:1 (dominant epistasis) 9:7 (duplicate recessive epistasis), 15:1 (duplicate dominant epistasis), and 13:3 (both dominant and recessive epistasis).

F. The action of a gene may affect more than one aspect of the phenotype. This phenomenon is called pleiotropy.

G. When individuals do not show a trait even though they have the appropriate genotype, the trait exhibits reduced or incomplete penetrance.

H. The term expressivity is used if a trait is not manifested uniformly among individuals that show it.

I. Continuous phenotypic variation can be explained by the combined effects of many genetic and environmental factors.

IMPORTANT TERMS

In the space allotted, concisely define each term.

complete or full dominance:

partial, incomplete, or semidominance:

codominance:

multiple alleles:

wild-type:

polymorphic:

null (amorphic):

mutation:

visible mutations :

sterile mutations:

lethal mutations:

incomplete penetrance:

expressivity:

epistasis:

pleiotropy:

IMPORTANT NAMES

In the space allotted, concisely state the major contribution made by the individual.

William Bateson:

TESTING YOUR KNOWLEDGE

*In this section, answer the questions, fill in the blanks, and solve the problems in the space allotted. Problems noted by an * are solved in the Approaches to Problem Solving section at the end of the chapter:*

1. The expression of both alleles of a heterozygous gene pair is called _____.

2. The percentage of individuals of a genotype that show the expected phenotype is called _____.

3. The degree to which a genotype is expressed in the phenotype is called _____.

4. The multiple phenotypic effect of a gene is called _____.

5. The suppression of the expression of a gene by another nonallelic gene is called _____.

6. How many different genotypes can occur in a population for a gene that exists in 6 different allelic forms?

7. Epistasis influences:

    (a) genotypic ratios
    (b) segregation of alleles
    (c) the kinds of gametes formed
    (d) the phenotypic ratio
    (e) none of the above

8. The phenomenon occurring when a single-gene heterozygous genotype results in a phenotype intermediate to that of the two respective homozygous genotypes is called _____.

9. When black mice of a true-breeding (homozygous) line are crossed with mice from a true-breeding white line, all the F_1s are black. The F_2 generation consists of about 9 black: 3 brown: 4 white. This is an example of _____ (be specific).

10. You are studying a disease in humans caused by a recessive allele and find that not all individuals who inherit two recessive alleles have the expected disease symptoms. Which genetic term is used to describe this phenomenon?

*11. A recessive trait (aa) in mice has a penetrance of 0.75. From the cross Aa X Aa, What is the expected proportion of normal to affected progeny?

*12. A mouse with the wild-type phenotype:

    (a) must be homozygous for recessive alleles at every locus
    (b) must be homozygous for dominant alleles at every locus
    (c) must have a dominant allele at each locus, but does not have to be homozygous
    (d) may be heterozygous at every locus and still be wild-type
    (e) none of the above

*13. You have obtained a series of mutants, all affecting the same phenotype, e.g., coat color. Each mutant, when crossed to a wild-type, genetically behaves as a single gene recessive mutation. Explain how you might determine whether these mutations are of the same gene or of more than one gene.

*14. In British cattle there are those phenotypes with normal long legs called Kerry and others with extremely shortened legs called Dexter. The Kerry cattle, when crossed to each other produce only progeny with the Kerry phenotype. Dexter cattle, when crossed with Kerry cattle produce the Kerry and Dexter phenotypes in a 1:1 ratio. It is not possible to obtain true breeding strains of the Dexter phenotype. When Dexter are crossed with Dexter, about 1/4 of the calves are the Kerry phenotype, 1/2 are the Dexter phenotype, and 1/4 are spontaneously aborted after about seven months of gestation. These calves are born dead and have extremely shortened legs and nose. Their superficial resemblance to a bulldog led to the name of bulldog calves. Interpret the genetics of the Kerry and Dexter phenotypes.

*15. An insect from a strain breeding true for white eyes was crossed to an insect from a strain breeding true for red eyes. The F_1s all had red eyes. The F_1s were crossed to produce an F_2 generation that consisted of 175 red eyed: 62 cream eyed: 81 white eyed.

(a) The results illustrate the phenomenon of _____.

(b) Provide genotypes for the parents, F_1s and F_2s of this cross.

(c) Illustrate a biochemical pathway that best explains the steps in pigment production, and indicate the step affected by each gene.

*16. Now assume when you crossed the red-eyed and white-eyed insects that the F_1s all had white eyes, and the F_2s consisted of 243 white-eyed: 58 red-eyed: 21 cream-eyed individuals.

(a) The results illustrate the phenomenon of _____.

(b) Provide genotypes for the parents, F_1s and F_2s of this cross.

(c) Illustrate a biochemical pathway that best explains the steps in pigment production, and indicate the step affected by each gene.

*17. In a particular species of fish, a region proximal on the tail may be unspotted or possess one spot, two spots; three spots, or a crescent marking. The following table lists a series of fish crosses and their offspring in regard to tail markings.

Cross	Parents	Offspring
A	one-spot X one-spot	95 one-spot
B	crescent X crescent	101 crescent
C	unspotted X unspotted	98 unspotted
D	one-spot X one-spot	74 one-spot, 26 unspotted
E	crescent X crescent	69 crescent, 22 unspotted
F	crescent X one-spot	49 crescent, 23 one-spot, 27 unspotted
G	crescent X two-spot	50 crescent, 24 two-spot, 25 unspotted
H	two-spot X one-spot	98 three-spot
I	three-spot X three-spot	55 three-spot, 27 one-spot, 28 two-spot

From the data listed above, indicate all possible genotypes for the following phenotypes (use logical allelic symbols of your choice).

unspotted   _____

one-spotted   _____

two-spotted   _____

three-spotted   _____

crescent   _____

*18. A geneticist crossed two true breeding lines of a pet bird. One line "chirped" and had maroon feathers. The other line "screeched" and had brown feathers. All the F_1s had maroon feathers and chirped. The F_1s were crossed and the F_2 data were distributed in the following ratio.

27/64 maroon feathers, chirper
12/64 brown feathers, chirper
9/64   maroon feathers, screecher
9/64   red feathers, chirper
4/64   brown feathers, screecher
3/64   red feathers, screecher

Provide a legend for the above inheritance and indicate the specific type of gene interaction (if any) that occurs in this cross.

*19. In wild sunflowers, populations occur that have yellow flowers with black centers (wild-type), and yellow flowers with yellow centers. The following summarizes a series of crosses involving true breeding sunflower plants:

Cross #1   P   wild-type X yellow-centered (plant 1)

              F_1 black-centered

              F_2 75 black-centered, 26 yellow-centered

Cross #2   P   wild-type X yellow-centered (plant 2)

              F_1 black-centered

              F_2 62 black-centered, 21 yellow-centered

Cross #3   P   yellow-centered (plant 1) X yellow-centered (plant 2)

              F_1 black-centered

              F_2 56 black-centered, 44 yellow-centered

a.) Using genetic symbols of your choice, provide a legend for the inheritance of color of flower centers in the sunflower.

b.) What specific kind of interaction is apparent from the results ?

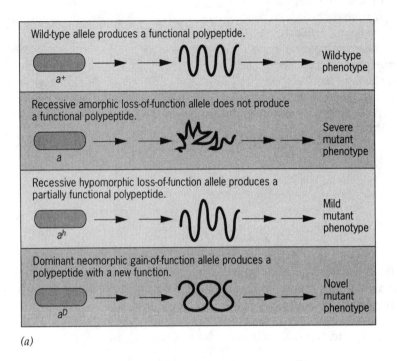

(a)

(b)

Figure 4.9. Differences between recessive loss-of-function mutations and dominant gain-of-function mutations. (a) Polypeptide products of recessive and dominant mutations. (b) Phenotypes of heterozygotes carrying a wild-type allele and different types of mutant alleles.

THOUGHT CHALLENGING EXERCISE

The himalayan phenotype in rabbits consists of a white body with dark extremities, i.e., the tips of the nose, ears and legs. This phenotype results from a single gene recessive mutation of a dominant wild-type allele. The wild-type allele determines a fully-colored rabbit. The himalayan mutation also occurs in Siamese cats and leads to the seal point phenotype of a lightly pigmented body and darkly pigmented extremities. It has been postulated that both the wild-type allele and the himalayan mutation produces an enzyme that catalyzes pigment formation, but that the himalayan allele produces a defective enzyme that is functionally temperature sensitive. How might you test the hypothesis that the expression of the himalayan allele is temperature sensitive?

SUMMARY OF KEY POINTS

Genes often have multiple alleles. Different alleles are created by the mutation of a wild-type allele. Mutant alleles may be dominant, recessive, incompletely dominant or codominant. The allelism of different recessive mutations can be tested by combining them in the same individual; if the individual has a mutant phenotype, then the mutants are alleles; if it has a wild phenotype, then they are not alleles. The function of most genes is to produce a polypeptide. Recessive mutations can cause a partial or complete loss of polypeptide activity. Dominant mutations can endow a polypeptide with new activity that may replace or interfere with the activity of the wild-type polypeptide.

Gene activity is affected by biological and physical factors in the environment. Two or more genes may determine a trait, and one of them may exert an overriding effect on it. A gene may affect many different phenotypes. Continuous variation in a trait can be explained by the combined action of many genes and environmental factors.

ANSWERS TO QUESTIONS AND PROBLEMS

1) codominance 2) penetrance 3) expressivity 4) pleiotropy 5) epistasis 6) 21 7) d
8) partial or incomplete dominance 9) recessive epistasis 10) reduced penetrance 11 to 19) see Approaches to Problem Solving section.

Some ideas concerning thought challenging exercise:

One way to test the hypothesis that the himalayan allele is a temperature sensitive mutant of the wild-type allele would be to shave some of the fur from the back of the cat and to put a cool pack on it for several days as the hair grows back. If the allele is temperature sensitive, the fur should grow back fully pigmented. Another test of the hypothesis would be to shave a patch of dark colored fur from the tail and cover the area with a thick patch to elevate the temperature. If the allele is temperature sensitive, it should not be expressed at the higher temperature and the fur should grow back lightly pigmented. Of course, more sophisticated experiments could be designed that involve cell extracts to measure the rate of conversion of a precursor compound to melanin pigment *in vitro*.

APPROACHES TO PROBLEM SOLVING

11. From the mating, Aa X Aa, there is a .25 probability of obtaining "aa". The "aa" genotype displays reduced penetrance (.75). Therefore, the probability of obtaining an individual expressing the trait is (.25)(.75) = .1875 or 3/16. The ratio of normal to affected progeny is expected to be 13/16: 3/16 = 13:3.

12. A phenotypically wild-type mouse will be expressing the most common, non-selected phenotype. Some of the wild-type traits will be determined by dominant alleles, others by recessive alleles. Therefore, the answer is (e) none of the above.

13. If all of the recessive mutant alleles are of the same gene, they should yield a mutant phenotype when combined in a hybrid. A dominant hybrid phenotype results from complementation of alleles of different genes. Additional evidence that the mutants are all of the same gene, is obtained if only monohybrid ratios are observed in F_2 progeny of crosses between the mutant individuals. If a dihybrid ratio is observed in the F_2 generation of a cross, then the parents must have differed at two gene pairs.

14. The cross of Dexter X Kerry gives a 1:1 ratio, suggesting that a heterozygous gene pair is segregating in one parent. Kerry X Kerry crosses always produce Kerry offspring. Dexter X Dexter crosses produce 1/4 Kerry: 1/2 Dexter: 1/4 bulldog calves. This indicates that the Dexter phenotype is heterozygous for the Dexter and Kerry alleles. Dexter is lethal in the homozygous condition resulting in bulldog calves, but acts like a dominant allele in the heterozygous condition.

15. The 175 red, 62 cream, and 81 white progeny in the F_2 reduces to a 9/16: 3/16: 4/16 ratio. The 16 in the denominator indicates that two gene pairs are segregating in the F_1, and therefore, the F_1 is a dihybrid. The F_1 can be arbitrarily given the genotype of $AaBb$. The phenotypes of the F_2 can be interpreted in terms of a modified dihybrid ratio:

$$9 \; A\_ \; B\_$$
$$3 \; A\_ \; b\,b$$
$$3 \; a\,a \; B\_$$
$$1 \; a\,a\,b\,b$$

The dominant alleles "A" and "B" interact to produce the red phenotype which is expected to occur in a frequency of 9/16. The white phenotype results from recessive epistasis of "aa", and comprises 4/16 of the progeny. A colored phenotype can occur only if the genotype is $A\_$. The cream allele "b" is a recessive mutant of the red allele "B". The cream phenotype results from the genotype $A\_ bb$ which occurs 3/16 of the time.

(a) The results illustrate the phenomenon of recessive epistasis (9:3:4 ratio).

(b) The genotypes for the parents, F_1s and F_2s of this cross are:

$$P \quad red \; (AABB) \; X \; white \; (aabb)$$

$$F_1 \qquad red \; (AaBb)$$

$$F_2 \qquad A\_ \; B\_ \; = red$$
$$A\_ \; b\,b \; = cream$$
$$a\,a \; B\_ \; = white$$
$$a\,a\,b\,b \; = white$$

(c) Illustrate a biochemical pathway that best explains the steps in pigment production, and indicate the step affected by each gene.

Dominant alleles code for enzymes which carry out specific steps in biochemical pathways. Recessive alleles are mutants of the dominant allele that code for a nonfunctional or partially functional enzyme or no enzyme at all. Homozygosity for a recessive allele may result in a block in a biochemical pathway because of production of a nonfunctional enzyme. With these concepts in mind, a biochemical pathway can be illustrated as follows;

$$\begin{array}{ccccc}
& \text{Gene A} & & \text{Gene B} & \\
& \Downarrow & & \Downarrow & \\
& \text{enzyme A} & & \text{enzyme B} & \\
\text{colorless} & \longrightarrow & \text{cream} & \longrightarrow & \text{red} \\
\text{precursor} & & \text{pigment} & & \text{pigment}
\end{array}$$

16. The 243 white: 58 red: 21 cream reduces to a 12/16: 3/16: 1/16 ratio or 12: 3: 1. The 16 in the denominator indicate that the F_1 is segregating for two gene pairs. It follows that the F_1 can be written as *Aa Bb*. The F_2 can be analyzed in respect to a modified 9:3:3:1 dihybrid ratio.

$$\begin{array}{cl}
9 & A\_\ B\_ \\
3 & A\_\ b\ b \\
3 & a\ a\ B\_ \\
1 & a\ a\ b\ b
\end{array}$$

If dominant epistasis is assigned to the white determining "*A*" allele, then any time "*A*" is present the eye color will be white, regardless of the genotype at the second gene pair. The insect must be "*aa*" for color to be expressed. The allele "*B*" results in red eyecolor only if the insect is "*aa*". The recessive cream eye color phenotype results if the insect is *aa bb*.

(a) The results illustrate the phenomenon of dominant epistasis (12:3:1 ratio).

(b) Provide genotypes for the parents, F_1s and F_2s of this cross.

$$\begin{array}{ll}
\text{P} & \text{red (}aa\ BB\text{)} \ \text{X} \ \text{white (}AAbb\text{)}
\end{array}$$

F_1 white (*Aa Bb*)

F_2
$$\begin{array}{ll}
A\_\_\_ & = \text{white} \\
a\ a\ B\_ & = \text{red} \\
a\ a\ b\ b & = \text{cream}
\end{array}$$

(c) Illustrate a biochemical pathway that best explains the steps in pigment production, and indicate the step affected by each gene.

A hypothetical biochemical pathway can be illustrated by having an enzyme, coded by allele "*B*", convert a cream colored precursor to the red pigment. Allele "*b*" does not code for a functional enzyme. In our model, an enzyme coded by allele "*A*" is much more active than the enzyme coded by allele "*B*", and when present all of the cream colored pigment is converted to a colorless compound and white eyes result. The alleles at the second gene pair are expressed only if the insect is "*aa*" at the first gene pair, assuming that the "*a*" allele does not code for a functional enzyme.

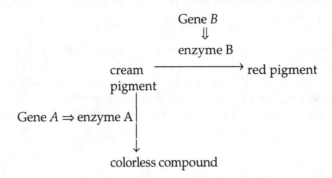

17. To approach this problem it is best to first look at the ratios of the crosses. In doing this it is seen that all data can be reduced to 3:1, 1:2:1, or 1:1 ratios, which all result from monohybrid crosses. Therefore, multiple alleles of a single gene are segregating in these crosses.

Crosses A, B, and C indicate that one-spot, crescent, and unspotted can be in the form of homozygous genotypes. This is concluded from the fact that no segregation of phenotypes is seen in these crosses.

Cross D indicates that the one-spot allele (arbitrarily designated as t^1) is dominant to the unspotted allele t^0 since a 3:1 ratio of one-spot to unspotted is observed among the progeny.

Cross E indicates that crescent (t^C) is dominant to unspotted.

Cross F suggests that the genotype of crescent is $t^C t^0$ and the genotype of one-spot is $t^1 t^0$. Since the crescent phenotype makes up 1/2 of the progeny, crescent must be dominant to both the unspotted and one-spot allele and one-spot is dominant to unspotted.

Cross G likewise indicates that the crescent fish is heterozygous for crescent and unspotted, and the two-spot fish is heterozygous for the two-spot and the unspotted alleles.
This cross, $t^C t^0$ X $t^2 t^0$ produces 1/2 crescent ($t^C t^2$ and $t^C t^0$), 1/4 two-spot ($t^2 t^0$) and 1/4 unspotted ($t^0 t^0$). The crescent allele is dominant to the two-spot allele. The two-spot allele is dominant to the unspotted allele.

Cross H suggests that three-spot results from the heterozygous condition for two-spot and one-spot.

Cross I confirms that the genotype of three-spot is $t^1 t^2$. Three spot ($t^1 t^2$) X three-spot ($t^1 t^2$) results in 1/2 three-spot ($t^1 t^2$), 1/4 one-spot ($t^1 t^1$) and 1/4 two-spot ($t^2 t^2$)

All possible genotypes are indicated for the following phenotypes:

unspotted $t^0 t^0$

one spotted $t^1 t^1$, $t^1 t^0$

two spotted $t^2 t^2$, $t^2 t^0$

three spotted $t^1 t^2$

crescent $t^c t^c$, $t^c t^0$, $t^c t^1$, $t^c t^2$

18. A superficial glance at the ratios with 64 in the denominator indicates that three gene pairs are segregating in this cross. It is easiest to analyze this cross by breaking it into simpler components. Therefore, first consider the inheritance of chirping and screeching.

$$P \qquad \text{chirper X screecher}$$

$$F_1 \qquad \text{chirper}$$

$$F_2 \qquad 27/64 + 12/64 + 9/64 = 48/64 \text{ chirpers}$$

$$9/64 + 4/16 + 3/64 = 12/64 \text{ screechers}$$

The 3:1 ratio indicates that chirper is dominant to screecher.

Next, consider the inheritance of maroon, red, and brown.

$$P \qquad \text{maroon feathers X brown feathers}$$

$$F_1 \qquad \text{maroon feathers}$$

$$F_2 \qquad 27/64 + 9/64 \qquad = 36/64 \text{ maroon}$$

$$9/64 + 3/64 \qquad = 12/64 \text{ red}$$

$$12/64 + 4/64 \qquad = 16/64 \text{ brown}$$

The 36:12:16 or 9:3:4 ratio indicates that two genes are involved in inheritance of feather color and also that recessive epistasis occurs for the brown allele.

The legend is:

chirper	$= C\_$	maroon	$= R\_ B\_$
screecher	$= c\,c$	red	$= r\,r\ B\_$
		brown	$= \_\_b\,b$

19. Crosses #1 and #2 indicate that the two yellow-center strains differ from the wild-type (black center) at one gene pair. Black is dominant to yellow as deduced from the 3:1 ratio in each cross. Further analysis indicates that the yellow-center alleles of cross #1 and cross # 2 are of two different gene pairs because they complement each other to give the wild-type in the F_1 hybrid. The dihybrid nature of this cross is seen when the F_2 data of 56 black-centered to 44 yellow-centered are reduced to a 9/16 to 7/16 ratio. This is a duplicate recessive epistatic ratio. Therefore, the yellow-centered alleles of each gene pair display recessive epistasis. The legend for this inheritance is:

$$B\_C\_ = \text{black center}$$
$$\_\_c\,c = \text{yellow center}$$
$$b\,b\_\_ = \text{yellow center}$$

5

The Chromosomal Basis of Mendelism

IMPORTANT CONCEPTS

   A. Each species has a characteristic chromosome number.
      1. A human has 23 pairs of chromosomes (2n = 46).
      2. A *Drosophila* has four pairs of chromosomes (2n = 8).
   B. The discovery of sex chromosomes paralleled the re-discovery of Mendel's work at the turn of the twentieth century.
      1. E. B. Wilson and others observed that the differences in the sexes were confined to the single pair of sex chromosomes, and that the behavior of these chromosomes during meiosis could account for the inheritance of sex.
         a. Such observations supported the emerging *chromosome theory of inheritance* which postulated that genes were found on chromosomes and the behavior of chromosomes during meiosis may explain Mendel's principles of segregation and independent assortment.
      2. In humans and *Drosophila*, females have two X chromosomes and males have an X and a Y chromosome.
         a. Chromosomes in the genome other than sex chromosomes are called autosomes.
   C. Thomas Hunt Morgan discovered that the white eye gene in *Drosophila* behaved genetically as if it were located on the X chromosome.
      1. Morgan's student, Calvin Bridges, proved that the X chromosome carried the white eye gene by demonstrating that exceptions to the pattern of X-linked inheritance were caused by nondisjunction of the X chromosomes during meiosis.
      2. Morgan's group discovered that many genes could be assigned to the X chromosome of *Drosophila*.

3. Morgan's group also identified many genes located on autosomes.

D. The early studies of *Drosophila* greatly strengthened the view that all genes were located on chromosomes and that Mendel's principles were explainable by the transmissional properties of chromosomes during meiosis.

E. The inheritance of X-linked traits conforms to the following pattern.
   1. A male progeny receives a Y chromosome from his father and an X chromosome from his mother. Therefore, an X-linked trait cannot be transmitted from father to son.
   2. A female receives one X chromosome from her mother and the other from her father.
   3. In pedigrees, an X-linked recessive trait is expressed more often in males than females. It typically skips generations; being passed from male to heterozygous daughters to about half of grandsons.
   4. For a female to express a sex-linked recessive trait, she must be homozygous; receiving an X chromosome carrying the recessive allele from both her mother and father.
   5. A sex-linked dominant trait will be passed from an affected male to all his daughters.
   6. A female, heterozygous for the dominant trait, will pass it on to about half of her progeny, regardless of sex.

F. In humans, hemophilia and color blindness are due to recessive mutations of X-linked genes.
   1. These traits are more prevalent in males than females because male are hemizygous for the X chromosome; affected males always inherit the mutant allele from their heterozygous mothers.
   2. For a female to be homozygous for a sex-linked recessive trait requires that her father and mother both carry the mutant allele.
   3. The fragile X syndrome is caused by an X-linked dominant allele with reduced penetrance.

G. Only a few Y-linked genes have been discovered in humans.
   1. A gene for the testis determining factor (*TDF*) which is required for male sexual differentiation is located on the Y chromosome; without this factor, a human embryo develops into a female.

H. Mechanisms of sex determination vary among organisms.
   1. In *Drosophila*, sex determination is based on the ratio of X chromosomes to sets of autosomes.
      a. A ratio of 1.0 or greater determines a female, a ratio of 0.5 or less results in a male, and a ratio between 0.5 and 1.0 leads to an intersex.
   2. In birds and moths females are the heterogametic sex (ZW) and males are the homogametic sex (ZZ).
   3. Diploid honeybees are females and haploids are males.
   4. In some insects, such as grasshoppers, the Y chromosome has been eliminated completely; females are XX and males are XO.

I. Dosage compensation is a phenomenon in which the activity of a gene is increased or decreased according to the number of copies of that gene in the cell.
   1. Dosage compensation for X-linked genes in *Drosophila* is achieved by hyperactivation of single X chromosome in males.
   2. In mammals, dosage compensation is achieved by randomly inactivating one of the two X chromosomes in females.
      a. The inactivated X chromosome takes the form of a heterochromatic Barr body and is late in replicating its DNA in the S phase of the cell cycle.
      b. In chromosomally abnormal mammals, including humans, with more than two X chromosomes, all the X chromosomes, except one, form a Barr body.

IMPORTANT TERMS

In the space allotted, concisely define each term.

chromatin:

euchromatin:

heterochromatin:

haploid:

haploid genome (n):

diploid (2n):

tetraploid (4n):

octaploid (8n):

X chromosome:

Y chromosome:

sex chromosomes:

autosomes:

hemizygote:

locus:

chromosome theory of heredity:

nondisjunction:

principle of segregation:

principle of independent assortment:

hemophilia:

color blindness:

fragile X syndrome:

pseudoautosomal genes:

testis determining factor (*TDF*):

sex determining region Y (*SRY*):

testicular feminization:

heterogametic:

homogametic:

hyperactivation:

genetic mosaics:

X-inactivation center (*XIC*):

Barr body:

IMPORTANT NAMES

In the space allotted, concisely state the major contribution made by each individual.

Thomas Hunt Morgan:

Edmund Beecher Wilson:

C. E. McClung:

N. M. Stevens:

W. S. Sutton:

Calvin B. Bridges:

Mary Lyon:

Murray Barr:

TESTING YOUR KNOWLEDGE

In this section, answer all questions, fill in the blanks, and solve the problems in the space allotted. Problems noted with an * are solved in the Approaches to Problem Solving Section at the end of the chapter.

1. The position a gene occupies on a chromosome is called a _____.

2. The X chromosome in mammals that is seen as a heterochromatin spot in an interphase nucleus is called a _____.

3. The person who first conclusively proved that a gene was located on a chromosome was _____.

4. In humans, about 1/500 phenotypically normal males have an extra Y chromosome , i.e., XYY. Account for the origin of the XYY karyotype.

5. A mammal with the 48 chromosomes, XXXY, would have _____ Barr bodies in somatic cell nuclei.

6. The person who discovered sex linkage in *Drosophila* was _____.

*7. A male is hemizygous for a sex linked dominant allele, causing a serious abnormality that shows 60% penetrance. What is the probability that if he has a daughter that she will be afflicted with the abnormality? What is the probability that he will have an afflicted son?

8. Why do nuclei of somatic cells of female, but not male, mammals possess a Barr body?

*9. A normal female has a father with a skin disorder called *ichthyosis*. This is caused by a sex linked recessive allele that results in severely-scaled skin.

   (a) What is the probability that she will have an affected son if she marries a normal male?

   (b) What is the probability that she will have an affected son if she marries a male with *ichthyosis*?

   (c) What is the probability that she will have an affected daughter if she marries a normal male?

   (d) What is the probability that she will have an affected daughter if she marries a male with *ichthyosis*?

*10. In *Drosophila*, waxy wings are inherited as a sex-linked recessive trait, and hairy body is inherited as an autosomal dominant trait. Indicate the F_1 and F_2 phenotypic ratios expected from a cross between a waxy wing male and a hairy body female.

*11. Provide a legend for the inheritances of the following cross of chickens:

P   white skin, non-barred feathers, female X yellow skin, barred feathers, male

F$_1$ both males and females have barred feathers and white skin

F$_2$ 3/16 barred feathers, white skin females
       1/16 barred feathers, yellow skin females
       3/16 non-barred feathers, white skin females
       1/16 non-barred feathers, yellow skin female
       6/16 barred feathers, white skin males
       2/16 barred feathers, yellow skin males

*12. What is the most probable mode of inheritance for the rare trait indicated in the following pedigree?

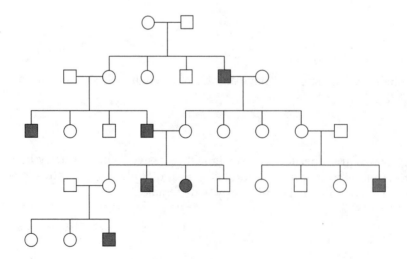

*13. What is the most probable mode of inheritance for the rare trait indicated in the following pedigree?

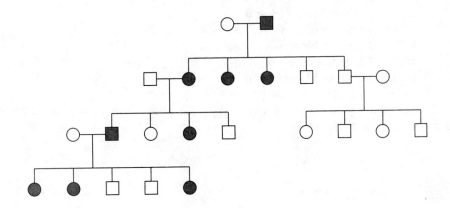

*14. What is the most probable mode of inheritance for the rare trait indicated in the following pedigree?

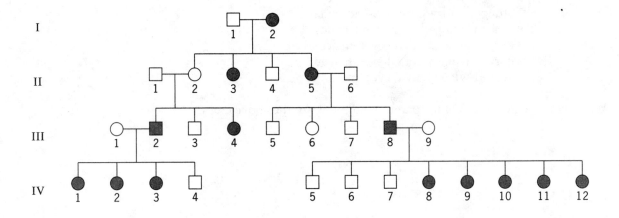

*15. A wildlife photographer who was also a geneticist observed one population of mountain sheep with blue eyes and horns, and another population with brown eyes and no horns. Several brown-eyed hornless females were mated with horned, blue-eyed males. The F_1 progeny consisted of only horned, brown-eyed males and hornless, brown-eyed females. Matings among these F_1s produced the following F_2 progeny:

3/16	horned, brown-eyed males
1/8	horned, brown-eyed females
1/16	hornless, brown-eyed males
3/8	hornless, brown-eyed females
3/16	horned, blue-eyed males
1/16	hornless, blue-eyed males

Using allelic symbols of your choice, provide a legend for the above inheritance. State the specific types of inheritance involved in the cross.

*16. In a *Drosophila* cross involving a straight-winged, round-eyed female and a curved-wing oval-eyed male, all the F_1 flies had straight wings and round eyes. The F_2 generation was as follows:

305	straight-winged, round-eyed males
607	straight-winged, round-eyed females
299	straight-winged, oval-eyed males
103	curved-winged, round-eyed males
201	curved-winged, round-eyed females
99	curved-winged, oval-eyed males

What were the genotypes of the parents and the F_1s of the above cross?

*17. You have selected a laboratory population of mice consisting of black and orange males and tortoise shell females. You make the observation that the individual tortoise shell females have mated with both black and orange mice in the cage. What type of litter would unambiguously demonstrate that a tortoise shell mouse actually conceived by more than one male?

KEY FIGURES

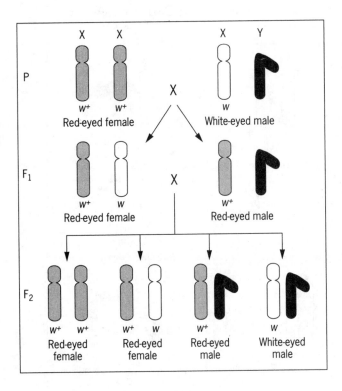

Figure 5.3 Morgan's experiment studying the inheritance of white eyes in *Drosophila*. The transmission of the mutant condition is associated with sex suggesting that the gene for eye color was present on the X chromosome but not on the Y chromosome.

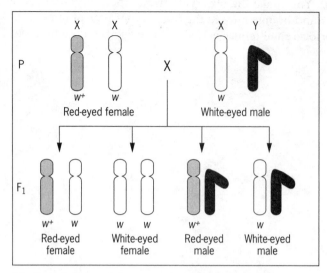

(a) Cross between a heterozygous female and a hemizygous mutant male.

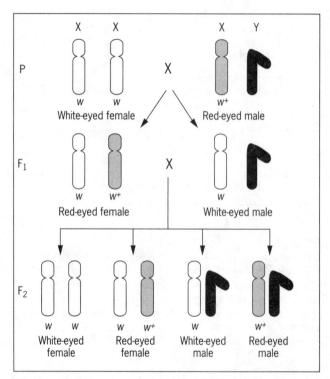

(b) Cross between a homozygous mutant female and a hemizygous wild-type male.

Figure 5.4 Experimental tests of Morgan's hypothesis that the gene for eye color in Drosophila is X-linked. In each experiment, eye color is inherited along with the X chromosome. Thus the results of these crosses supported Morgan's hypothesis that the gene for eye color is X-linked.

THOUGHT CHALLENGING EXERCISE

A lawyer friend asks your opinion on whether or not he should pursue a legal case requested of a client. He explains to you that a woman has given birth to a son that has been medically diagnosed with Lesch-Nyhan syndrome. She claims that her husband worked in a chemical plant for ten years before they met and was chronically exposed to vinyl chloride (a suspected mutagen). The couple has it set in their minds that the exposure to vinyl chloride induced a mutation in the man that in turn was passed on to their son. Based upon genetic knowledge, what would be your advice to the lawyer?

SUMMARY OF KEY POINTS

Chromosomes emerge from a diffuse network of chromatin fibers during cell division. Diploid somatic cells have twice as many chromosomes as haploid gametes. In some species, the X and Y sex chromosomes distinguish the cells of males and females — XY in males and XX in females. In other species, the female has two X chromosomes and the male has only one X chromosome and no Y chromosome.

Genes are located on chromosomes. The disjunction of chromosomes during meiosis is responsible for the segregation and independent assortment of genes.

Disorders caused by recessive X-linked mutations such as hemophilia and colorblindness are more common in males than females. In humans, the Y chromosome carries few genes. Some of these genes are also carried by the X chromosome.

In humans the testis determining factor (TDF) on the Y chromosome causes an embryo to develop into a male. Without this factor, an embryo develops into a female. In *Drosophila*, sex is determined by the ratio of X chromosomes to sets of autosomes (X:A). For X:A < 0.5, the fly develops as a male, for X:A > 1.0, it develops as a female, and for 0.5 < X:A < 1.0, it develops as an intersex.

In *Drosophila*, dosage compensation for X-linked genes is achieved by hyperactivating the single X chromosome in males. In mammals, dosage compensation is achieved by inactivating one of the two X chromosomes in females.

ANSWERS TO QUESTIONS AND PROBLEMS

1) locus 2) Barr body 3) Calvin Bridges 4) nondisjunction of the Y chromosomes in the second meiotic division of males to yield an YY sperm 5) 2 6) Morgan 7) 0.60, 0 8) it is a means of dosage compensation in females that have 2 X chromosomes 9) a. 1/2; b. 1/2; c. 0; d. 1/2 10) F_1 - both males and females are hairy, F_2 - 3/16 hairy males, 1/16 normal males, 3/16 waxy hairy males. 1/16 waxy males, 3/8 hairy females, 1/8 normal females. 11) W _ = white, ww = yellow for both males and females (white is dominant to yellow, autosomally inherited); $Z^B Z^B$ and $Z^B Z^b$ = barred males, $Z^b Z^b$ = non-barred male, $Z^B W$ = barred female, $Z^b W$ = nonbarred female (barred is a Z-

linked dominant allele) **12)** sex-linked recessive **13)** sex-linked dominant **14)** sex-linked dominant with reduced penetrance **15)** HH = horned males and females, Hh = horned males and hornless females, hh = hornless males and females (horned is dominant in males and recessive in females, or sex influenced); $X^B X^B$, $X^B X^b$ = brown-eyed female, $X^b X^b$ = blue-eyed female, $X^B Y$ = brown-eyed male, $X^b Y$ = blue-eyed male (brown is dominant to blue and sex-linked) **16)** parents - $c^+ c^+$ $o^+ o^+$ female X cc o Y male; F$_1$s, - $c^+ c$ $o^+ o$ females and $c^+ c$ $o^+ Y$ males (curved is an autosomal recessive allele; oval is a sex-linked recessive allele), **17)** a litter possessing a black female and also an orange female.

Ideas concerning the thought challenging exercise:

Lesch-Nyhan syndrome is caused by a sex-linked recessive allele. You would tell the lawyer that the son inherited the mutant allele from his heterozygous mother. It would be best to diagram the inheritance of a sex-linked trait. Indicate that males receive their X chromosome only from their mothers. Show why the son could not have received a mutant allele on his X chromosome from his father. The lawyer would then realize that there were no grounds for the claim that the vinyl chloride induced the mutation that the son inherited, because he inherited it from his mother.

APPROACHES TO PROBLEM SOLVING

7. A male will pass his X chromosome to all his daughters. Therefore, if the gene shows 60% penetrance, 0.60 of the females should be afflicted with the abnormality. None of the sons will be afflicted because they inherit their X chromosome from their mother.

9. Approach this problem by characterizing the normal female. Since her father had *ichthyosis*, he was of the genotype i Y. The normal daughter must be heterozygous, I i, because she inherited one of her X chromosomes from her father.

   (a) Males receive their X chromosome from their mother. Therefore, the normal heterozygous female, Ii, would have a 1/2 or 50% chance that each male child would be hemizygous, i Y, for the recessive allele.
   (b) Since it is the female parent that contributes the X chromosome to her sons, the genotype of the father is inconsequential in this case. The probability is still 1/2.
   (c) For a daughter to be homozygous for a sex-linked recessive allele, her father must have been hemizygous and her mother at least heterozygous for the allele. Since the normal male can not carry a sex-linked recessive allele, assuming full penetrance, no affected daughters will result from this mating.
   (d) If the heterozygous woman, I i, marries a male with *ichthyosis*, i Y, then there is a 50% chance that each daughter will be affected.

10. Approach this problem by first denoting the genotypes at both genes of the parents.

    P   waxy male X hairy female
       wY $h^+ h^+$ $w^+ w^+$ HH

    Then, proceed to the F$_1$ generation. Remember males get their X chromosome from their mother and their Y chromosome from their father, and females get an X chromosome from both their mother and father.

    F$_1$ w^+Y Hh^+ males $w^+ w$ Hh^+ females

Next, indicate the F_2 generalized genotypes that will result in each phenotype. Calculate the probability of obtaining the genotype at each gene pair and multiply the individual probabilities. This gives the expected ratio of the corresponding phenotype as indicated below.

F_2	$w^+\_\, H\, \_$	probability = $1/2 \times 3/4 = 3/8$	hairy females
	$w^+\_\, h^+h^+$	probability = $1/2 \times 1/4 = 1/8$	wild-type females
	$w^+Y\, H\, \_$	probability = $1/4 \times 3/4 = 3/16$	hairy males
	$w^+Y\, h^+h^+$	probability = $1/4 \times 1/4 = 1/16$	wild-type males
	$wY\, H\, \_$	probability = $1/4 \times 3/4 = 3/16$	waxy, hairy males
	$wY\, h^+h^+$	probability = $1/4 \times 1/4 = 1/16$	waxy males

11. This problem is approached by the typical method of analyzing one set of traits at a time. First analyze the inheritance of white and yellow skin.

P   white skin   X   yellow skin

F_1 All white skin

F_2 white skin females, $3/16 + 3/16 = 6/16$
yellow skin females, $1/16 + 1/16 = 2/16$ 3:1 white:yellow

white skin males, $6/16$
yellow skin males, $2/16$ 3:1 white:yellow male

Since white skin and yellow skin segregate in a 3:1 ratio in both sexes, one autosomal gene is involved with the white allele (W) being dominant to yellow (w).

Next, analyze the inheritance of barred and non-barred.

P   non-barred female   X   barred male

F_1 All barred feathered

F_2 barred females, $3/16 + 1/16 = 4/16$
non-barred females, $3/16 + 1/16 = 4/16$ 1:1 barred: non-barred

barred males, $6/16 + 2/16 = 8/16$
non-barred males, none

The 1:1 segregation ratio of barred and non-barred in the F_2 females, in addition to the F_2 males being all barred, indicate a sex-linked trait. Barred is dominant to non-barred. The females show the 1:1 F_2 ratio and therefore are the heterogametic sex. This is a Z Z and Z W sex determining mechanism.

The legend is as follows:

$W\, \_$ = white skin $Z^B\, W$ = barred feathered female
$w\, w$ = yellow skin $Z^b\, W$ = nonbarred feathered female
                     $Z^B\, Z^B, Z^B\, Z^b$ = barred feathered male
                     $Z^b\, Z^b$ = non-barred feathered male

12. Since the trait occurs mostly in males and generally skips a generation, thereby appearing in grandfathers and grandsons, it fits the pattern of a sex-linked recessive trait. The only female in

the pedigree with the trait had an affected father and a mother who was a carrier (heterozygous) for the trait.

13. Notice that this trait does not skip a generation when proceeding from an affected individual on the bottom and ascending to the top of the pedigree. Therefore, it must be determined by a dominant allele. Next, it must be determined whether it is sex-linked or autosomal. The definitive parts of the pedigree are that all affected males pass the trait to all of their daughters, and to none of their sons. This is the expected pattern of inheritance for a dominant sex-linked trait.

14. This pedigree is similar to the one in problem 13, in that males pass the trait on to their daughters. There is an exception in individual II-2. The simplest interpretation is that this trait is due to a sex-linked dominant trait with incomplete penetrance. That assumption is supported by the fact that individual II-2 has affected children.

15. The cross is analyzed one set of traits at a time as follows:

P   brown-eyed females X blue-eyed males

F_1 all brown-eyed

F_2 brown-eyed females, $1/8 + 3/8 = 4/8$ ·

brown-eyed males, $3/16 + 1/16 = 4/16$
blue-eyed males, $3/16 + 1/16 = 4/16$ 1:1 brown:blue

All F_2 females are brown-eyed, but the F_2 males segregate brown and blue eyes in a 1:1 ratio. These F_2 ratios are expected of a sex-linked trait with brown being dominant to blue, and the male being the heterogametic sex.

P   hornless female X horned males

F_1 all males horned, all females hornless

F_2 horned males, $3/16 + 3/16 = 6/16$
hornless males, $1/16 + 1/16 = 2/16$ 3:1 horned:hornless

horned females, $1/8$
hornless females, $3/8$ 1:3 horned:hornless

The 3:1 F_2 ratio in each sex suggests that a single gene pair is segregating with simple dominance of one allele. However, the reversed 3:1 ratios indicate that horned is dominant in males and hornless is dominant in females. Alleles showing reversal of dominance related to the sex of the individual are called *sex influenced*.

A legend is: <u>females</u>   <u>males</u>

HH horned HH, Hh
Hh, hh hornless hh

$X^B X^B, X^B X^b$ Brown $X^B Y$
$X^b X^b$ blue $X^b Y$

16. For this problem, we again use the approach of breaking the more complex set of data into simpler components. Let's arbitrarily start with the inheritance of straight and curved wings.

> P   straight-winged female X curved-winged male
>
> F_1 all straight-winged
>
> F_2 straight-winged males, $305 + 299 = 604$
> curved-winged males, $103 + 99 = 202$ 3:1 straight:curved
>
> straight-winged females, 607
> curved-winged females, 201   3:1 straight:curved

The 3:1 ratio of straight to curved in the F_2 generation indicates that one gene is involved in the inheritance of straight and curved. The straight allele (c^+) is dominant to the curved allele (c).

> P   round-eyed female X oval-eyed male
>
> F_1 all round-eyed
>
> F_2 round-eyed males, $305 + 103 = 408$
> oval-eyed males, $299 + 99 = 398$ 1:1 round:oval
>
> round-eyed females, $607 + 201 = 808$
> oval-eyed females, none

The F_2 1:1 ratio of round to oval eyes in males and only round eyes in F_2 females suggest that the oval eye locus is sex-linked with round ($o+$) dominant to oval (o).

The genotypes of the parents were: female, $c^+c^+ o^+o^+$; male, $cc\ o\ Y$.

The genotypes of the F_1s were: females, $c^+c\ o^+o$; males, $c^+c\ o^+Y$.

17. A tortoise shell mouse is always a female and heterozygous for the sex-linked alleles, orange (c^o) and black (c^b). The random lyonization of one of the X chromosomes in each cell at an early stage of embryonic development and the retaining of the inactivation of the same X chromosome in all cells of a lineage result in a mosaic of patches of orange and black fur. If the tortoise shell mouse (c^oc^b) conceived only by a black male ($c^b\ Y$), then the female progeny could be tortoise shell (c^oc^b) or black (c^bc^b). If she conceived only by an orange male ($c^o\ Y$), then the female progeny could be either tortoise shell (c^oc^b) or orange (c^oc^o). Therefore, a litter having both an orange female and a black female would indicate that the female tortoise shell mouse had conceived by two different males.

6

Variation in Chromosome Number and Structure

IMPORTANT CONCEPTS

A. Cytogenetic analysis of stained mitotic chromosomes is generally performed at late prophase or metaphase.
 1. Specific treatments with stains such as quinicrine and Giemsa create a characteristic banding pattern along the chromosomes.
 2. The human karyotype consists of 23 pairs of chromosomes, numbered (approximately) from the largest to smallest.
 3. Each chromosome is characterized by a long (q) and a short (p) arm.
 4. Cytological analysis can reveal variation in chromosome number and structure.

B. Polyploidy is the presence of chromosome sets above the diploid (2n) level, e.g., triploidy (3n), tetraploidy (4n), hexaploidy (6n).
 1. Many polyploids are sterile or only show partial fertility because their chromosomes segregate irregularly during meiosis.
  a. The presence of incomplete sets of chromosomes upsets the genetic balance and causes inviability of gametophytes in plants, and inviability of zygotes or abnormal development of embryos in animals.
 2. Alloploids, produced by doubling the chromosome number of chromosomally sterile hybrids are often fertile because each chromosome has a homolog with which to pair.
 3. Cytogeneticists have been able to induce polyploidy by treating organisms with the spindle-poisoning drug colchicine.

C. A special form of polyploid tissue called polyteny occurs in salivary glands of *Drosophila* larvae.

1. In polytene nuclei, sister chromatids remain together through many rounds of DNA synthesis, forming large, banded, polytene chromosomes.
   a. These interphase chromosomes are excellent materials for cytogenetic analysis.
D. Aneuploidy involves the over- or under-replication of particular chromosomes or chromosome segments.
   1. Aneuploidy in animals was first detected in *Drosophila* and involved the sex chromosomes.
   2. Aneuploidy in plants was originally discovered in mutant strains of *Datura*.
   3. Aneuploidy also occurs in humans.
      a. The most common aneuploidy is trisomy for chromosome 21 (Down syndrome).
      b. Triplo-X and XXY (Klinefelter syndrome) are other viable human trisomics.
      c. The only viable human monosome is the 45, X karyotype (XO or Turner syndrome).
   4. Deletions or duplications of chromosome segments may also cause aneuploidy.
      a. A deletion of the short arm of human chromosome 5 (5p⁻) causes the *cri-du-chat* syndrome.
      b. In *Drosophila*, a duplication of a segment of the X chromosome causes the Bar eyes phenotype.
E. The gross structure of chromosomes may be altered by inversions and translocations.
   1. Inversions reverse the order of genes within a segment of a chromosome.
      a. An inversion that includes the centromere is called pericentric.
      b. Inversions in a chromosome arm that don't involve the centromere are called paracentric.
   2. Reciprocal translocations interchange segments between two nonhomologous chromosomes.
      a. In translocation heterozygotes, chromosome disjunction during meiosis may produce aneuploid gametes that cause reduced fertility.
   3. Compound chromosomes, such as attached-X, are formed by the fusion of homologous chromosome segments.
   4. Robertsonian translocations are formed by fusions between the centromeres of nonhomologous chromosomes.
   5. Genes near the breakpoints of inversions and translocation may exhibit position effect on expression leading to a variegated pattern of expression.
      a. The white-mottled mutation of *Drosophila* is a classic example of a position effect.
   6. Phylogenetic studies indicate that chromosome rearrangements have played a role in evolution.

IMPORTANT TERMS

In the space allotted, concisely define each term.

cytogenetics:

Q banding:

G banding:

R banding:

C banding:

karyotype:

univalent:

trivalent:

allopolyploids:

autopolyploids:

amphidiploid:

colchicine:

endomitosis:

polytene:

chromocenter:

aneuploidy:

trisomies:

hypoploid:

hyperploid:

Down syndrome:

Klinefelter syndrome:

monosomy:

amniocentesis:

chorionic biopsy:

Patau syndrome:

Edward syndrome:

Turner syndrome:

cri-du-chat syndrome:

deletion (deficiency):

duplication:

pericentric inversion:

paracentric inversion:

translocation:

reciprocal translocation:

adjacent disjunction:

alternate disjunction:

compound chromosome:

isochromosome:

attached-X chromosome:

Robertsonian translocation:

position effect:

IMPORTANT NAMES

In the space allotted, concisely state the major contribution made by the individual.

J. O. Beasley:

E. G. Balbiani:

Theophilus Painter:

C. B. Bridges:

Albert Blakeslee:

John Belling:

Lillian Morgan:

F. W. Robertson:

TESTING YOUR KNOWLEDGE

*In this section, answer the questions, fill in the blanks, or solve the problem in the space allotted. Problems noted with an * are solved in the Approaches to Problem Solving section at the end of the chapter.*

1. Trisomy for chromosome 21 in humans results in _____ syndrome.

2. An inversion that does not include the centromere is called _____.

3. An aberrant chromosome number in which a normally diploid cell has three copies of one chromosome and two copies of all the others is called _____.

4. An organism with four sets of homologous chromosomes is called an _____.

5. A reciprocal exchange of non-homologous chromosomes is called a _____.

6. A phenotypically normal man or woman with only 45 chromosomes and carrying a 14/21 translocation has about a 10% to 15% chance that each child will be afflicted with _____ syndrome.

7. An aberrant chromosomal condition in which a normally diploid cell has one copy of a chromosome and two copies of all others is called _____.

8. An inversion that includes the centromere is called _____.

9. An organism with four sets of chromosomes, two of one genome and two of a different genome, is called an _____ .

10. A child trisomic for chromosome 18 has _____ syndrome.

11. An individual possessing three sets of chromosomes is called _____.

12. How many chromosomes would be found in a root tip nucleus of a hexaploid plant, derived from two plants, a diploid with 2n = 14 , and a tetraploid with 4x = 2n = 30?

13. Which chromosome aberration in the heterozygous condition may lead to an acentric fragment and a bridge at anaphase I of meiosis?

14. A child trisomic for chromosome 13 has _____ syndrome.

15. A deletion of the short arm of one member of chromosome 5 results in _____ syndrome.

16. If a female human has only one X chromosome, an abnormal phenotype results called _____ syndrome.

17. If a human male has two X chromosomes and a Y chromosome, an abnormal phenotype results that is called _____syndrome.

*18. A couple gives birth to an extremely mentally and physically retarded child. Upon karyotyping of the parents, it is discovered that the man is heterozygous for a normal chromosome and a chromosome carrying a large inversion.

(a) What is the likely cause of the phenotypically abnormal child?

(b) What advice would you give the parents on the probability of them having a second abnormal child?

*19. Two wild bluebell species, when crossed, produced a vigorous but sterile hybrid with large, horticulturally-desirable flowers. You want to commercially develop a plant having the showy flowers of the hybrid. Unfortunately, the wild species grow so poorly under cultivated conditions that mass production of hybrid seed is unfeasible. Upon chromosomal analysis of the wild species you find that one has a somatic chromosome number of 2n = 10, and the other has a somatic chromosome of 2n = 6. Diagram the easiest procedure by which you may potentially derive a fertile, large-flowered variety for commercial seed production. Indicate the chromosome composition of the derived variety.

*20. Strains 1 and 2 are homozygous for the following respective chromosomal arrangement and alleles.

1. L M N   O P Q   2. l m p o   n q

   L M N   O P Q   l m p o   n q

(a) Diagram pairing of the chromosomes during meiosis of the hybrid obtained from crossing strains 1 and 2. Indicate a chiasma in the region between the O and P loci.

(b) Indicate the genotype of the meiotic products formed by the hybrid, if a single crossover occurred between the O and N loci.

(c) Which meiotic products would result in inviable gametophytes if the organism were a plant, or abnormal or inviable zygotes upon fertilization if the organism were an animal? Why?

*21. The New World allotetraploid cotton *Gossypium hirsutum* (4x = 2n = 52) has 13 relatively large pairs and 13 smaller pairs of chromosomes. Thirteen of the chromosome pairs apparently were derived from the Old World *G. herbaceum*, which possesses 2n = 26 large chromosomes. The other thirteen pairs were apparently derived from the New World *G. raimondii*, or a very similar species, which has 2n = 26 small chromosomes.

(a) If *G. hirsutum* were crossed with *G. herbaceum*, how many bivalents and univalents and their sizes would you expect to see in the hybrid at metaphase I of meiosis?

(b) If *G. hirsutum* were crossed with *G. raimondii*, how many bivalents and univalents and their sizes would you expect to see in the hybrid at metaphase I: of meiosis?

(c) How many bivalents would you expect to observe at metaphase I of meiosis in a hybrid between *G. raimondii* and *G. herbaceum*? How many univalents?

*22. In the Jimson weed, trisomic plants can be found for each of the twelve chromosomes. From various crosses of plants (female is always written first), disomic or trisomic for the chromosome on which the locus for purple (p^+) and white (p) flower color resides, indicate the expected proportions of purple and white flower plants among the progeny. Female gametes may carry an extra chromosome, but pollen possessing an extra chromosome is inviable.

(a) $p^+p^+p \times p\,p$

(b) p^+p^+p X p^+p^+p

(c) p^+p^+p X p^+p

(d) $p^+p\,p$ X p^+p

(e) $p^+p\,p$ X $p^+p\,p$

*23. The following diagram depicts a pair of mouse chromosomes.

A B C D E F G

A B C D C D E F G

(a) What type of chromosomal aberration is depicted in this heterozygous chromosome pair?

(b) Diagram pairing of these chromosomes at pachynema.

*24. Show how two nonhomologous chromosomes can be rearranged so that genes, formerly on different chromosomes, share the same chromosome. Assume that the resulting individual is fully fertile and has only bivalents formed in meiosis.

*25. What is the expected phenotypic ratio in the progeny of a cross involving two autotetraploids of the genotype AaAa (assume random chromosome segregation)?

*26. What is the expected phenotypic ratio of a cross between two allotetraploids, $A_1a_1 A_2a_2$ X $a_1a_1 a_2a_2$ (the subscripts designate chromosomes of the two different genomes)?

*27. What is the expected phenotypic and genotypic ratios of a cross between two allotetraploids, $A_1A_1 a_2a_2$ X $A_1A_1 a_2a_2$ (the subscripts designate chromosomes of the two different genomes)?

*28. Illustrate how you can get an *aa* gamete from an autotetraploid of the genotype *AAAa*.

THOUGHT CHALLENGING EXERCISE

In some crop plants such as tomato (2n = 24), plants can be maintained that are individually trisomic for one of the twelve chromosomes. The twelve different trisomics can be easily identified by their specific abnormal effect on the phenotype. Describe a procedure, using trisomic plants, by which you could assign the newly discovered mutant gene in tomato to a specific chromosome.

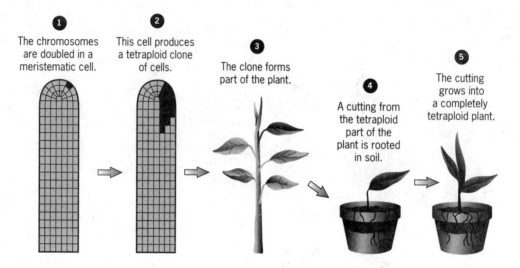

(a) Chromosome doubling in meristematic tissue.

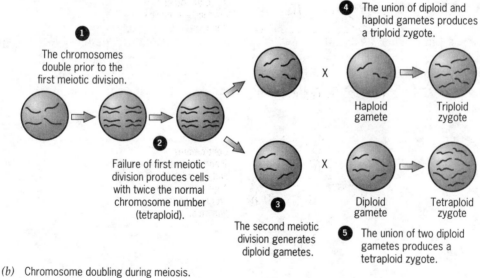

(b) Chromosome doubling during meiosis.

Figure 6.11 Chromosome doubling and the origin of polyploids.

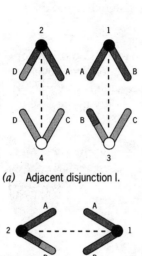

(a) Adjacent disjunction I.

Centromeres 1 and 2 go to one pole and centromeres 3 and 4 go to the other pole, producing aneuploid gametes.

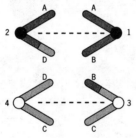

(b) Adjacent disjunction II.

Centromeres 1 and 3 go to one pole and centromeres 2 and 4 go to the other pole, producing aneuploid gametes.

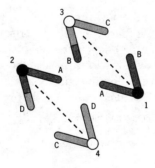

(c) Alternate disjunction.

Centromeres 2 and 3 go to one pole and centromeres 1 and 4 go to the other pole, producing euploid gametes.

Figure 6.27 Types of disjunction in a translocation heterozygote.

SUMMARY OF KEY POINTS

Chromosomes are most easily analyzed in dividing cells. Among the various dyes used to stain chromosomes, quinacrine and Giemsa produce a banding pattern that is useful in identifying individual chromosomes within a karyotype.

Polyploidy involves the presence of extra sets of chromosomes. Many polyploid species are sterile because their multiple sets of chromosomes segregate irregularly in meiosis. However, polyploids produced by chromosome doubling in interspecific hybrids may be fertile if their constituent genomes segregate independently. In some polyploid tissues, sister chromatids remain together, forming a large, polytene chromosome.

Aneuploidy involves the under- or over-representation of a chromosome or chromosome segment. In a trisomy, such as Down syndrome in human beings, a chromosome is over-represented; in a monosomy, such as Turner Syndrome, a chromosome is underrepresented. Deletions and duplications of particular chromosome segments also cause aneuploidy — hypoploidy in the case of a deletion and hyperploidy in the case of a duplication.

An inversion reverses the order of a segment within a single chromosome; a translocation interchanges segments between two nonhomologous chromosomes. During meiosis, the chromosomes in an inversion heterozygote pair by forming a loop; in a translocation heterozygote, they pair by forming a cross. Chromosome rearrangements may alter the expression of genes by moving them to new chromosomal positions.

ANSWERS TO QUESTIONS AND PROBLEMS

1) Down 2) paracentric 3) trisomic 4) autotetraploid 5) reciprocal translocation 6) Familial Down 7) monosomic 8) pericentric 9) allotetraploid 10) Edward 11) triploid 12) 44 13) paracentric inversion 14) Patau 15) Cri-du-chat 16) Turner 17) Klinefelter 18-28) see *Approaches to Problem Solving* section.

Ideas concerning thought challenging exercise:

A procedure to determine the chromosome on which the new mutant resides follows: 1) cross the mutant plant as a male parent to each of the twelve trisomic plants; 2) select a trisomic hybrid plant from the progeny of each cross; 3) testcross each trisomic plant; 4) determine the phenotypic ratio of mutant to normal in the testcross progeny. For eleven of the twelve trisomics selected, a 1:1 testcross ratio indicates disomic inheritance and the gene is not located on the chromosome that was trisomic. For one of the trisomics a distorted testcross ratio (2:1 to 5:1) will be observed that results from segregation of the heterozygous trisomic condition. This indicates that the gene is located on the chromosome that is trisomic in this cross.

APPROACHES TO PROBLEM SOLVING

18. (a) The key to the abnormal child resides in the fact that crossing over within the inversion loop of an inversion heterozygote will generate recombinant chromatids with terminal duplications and deficiencies. The duplicated/deficient chromosomes will cause genetic imbalance and hence abnormal development.

    (b) The parents should be advised that there is an increased risk of having a second abnormal child. However, a numerical probability cannot be assigned without knowing the frequency of recombination within the inversion loop.

19. The key to solving this problem is the sterility of the hybrid and the fact that the two species differ in chromosome number. This suggests that the sterility of the hybrid may result from lack of chromosome pairing due to divergence of the chromosomes. The doubling of the chromosome number by treatment with colchicine may lead to restoration of fertility because all chromosomes would now have a homolog. If so, then the resulting allotetraploid (4x = 2n = 16) could be mass produced from seeds.

20. (a) By analyzing the arrangement of chromosome regions in the two strains, it is seen that the sequence N to P is inverted in strain 2 relative to strain 1. Maximizing of pairing of homologous chromosome regions will lead to the formation of an inversion loop as depicted below at pachynema.

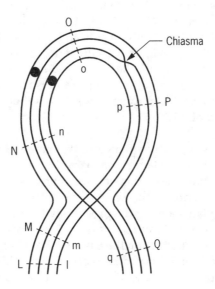

(b) The crossover between O and P will lead to one chromatid of each chromosome having a duplicated and deficient end. At the end of meiosis, there will be four meiotic products; one with a normally arranged chromosome, one with the chromosome carrying the inversion, and two carrying a duplicated/deficient chromosome as follows:

$$\underline{L\ M\ N\ O\ P\ Q}\qquad \underline{L\ M\ N\ O\ p\ m\ l}\qquad \underline{q\ n\ o\ P\ Q}\qquad \underline{l\ m\ p\ o\ n\ q}$$

Normal	Duplicated for m and l	Duplicated for Q	Inversion
	Deficient for Q	Deficient for l and m	

(c) Due to the genetic imbalance, the meiotic products with duplicated/deficient chromosomes would result in inviable gametophytes in plants and inviable zygotes or abnormal embryonic development in animals.

21. (a) Since *G. hirsutum* contains 13 chromosomes similar to those of *G. herbaceum*, these chromosomes will pair in the triploid hybrid. Therefore, in the hybrid 13 larger bivalents and 13 smaller univalents would be seen at metaphase I of meiosis.

(b) *G. hirsutum* also has 13 chromosomes similar to those of *G. raimondii*. These would pair with the 13 chromosomes of *G. raimondii* in the hybrid forming 13 smaller bivalents; 13 larger univalents would also be observed.

(c) If no multivalents are observed in normal *G. hirsutum* (as is the case), and no trivalents are formed in the hybrids between *G. hirsutum* and the diploids, *G. raimondii* and *G. herbaceum*, then no pairing would be expected in the *G. raimondii* X *G. herbaceum* hybrid; 26 univalents and no bivalents should be observed.

22. To approach this problem, first determine the various genotypes of meiotic products formed by the random (two chromosomes to one pole and one chromosome to the other pole) segregation of the trisomic chromosomes. Also note that male gametophytes containing an extra chromosome grow slower and generally are not involved in fertilization of the egg. Therefore, the extra chromosome is transmitted only through the female parent.

(a) Let's consider the random segregation of chromosomes from the trisome p^+p^+p. To make it easier to follow, the three chromosomes will be denoted as 1, 2, and 3, i.e., $_1p^+{}_2p^+{}_3p$. Three random types of segregation occur during meiosis: 1 and 2 to one pole and 3 to the other pole; 2 and 3 to one pole and 1 to the other; 1 and 3 to one pole and 2 to the other. These segregations result in the four genotypes:, p^+p^+, p^+p, p^+ and p in a 1:2:2:1 ratio. Since 1/6 of the gametes from this trisomic plant carry only p there should be a distorted testcross ratio of 5:1 purple to white. Since all female gametophytes carrying an extra chromosome may not survive and since trivalent association of the trisomic chromosomes does not occur all the time, the ratio is generally about 4:1 instead of 5:1.

(b) The female trisomic plant will produce gametophytes containing just p about 1/6 of the time. Since the male meiotic products containing an extra chromosome do not function in fertilization, the frequency of chromosomally normal male gametophytes containing just p will be 1/3. Therefore, in this cross the frequency of white flower plants will be $(1/6)(1/3) = 1/18$. The ratio of purple to white flower plants among the progeny will be 17:1.

(c) The frequency of p meiotic products from the trisomic plant is 1/6. The heterozygous disomic male, p^+p, produces 1/2 of the meiotic products carrying p. Therefore, the expected frequency of pp (white) offspring is expected to be 1/12. The ratio of purple to white is 11:1.

(d) The random segregation from the trisomic female p^+pp will produce the meiotic products pp, p^+p, p, and p^+, in a 1:2:2:1 ratio. Half of the meiotic products are of the genotype pp or p. The male disomic parent produces p^+ and p meiotic products in a 1:1 ratio. This cross segregates purple to white in a 3:1 ratio.

(e) Half of the meiotic products of the female parent are p or pp. The meiotic products with an extra chromosome do not function as viable gametophytes in the male parent. Therefore the male parent, p^+pp, will produce gametophytes carrying p and p^+ in a 2:1 ratio, and the expected ratio of purple to white from this cross is 2:1.

23. (a) An analysis of this pair of chromosomes reveals that the region C - D is represented twice in one of the chromosomes. Therefore, a duplication occurred in one chromosome.

(b) The chromosomes will pair as homologously as possible. However the duplicated region will form a buckle as indicated below:

24. A translocation between nonhomologous chromosomes initially results in a translocation heterozygote as illustrated below:

```
A | A |     W | W |                              A | W |     A | W |
  |   |       |   |                                |   |       |   |
B | B |     X | X |       ────────────────────▶   B | X |     B | X |
  |   |       |   |                                |   |       |   |
C | C |     Y | Y |    reciprocal translocation   C | C |     Y | Y |
  |   |       |   |    occurs between two          |   |       |   |
D | D |     Z | Z |    nonhomologous              D | D |     Z | Z |
                      chromosomes
```

Alternate segregation from the cross-shaped quadrivalent that results from homologous pairing of the two normal and two translocated chromosomes will produce meiotic products containing either the two normal chromosomes or the two translocation chromosomes. These meiotic products are genetically balanced and have all the regions of the chromosomes present twice as in the normal cells, except that the linkage relationships have changed in the gametes that get the two translocated chromosomes. These are indicated below:

```
A | W |                          W | A |
  |   |                            |   |
B | X |                          X | B |
  |   |                            |   |
C | Y |                          C | Y |
  |   |                            |   |
D | Z |                          D | Z |

normal chromosomes              translocated chromosomes
in half of the gametes          in half of the gametes
resulting from adjacent         resulting from adjacent
segregation                     segregation
```

Fusion of two gametes containing the two translocated chromosomes will result in a zygote homozygous for both of the translocated chromosomes that form two separate bivalents during meiosis. The chromosomal constitution of such a zygote is shown below:

```
W | W |     A | A |
  |   |       |   |
X | X |     B | B |
  |   |       |   |
C | C |     Y | Y |
  |   |       |   |
D | D |     Z | Z |
```

The translocation homozygote has two new linkage groups that have brought together genes formerly on different chromosomes. Since it is homozygous for these rearranged chromosomes, two bivalents will form in meiosis and the plant will be fully fertile.

25. To start solving this problem, number the four homologous chromosomes of the autotetraploid that carry the A or a alleles as follows:

$$
\begin{array}{cccc}
1 & 2 & 3 & 4 \\
A & a & A & a
\end{array}
$$

Next, determine the genotypes of the meiotic products resulting from random, two-and-two segregation of the four homologous chromosomes. These are indicated below:

$$1 \quad 2 \quad \Leftrightarrow \quad 3 \quad 4 \qquad 1 \quad 3 \quad \Leftrightarrow \quad 2 \quad 4 \qquad 1 \quad 4 \quad \Leftrightarrow \quad 2 \quad 3$$
$$A \quad a \qquad A \quad a \qquad A \quad A \qquad a \quad a \qquad A \quad a \qquad a \quad A$$

As evident, the *aa* gametes will occur in a frequency of 1/6. The homozygous recessive genotype will occur in a frequency of $(1/6)(1/6) = 1/36$. Therefore, the phenotypic ratio of the dominant phenotype $A \_ \_ \_$ to the recessive phenotype *aaaa* is 35:1.

26. Since this plant is an allotetraploid, the chromosomes designated 1 and 2 will not pair with each other. The two chromosomes designated 1 will pair and the two designated 2 will pair. Therefore, these chromosome pairs will undergo independent assortment. The $a_1 a_2$ gamete is expected to occur in a frequency of 0.25. From the testcross indicated, the frequency of the $a_1a_1 a_2a_2$ progeny is 0.25. The expected phenotypic ratio is 3:1 dominant ($A\_ \_ \_$) to recessive (*aaaa*).

27. Because the chromosome pair designated 1 is homozygous for the dominant allele A_1 in both parents, and the second chromosome pair is homozygous for a_2, all gametes from each parent are $A_1 a_2$. All progeny are $A_1 A_1 a_2 a_2$.

28. To obtain an *aa* gamete from an autotetraploid of the genotype *AAAa* requires crossing over between the gene and the centromere with both recombinant chromosomes going to the same pole at anaphase I of meiosis. This secondary meiocyte will contain two recombinant chromosomes of the genotype *Aa* and *Aa*. This secondary meiocyte will proceed through the second meiotic division to produce two meiotic products. These meiotic products would be expected to be of the genotypes *Aa* and *Aa* 50% of the time and *AA* and *aa* 50% of the time. This is illustrated below.

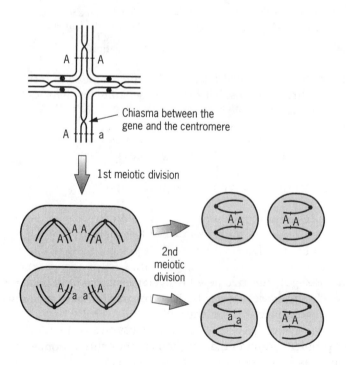

7

Linkage, Crossing Over, and Chromosome Mapping in Eukaryotes

IMPORTANT CONCEPTS

A. Genes that are located on the same chromosome are called linked.
1. Linked genes are an exception to the Principle of Independent Assortment.
2. The intensity of linkage is measured by the frequency of recombination, which has a maximum value of 50%.
   a. The less often recombination occurs, the tighter is the linkage.
      (1) No recombination between linked genes results in complete linkage.
3. Recombination is caused by crossing over between homologous chromosomes after the chromosomes have replicated.
   a. Crossing over apparently occurs during pachynema of meiosis.
   b. Each crossover event is a physical exchange between two of the four chromatids in a tetrad.
   c. Multiple crossovers involving different combinations of chromatids may occur along the length of a tetrad.
   d. Crossovers become visible as chiasmata from diplonema (pachynema in cytologically favorable material) through metaphase I of meiosis.

B. The average number of crossovers on chromosome pairs during meiosis is a measure of genetic distance, calibrated in centiMorgans (cM), and is the basis for constructing chromosome maps.
1. Map distance can be estimated by counting chiasmata in cytological preparations or by calculating the frequency of recombination observed between genes in experimental crosses, e.g., in two- or three-point testcrosses.
2. One chiasmata is equivalent to a distance of 50 cM and one percent recombination is equivalent to one cM, providing that multiple crossovers do not occur.
3. Due to formation of multiple crossovers, recombination frequencies greater than 20 - 25% usually underestimates genetic distance.
   a. Multiple crossovers are rare over short genetic distances. This is due to a phenomenon called interference.

(1) Interference is crossing-over at one point reducing the chance of another crossover occurring nearby.
(2) The amount of interference is measured by the coefficient of coincidence.
4. A recombination map showing the positions of genes on a chromosome is colinear with the physical map of the chromosome; however, recombination distances between genes are not proportional to physical distances.
C. Recombination generates new combinations of alleles, some of which are favorable, upon which selection may act.
1. Recombination and the recovery of recombinant chromosomes is suppressed by heterozygosity for chromosomal rearrangements, especially inversions.

IMPORTANT TERMS

In the space allotted, concisely define each term.

linkage:

chiasma:

crossing over:

linkage phase:

coupling:

repulsion:

tetrad:

map unit:

centiMorgan (cM):

two-point testcross:

three-point testcross:

interference:

coefficient of coincidence:

balancers:

IMPORTANT NAMES

In the space allotted, concisely state the major contribution made by the individual or group.

Thomas Hunt Morgan:

Alfred H. Sturtevant:

W. Bateson and R. C. Punnett:

Curt Stern:

Harriet Creighton and Barbara McClintock:

Calvin B. Bridges and T. M. Olbrycht:

TESTING YOUR KNOWLEDGE

In this section, answer the questions, fill in the blanks, or solve the problems in the space allotted. Problems noted with an * are solved in the Approaches to Problem Solving section at the end of the chapter.

1. The linkage phase in which the dominant alleles of two genes are on one homologous chromosome and the two respective recessive alleles are on the other homolog is called
    _____.

2. The phenomenon of a crossover at one point inhibiting the formation of another crossover in an adjacent region is called _____.

3. The term describing genes located on the same chromosome is _____.

4. The linkage phase in which a dominate allele of one gene and a recessive allele of a second gene are found on one homolog and the respective recessive and dominant alleles are on the other homolog is called _____.

5. If in the hybrid, *E f/e F*, <u>crossing over</u> occurs 40% of the time, then the frequency of the *E F* gamete will be _____.

*6. A dihybrid plant, *Cc Dd*, is self-fertilized and the double recessive, *cc dd*, phenotype occurs in 4% of the progeny.

    (a) What is the expected frequency of the *c D* gamete?

    (b) What is the linkage phase of this dihybrid plant?

*7. It is observed that two heterochromatic knobs on chromosome 6 of maize have a chiasma between them in 20% of the meiocytes studied. What is the genetic map distance between these knobs?

*8. A mouse has a chromosome number 2n = 40. How many linkage groups does the mouse have?

9. A dihybrid plant, *Cc Dd*, is self-fertilized and the double recessive, *cc dd*, phenotype occurs in 16% of the progeny.

    (a) What is the expected frequency of the *c D* gamete?

    (b) What is the linkage phase of this dihybrid plant?

*10. In Upland cotton, the laciniate leaf and brown lint loci are 34 map units apart. What percentage of the meioses is expected to have a chiasma involving the region of the bivalent between these loci?

*11. The following are progeny of a heterozygous tomato plant, *R r Wf wf* (*R* = red fruit, *r* = yellow fruit; and *Wf* = yellow flowers, *wf* = white flowers) testcrossed to a *rr wfwf* plant: 77 red fruit, yellow flowers; 73 yellow fruit, white flowers; 28 red fruit, white flowers; 32 yellow fruit, yellow flowers.

    (a) What is the recombination frequency between the fruit color and flower color loci?

    (b) What is the linkage phase of the alleles in the dihybrid parent?

*12. In cotton, genes located on chromosome 15 give the following recombination frequencies as determined by two-point testcross mapping: L^0-Lg = 50%; L^0-xl = 3%, s -cr = 0.2%; vf -s = 1%; sxl - Lg = 44%; Lg - vf = 5%; and Lg - cr = 6%. What is the relative order of these genes?

*13. A large inversion has been detected in pachytene nuclei of corn that involves about 25% of chromosome 2. A detailed analysis of pachytene chromosomes of plants heterozygous for this inversion showed that an inversion loop forms essentially in all of the meioses. Furthermore, a chiasma occurs in 54% of the inversion loops. How large is this inversion measured in cM map units?

*14. In cotton, the dwarf red mutation (*Rd* = dwarf red, *rd* = green) and the virescent mutation (*v* = virescent, *V* = nonvirescent) mapped about 10 cM apart. List the phenotypes and their frequencies if a *Rd V / rd v* plant is allowed to self-pollinate.

*15. In *Drosophila* , rose eyes (*rs*), dichaete wings (*Dt*) and curled wings (*cu*) reside at the following positions on chromosome 3.

rs	*Dt*	*cu*
35	40	50

List all gametes and their frequencies that will be produced by a female fly of

the genotype $\dfrac{rs^+ \quad\quad Dt^{+} \quad cu^+}{rs \quad\quad Dt \quad cu}$ assuming a coefficient of coincidence of 0.90.

*16. Rabbits fully heterozygous at three loci, $B\,b\;c^{ch}\,c^h\;S\,s$ (B = black, b = brown; c^{ch} = chinchilla, c^h = himalayan; S = long fur, s = short fur) are testcrossed and the 200 progeny are tabulated.

Phenotype			Number
black	himalayan	long	35
brown,	chinchilla	long	36
black	himalayan	short	35
brown	chinchilla	short	37
black	chinchilla	long	13
brown	himalayan	long	14
black	chinchilla	short	13
brown	himalayan	short	17

(a) Which genes are linked? Which gene, if any, is not linked to the others?

(b) Indicate the map distance between all linked genes.

*17. The progeny of a *Drosophila* female (heterozygous at three loci, $y^+y\;ct^+ct\;w^+w$) crossed to a wild-type male are listed as follows:

	Phenotype			Number
females:	y^+	ct^+	w^+	2000
males:	y^+	ct^+	w	773
	y	ct	w^+	782
	y	ct^+	w^+	201
	y^+	ct	w	209
	y^+	ct	w^+	15
	y	ct^+	w	16
	y	ct	w	3
	y^+	ct^+	w^+	1

(a) On which chromosome do these genes reside?

(b) What is the gene order and gene arrangement in the trihybrid female parent?

(c) What are the map distances between these genes?

(d) Calculate the coefficient of coincidence and the value of interference for this cross.

*18. In tomato the genes for jointless (*j*), hairless (*hl*)and leafy (*lf*) are known to be linked on chromosome 5. Plants heterozygous for these three gene pairs, j^+j hl^+hl lf^+lf were testcrossed and the progeny listed:

Phenotypes of progeny			Number
jointed	hairless	normal	142
jointless	hairy	leafy	139
jointed	hairy	leafy	9
jointless	hairless	normal	10
jointed	hairy	normal	96
jointless	hairless	leafy	103
jointless	hairy	normal	1
			500

(a) What is the correct sequence of genes in the linkage group?

(b) What is the correct linkage phase of the trihybrid parent?

(c) What are the map distances between the loci?

(d) What is the coefficient of coincidence and the value of interference for this cross?

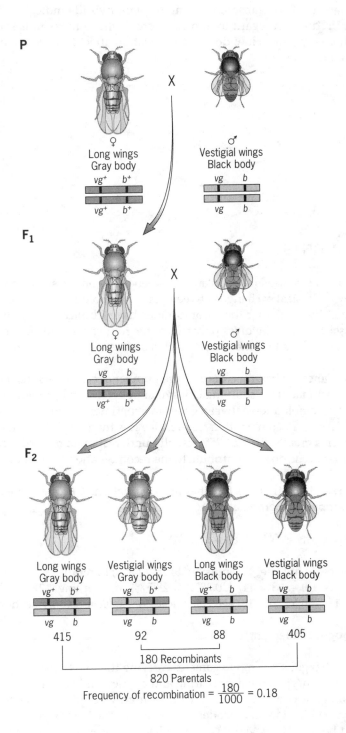

Figure 7.14 An experiment involving two linked genes, *vg* (vestigial wings) and *b* (black), in
Drosophila. The frequency of recombination between the linked genes is determined by
analyzing the testcross progeny and dividing the number of recombinants by the total
number of flies.

THOUGHT CHALLENGING EXERCISE

Meiotic cytogenetic analysis of chromosome pairing of plant hybrids indicates that different populations and species often differ in regard to translocations in the homozygous state. Explain how the translocation became homozygous. How does a translocation affect linkage relationships? How might these be of selective advantage?

SUMMARY OF KEY POINTS

The frequency of recombination is a measure of linkage between genes on the same chromosome. Recombination is caused by a physical exchange between paired homologous chromosomes early in prophase of the first meiotic division, after the chromosomes have duplicated. At any one point, the process of exchange — crossing over —involves only two of the four chromatids in a meiotic tetrad. Later in prophase I, these crossovers become visible as chiasmata.

The genetic maps of chromosomes are based on the average number of crossovers that occur during meiosis. Map distances are estimated by counting chiasmata in cytological preparations or by calculating the frequency of recombination between genes in experimental crosses. One chiasma is equivalent to a distance of 50 cM. Recombination frequencies less than 20-25% estimate map distance directly. However, frequencies greater than 20-25% usually underestimate map distance because some of the multiple crossovers that occur do not contribute to the recombination frequency.

Recombination can bring favorable mutations together in a population. Chromosome rearrangements, especially inversions, can suppress recombination.

ANSWERS TO PROBLEMS AND PROBLEMS

1) coupling 2) interference 3) linkage 4) repulsion 5) 10 6) a. 3; b. repulsion or *trans* 7) 10 cM 8) 20 9) a. 0.1, b. coupling or *cis* 10) $\cong 68\%$ 11) a. 0.2857 or 28.57%, b. coupling 12) L^0-*sxl* - *cr* - *s* - *vf* - *Lg* or L^0- *sxl* - *Lg* - *wf* - *s* - *cr* 13. 27 cM 14 - 18) see *Approaches to Problem Solving* section.

Some ideas concerning thought challenging exercise:

A translocation becomes homozygous when two gametes carrying the two translocated chromosomes fuse. These chromosomes show normal bivalent pairing and are new linkage groups. The original translocation in the heterozygous condition may have kept highly adaptive allele combinations on the chromosomes from being recombined by independent assortment. The heterozygous plant displays semisterility (which would be disadvantageous) due to adjacent segregation of the translocation quadrivalents. The plants that are homozygous for the translocation would retain the advantageous alleles and would be fully fertile. Geographic isolation from the parental population would be favorable to maintain the translocation homozygosity, because it minimizes hybridization with the normal plants that would generate translocation heterozygotes.

APPROACHES TO PROBLEM SOLVING

6. (a) The homozygous recessive phenotype (*cc dd*) occurred in a frequency of 4% or 0.04. Since the *cc dd* genotype results from the fusion of two *c d* gametes, the frequency of the *c d* gamete must be 0.2, $(0.2 \times 0.2 = 0.04)$ This frequency is less than 0.25 and thus *c d* is a recombinant. The other recombinant gamete *C D* also should occur in a frequency of 0.2. This leaves 0.6 of the gametes as nonrecombinants. *c D* is one of the two nonrecombinant types. Its frequency is 0.3.

   (b) From the calculations and reasoning involved in part (a), we can assign the linkage phase of the dihybrid parent as *c D/C d*. This is the repulsed or *trans* linkage phase.

7. Chiasmata represent crossovers. Because only two of the four chromatids are involved in crossing over at any one point, the amount of recombination is 1/2 the frequency of crossing-over. One map unit (cM) equals 1% recombination. Therefore, the map distance between the knobs is 10% or 10 cM.

8. The number of linkage groups is determined by the number of different chromosomes. The mouse has (2 n = 40; 1 n = 20) 20 different chromosomes or 20 linkage groups.

10. The 34 map units is 34% recombination. Since recombination is 50% of the frequency of crossing-over, we would expect to see 68% of the meioses with a single chiasma between the laciniate leaf and brown lint loci.

11. When the data are analyzed it is observed that yellow (73 + 32 = 105) and red (77 + 28 = 105) fruit segregate in a 1:1 ratio, and yellow (77 + 32 = 109) and white (73 + 28 = 111) flowers segregate in a 1:1 ratio. Therefore the deviation away from a 1:1:1:1 testcross ratio is not due to inviability of certain genotypes or factors that interfere with segregation of the genes, but is caused by linkage.

   (a) To calculate the recombination ratio between these genes first add the individuals to obtain a total (77 + 73 + 28 +32 = 210). The two most abundant phenotypes (77 red fruit, yellow flowers; 73 yellow fruit white flowers) are the parental types or nonrecombinants. The two less frequent phenotypes (28 red fruit, white flowers; 32 yellow fruit, yellow flowers) are the recombinants. The recombination frequency is calculated by dividing the number of recombinants (28 + 32 = 60) by the total number of progeny (210). This is 0.2857 or 28.57%.

   (b) The nonrecombinant chromosomes from the dihybrid parent were R Wf and r wf. Therefore, the genotype of the dihybrid was R Wf and the linkage phase was coupling.
   r wf

12. To work this problem it is important to consider the additive nature of the map distances. One way to start the problem is to consider the two genes that are farthest apart in terms of map recombination distance, i.e., L^0 -Lg. = 50%. But 50% is the maximum amount of recombination and is not definitive by itself. Next consider *sxl - Lg* as a starting point. Add L^0 to the map. It will be 3 units to the left of *sxl*, so as to lengthen the distance between *Lg* and L^0. Bring in the additional markers, again trying to arrange them in an additive fashion.

sxl ———————————— 44 ————————————— *Lg*

L^0 3 *sxl*   *cr*   6   *Lg*

                                               cr .2 *s*   *vf*   5   *Lg*

                                                    s 1 *vf*

Not enough data are provided to assemble an unambiguous sequence. The map order is likely either

$$L^0\text{ -}sxl - cr - s - vf - Lg \text{ or } L^0 - sxl -Lg - vf - s - cr$$

13. Since a chiasma occurs within the inversion loop in 54% of the meioses, this represents 54% crossing-over within the inversion loop. Since the amount of recombination is 1/2 the frequency of crossing-over, this equates to 27% recombination. The inversion loop, therefore, covers 27 cM.

14. Since the dihybrid is in the coupled phase, the recombinant chromosomes will be $\underline{Rd\ \ v}$ and $\underline{rd\ \ V}$. The loci are 10 cM apart. Therefore, each recombinant will occur in a frequency of 0.05. The parental types will make up 90% of the progeny. The $\underline{Rd\ \ V}$ and $\underline{rd\ \ v}$ gametes each will occur at a frequency of 0.45. The phenotypic frequency of the progeny from selfing the $\underline{Rd\ \ V}$ plant can be determined by using a Punnett square.

	$\underline{Rd\ \ V}$ (0.45)	$\underline{rd\ \ v}$ (0.45)	$\underline{Rd\ \ v}$ (0.05)	$\underline{rd\ \ V}$ (0.05)
$\underline{Rd\ \ V}$ (0.45)	RdRd VV (.2025)	Rdrd Vv (.2025)	RdRd Vv (.0225)	Rdrd VV (.0225)
$\underline{rd\ \ v}$ (0.45)	Rdrd Vv (.2025)	rdrd vv (.2025)	Rdrd vv (.0225)	rdrd Vv (.0225)
$\underline{Rd\ \ v}$ (0.05)	RdRd Vv (.0225)	Rdrd vv (.0225)	RdRd vv (.0025)	Rdrd Vv (.0025)
$\underline{rd\ \ V}$ (0.05)	Rdrd VV (.0225)	rdrd Vv (.0225)	Rdrd Vv (.0025)	rdrd VV (.0025)

The phenotypes can then be summed as follows.

Red, nonvirescent ($Rd\_\ V\_$): .2025 + .2025 + .0225 + .0225 + .2025 + .0225 + .0025 + .0225 + .0025 = .7025

Red, virescent ($Rd\ \_\ vv$): .0225 + .0225 + .0025 = .0475

green, nonvirescent ($rdrd\ V\_$): .0225 + .0225 +.0025 = .0475

green, virescent ($rdrd\ vv$): .2025

15. To start solving this problem first calculate the frequency of the double recombinant gametes. These are expected to occur at a frequency of recombination between rs and Dt (0.05) X frequency of recombination between Dt and cu (0.10) X the coefficient of coincidence (0.9). The frequency of the two double recombinants should total 0.0045.

rs^+ Dt $cu^+ = .00225$

rs Dt^+ $cu = .00225$

The recombination between the rose and dichaete locus is .05. Since the map distance includes the double recombinants, we must subtract the frequency of the double recombinants (.0045) from .05. The frequency of these single recombinants total .0455.

rs^+ Dt $cu = .02275$

rs Dt^+ $cu^+ = .02275$

The recombination between the dichaete and curled loci is .10. Again, subtract the frequency of the double recombinants to obtain a frequency of the single recombinant gametes of .0955.

rs^+ Dt^+ cu = .04775

rs Dt cu^+ = .04775

The remaining gametes will be the nonrecombinants. These are obtained by subtracting the total number of recombinants from one. NR = (1 - [.0045 + .0455 + .0955]) = .8545.

rs^+ Dt^+ cu^+ = .42725

rs Dt cu = .42725

16. (a) When you look at the data it is seen that the eight phenotypic classes are distributed in two groups of four. This suggests that two genes are linked and a third gene is segregating independently. The key to solving the problem is to detect the linked genes. It may be easier to convert the phenotypes to allelic symbols.

$$
\begin{array}{cccc}
B & c^h & S & 35 \\
b & c^{ch} & S & 36 \\
B & c^h & s & 35 \\
b & c^{ch} & s & 37 \\
B & c^{ch} & S & 13 \\
b & c^h & S & 14 \\
B & c^{ch} & s & 13 \\
b & c^h & s & 17 \\
\end{array}
$$

The linked genes are B and c. S segregates independently. This is apparent if the data are rearranged as follows:

$\underline{B \quad c^h}$ S 35	$\underline{b \quad c^{ch}}$ S 36	$\underline{B \quad c^{ch}}$ S 13	$\underline{b \quad c^h}$ S 14
$\underline{B \quad c^h}$ s 35	$\underline{b \quad c^{ch}}$ s 37	$\underline{B \quad c^{ch}}$ s 13	$\underline{b \quad c^h}$ s 17

(b) The nonrecombinants are $\underline{B \quad c^h}$ and $\underline{b \quad c^{ch}}$ and the recombinants are $\underline{B \quad c^{ch}}$ and $\underline{b \quad c^h}$. The recombinants total 57. There are 198 total progeny. The amount of recombination = 57/198 = .2879. The map distance between these linked genes is 28.79 cM.

17. (a) This problem involves sex-linked genes. This is apparent from the trihybrid female X wild-type male producing all wild-type female progeny, but various phenotypic combinations among male progeny.

(b) To determine gene order requires recognizing nonrecombinant and double recombinant phenotypes among the male progeny. The 2,000 male progeny are distributed in eight phenotypes, the two most frequent (y^+ ct^+ w and y ct w^+) representing the parental or nonrecombinant chromosomes, and the two least frequent (y ct w and y^+ ct^+ w^+) the double recombinant chromosomes that came from the trihybrid female. The genotype of the trihybrid parent in terms of the linkage phases is:

$$\frac{y^+ \quad ct^+ \quad w}{y \quad ct \quad w^+}$$

A simple way to determine the order of the genes is to look at the genotypes (in regard to the chromosome received from the trihybrid) of the double recombinant phenotypes among the progeny.

$$\underline{y \quad ct \quad w}$$

and

$$\underline{y^+ \quad ct^+ \quad w^+}$$

Since a double crossover switches the middle gene pair, look for the gene pair that would have switched in the trihybrid designated above to generate the double recombinants. This gene pair is w/w^+. Therefore the white locus is positioned between yellow and cut. The correct gene order and genotype of the trihybrid female follows:

$$\frac{y \qquad w^+ \qquad ct}{y^+ \qquad w \qquad ct^+}$$

(c) The map distances are determined by first summing the phenotypes representing single recombinants between the loci. Remember that double recombinants result from recombination in both regions simultaneously and must be included. The single recombinants between the yellow and white loci are $y \ w \ ct^+$ (16) and $y^+ \ w^+ \ ct$ (15), and the single recombinants between the white and cut loci are $y \ w^+ \ ct^+$ (201) and $y^+ w \ ct$ (209).

$$\frac{y \qquad w^+ \qquad ct}{y^+ \qquad w \qquad ct^+}$$

15	201
16	209
3	3
1	.1
35	414

The map distance between the yellow and white loci is $(35/2{,}000)(100) = 1.75$ cM.

The map distance between the white and cut loci is $(414/2{,}000)(100) = 20.7$ cM.

(d) The coefficient of coincidence (CC) is calculated by the formula

$$CC = \frac{\text{observed number of double recombinants}}{\text{expected number of double recombinants}} = \frac{.002}{(.207)(.0175)} = \frac{.002}{.0036} = .56$$

Interference = 1 - CC = 1 - .56 = .44

18. To solve this problem it is beneficial to substitute allelic symbols for the phenotypes. Also remember that in a testcross the phenotypes of the progeny indicate the alleles that came from the testcross parent (trihybrid in this case). It follows that the following chromosomes in the progeny were from the trihybrid parent:

j^+	h l	l f$^+$	142
j	h l$^+$	l f	139
j^+	h l$^+$	l f	9
j	h l	l f$^+$	10
j^+	h l$^+$	l f$^+$	96
j	h l	l f	103
j	h l$^+$	l f$^+$	1
j^+	h l	l f	0
			500

One genotype did not occur. This was one of the double recombinants and so it was put in the table with a frequency of zero. The two most frequent phenotypes represent the parental types. These can be put together to reconstruct the genotype of the trihybrid as indicated below:

$$\frac{j^+ \quad h\,l \quad l\,f^+}{j \quad h\,l^+ \quad l\,f}$$

The gene order can be immediately determined by comparing the trihybrid indicated above with the genotype of the double recombinants. The double recombinants, the least frequent phenotypes are:

$$\underline{j \quad h\,l^{+-} \quad l\,f^+}$$

and

$$\underline{j^+ \quad h\,l \quad l\,f}$$

The gene pair that was switched to produce the the genotypes of the double recombinants from the trihybrid was lf^+/lf. Therefore, the leafy gene must be the middle gene.

(a) The correct sequence of genes must be $j \quad \underline{lf} \quad \underline{hl}$.

(b) The correct linkage phase is $\dfrac{j^+ \quad\quad lf^+ \quad\quad hl}{j \quad\quad lf \quad\quad hl^+}$

(c) The single recombinants between the jointless and leafy loci are $\underline{j \quad lf^+ \quad hl}$ (10) and $\underline{j^+ \quad lf \quad hl^+}$ (9). The double recombinants (1 and 0) also are added to make the total number of recombinants between these loci 20. The frequency of recombination in this region is 20/500 = .04 or 4 cM.

The single recombinants between the leafy and the hairless loci are $\underline{j^+ \quad lf^+ \quad hl^+}$ (96) and $\underline{j \quad lf \quad hl}$ (103). The double recombinants (1 and 0) also are added to make the total number of recombinants between these loci 200. The frequency of recombination in this region is 200/500 = .40 or 40 cM.

(d) The CC = $\dfrac{.002}{(.40)(.04)}$ = .125.   Interference = 1 - CC = 1 - .125 = .875.

8

Advanced Linkage Analysis

DETECTION OF LINKAGE IN EXPERIMENTAL ORGANISMS
    Testcross Analysis to Detect Linkage in Fungi
    Balancer Chromosome Technique to Assign a Gene to a Chromosome in *Drosophila*

SPECIALIZED MAPPING TECHNIQUES
    Centromere Mapping with Ordered Tetrads in *Neurospora*
    Cytogenetic Mapping with Deletions and Duplications in *Drosophila*

LINKAGE ANALYSIS IN HUMANS
    Detection of Linked Loci by Pedigree Analysis
    Somatic Cell Genetics: An Alternative Approach to Gene Mapping
    The Human Gene Map

IMPORTANT CONCEPTS

A. Linkage can be readily detected in organisms amenable to tetrad analysis.
1. In two-point crosses in yeast, for example, linkage is detected when parental ditype tetrads are more frequent than nonparental ditype tetrads.
2. The ordered tetrads of *Neurospora* make it possible to map centromeres relative to genes.
    a. The centromere-to-gene distance is one-half the frequency of the second division segregation tetrads, because only half of the spores of a tetrad arrangement resulting from crossing over are recombinant.

B. Several procedures are available to study linkage in *Drosophila*.
1. Balancer chromosome stocks are used to determine which chromosome carries a particular gene.
2. Genes can be localized on the polytene chromosome maps by testing recessive mutations against cytologically defined deletions and duplication.
    a. A deletion will reveal the phenotype of a recessive mutation located between its endpoints, whereas a duplication will conceal the mutant phenotype.

C. Until recently, human gene mapping has been difficult because of uninformative pedigrees and small number of progeny.
1. Pedigree analysis has on occasion yielded information about the linkage between two genes.
    a. A statistical procedure called the *lod score method* assesses data from pedigrees and estimates the likelihood that two genes are linked.
    b. Interspecific somatic cell hybridization techniques, developed during the 1960s, allow researchers to link a gene to a specific autosome.
        (1) Using chromosome rearrangements such as translocations and large deletions, in cell hybrid lines, researchers can identify the specific chromosome and region of the chromosome that carries a gene.
        (2) In deletion mapping, a gene's position can be determined by correlating absence of a gene product with the absence of a chromosome segment.
        (3) In duplication mapping, a gene's position can be determined by correlating excess copies of a gene with a duplicated chromosome segment.

D. The Human Genome Project is a massive international effort to map and sequence all the genes in the human genome.

IMPORTANT TERMS

In the space allotted, concisely define each term.

Huntington's disease:

parental ditype:

nonparental ditype:

tetratype:

ordered tetrads:

first division segregation:

second division segregation:

deletion mapping:

duplication mapping:

lod score method:

somatic cell hybridization (cell fusion):

HAT medium:

gel electrophoresis:

Duchenne's muscular dystrophy (DMD):

Human Genome Project:

IMPORTANT NAMES

In the space allotted, concisely state the major contribution made by the individual or group.

George Beadle:

Burke Judd, Margaret Shen, and Thomas Kaufman:

J. H. Renwick and S. D. Lawler:

N. E. Morton:

G. Barski and B. Ephrussi:

Oscar J. Miller:

Louis Kunkel:

Ron Worton:

TESTING YOUR KNOWLEDGE

*In this section, answer the questions, fill in the blanks, and solve the problems in the space allotted. Problems noted with an * are solved in the Approaches to Problem Solving section at the end of the chapter.*

1. The occurrence of two or more loci on the same chromosome, without regard to the distance between them is called _____.

2. In fungi, what is the designation for a tetrad of spores that contains four different types, e.g., *AB, Ab, aB, ab* ?

3. In fungi, a tetrad containing two kinds of spores, e.g., 2 *AB* and 2 *ab*, is called a _____.

4. In analysis of linkage by pedigree analysis in humans is it easier to detect linkage of sex-linked genes or autosomally inherited genes?

5. Which of the following genes or DNA sequences can <u>not</u> be easily detected and mapped following somatic cell hybridization between human and mouse cells?

    (a) gene coding an enzyme that has an assay to visualize its electrophoretic migration distance.
    (b) electrophoretically detectable DNA sequence that shows polymorphisms between the human and mouse genomes.
    (c) DNA sequence that has not diverged between the human and mouse genomes.
    (d) genes coding for enzymes that show identical electrophoretic migration rates.
    (e) both c and d.
    (f) none of the above; all can be used with equal efficacy in somatic cell genetics.

*6. Suppose that you discover a new yellow eye color mutation in *Drosophila*. You then cross a recessive yellow-eyed female to a fly from a *Cy/Pm; Tb/Sb* tester stock. The male, but not female, progeny had yellow eyes. On which chromosome is the yellow eye mutant located?

*7. Suppose that you discover another new recessive mutant, this one causing blue eyes in *Drosophila*. You cross a blue-eyed female fly with a male from the $Cy/Pm; Tb/Sb$ tester stock . The F_1 are all wild -type for eye color. Curly, tubby F_1 flies are selected and backcrossed as the male parent to the blue-eyed female. The following progeny are obtained:

> 98 curly, tubby
> 93 curly, blue
> 99 tubby
> 95 blue

On which chromosome is the blue-eyed gene located?

*8. The following pedigree is for a family in which Duchenne's muscular dystrophy (DMD) and colorblindness are segregating. For this problem assume that the DMD and colorblindness loci are 10 cM apart on the X chromosome.

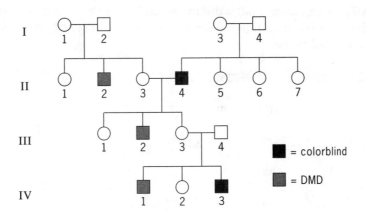

= colorblind

= DMD

(a) What is the probability if III-3 and III-4 have another son that he will have DMD?

(b) What is the probability if III-3 and III-4 have another son that he will have both colorblindness and DMD?

*9. A geneticist is researching the genetics of a previously unstudied dipteran insect. Recessive alleles of six different loci have been found that map in the following relative order on an unidentified chromosome: <u>eb   p   m   c   sl   ov</u> (eb = ebony body, p = pink eyes, m = miniature wings, c = curly wings, sl = short legs, ov =oval eyes). Two other dominant mutants, (Pm = plumb eye color, and Wl = wingless) were detected that are lethal when homozygous. A study of polytene chromosomes identified Pm as a deletion of bands 87 - 95 on chromosome 2, and Wl as a deletion of bands 22 - 30 on the same chromosome (band 1 is near one end of the chromosome). The geneticist crosses a short-legged fly with a plum-eyed fly and one-half of the progeny have plum eyes and short legs. The other one-half of the progeny are wild-type. A cross between a pink eye and a wingless fly produce two types of progeny, one-half wingless pink and one-half wild type. What do these crosses uncover about the chromosomal location and orientation of the linkage group, eb — ov?

*10. A new spore color mutant, maroon (m) is discovered in *Neurospora*. A maroon-spored fungus is crossed to a normal gray spore color (m⁺) fungus. The segregation patterns of the spores in individual asci are listed below:

<u>Number</u>	<u>Genotype of spore pairs</u>			
39	m	m	m^+	m^+
40	m^+	m^+	m	m
9	m^+	m	m^+	m
8	m	m^+	m	m^+
7	m^+	m	m	m^+
6	m	m^+	m^+	m

What is the distance in cM between the centromere and the gene determining maroon and gray spores?

*11. The following data are from a cross of a wild-type *Neurospora* with a strain that has purple spores (*p*) and shiny spore coats (*s*). The following ordered tetrads of sexual spores resulted:

	Spore pairs			Number
1	2	3	4	of asci
p +	*p* +	+ *s*	+ *s*	15
p s	*p s*	+ +	+ +	99
p s	*p* +	+ *s*	+ +	64
p s	+ *s*	*p* +	+ +	3
p s	+ +	*p s*	+ +	17
p +	+ *s*	*p* +	+ *s*	2
p s	+ +	+ *s*	*p* +	4
				204

(a) What is the map distance between purple and shiny?

(b) What is the distance between purple and the centromere?

(c) What is the map distance between shiny and the centromere?

*12. The following table summarizes the human enzymes and human chromosomes that are present in stabilized clones derived from human/mouse somatic cell hybrids.

	Clone designation				
Human enzymes	A	B	C	D	E
ADK	+	+	+	-	+
CAT	+	-	-	-	-
ESD	-	-	+	+	-
Human chromosomes					
16	-	+	-	-	+
10	+	+	+	-	+
11	+	-	-	-	-
13	-	-	+	+	-

Assign the gene coding for each of the enzymes to a specific human chromosome.

*13. The following table summarizes the human enzymes and human chromosomes that are present in stabilized clones derived from human/mouse somatic cell hybrids.

	Clone designation					
Human enzymes	A	B	C	D	E	F
hepatic lipase	+	-	+	+	-	+
esterase D	+	+	-	-	-	-
catalase	-	-	+	+	+	+
alkaline phosphatase	+	+	+	-	+	+

Human chromosomes

2	q	q	+	p	+	+
11	-	q	+	+	+	p
13	q	+	-	-	-	p
14	+	+	+	-	-	-
15	q	-	+	+	-	+

+ = the presence of an enzyme or chromosome; - = the absence of an enzyme or chromosome. p = short arm of chromosome only; q = long arm of chromosome only.

Indicate the chromosome and arm (if possible) that house the gene coding for each of the above enzymes.

THOUGHT CHALLENGING EXERCISE

A student is studying the inheritance of a newly induced recessive mutation of *Drosophila*, and would like to quickly and unambiguously assign the gene to a linkage group and chromosome. Balancer chromosome stocks to carry out the analysis are not readily available and the student does not want to waste the time necessary to order and have them mailed. However, mutant stocks representing one and only one gene on each of the three autosomes are maintained in the laboratory. How could the new mutant be unambiguously assigned to a chromosome?

KEY FIGURE

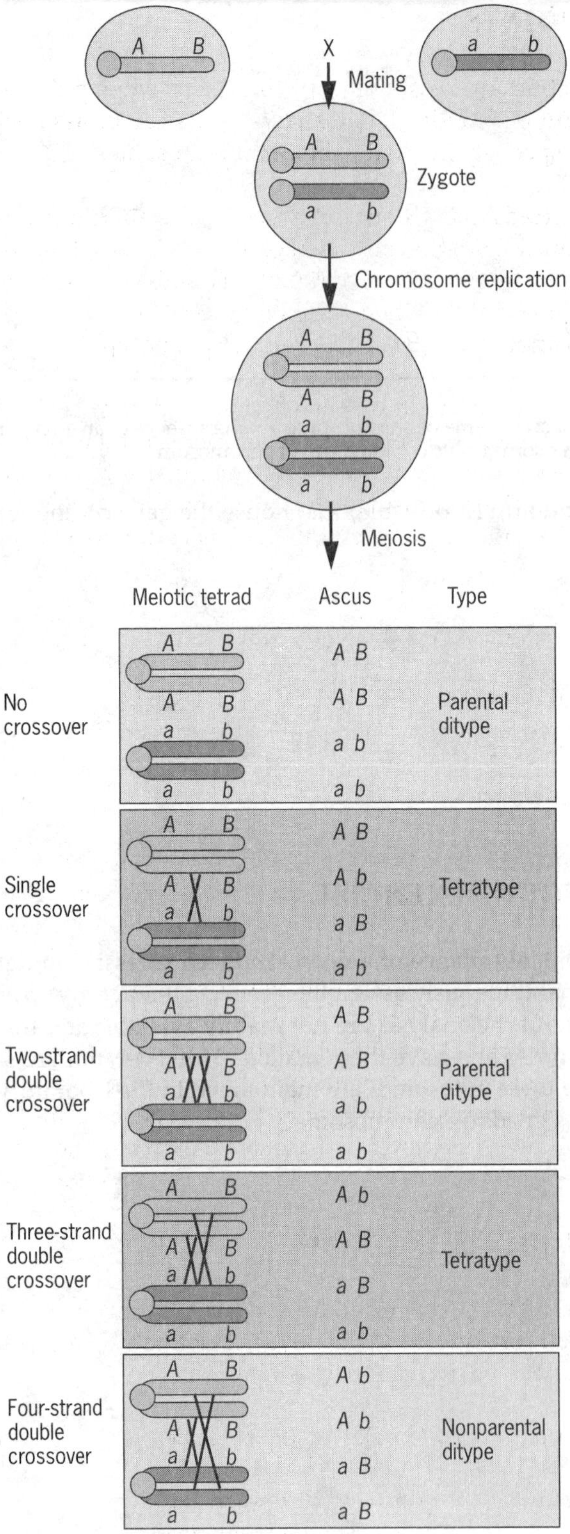

Figure 8.3 Cross with linked yeast mutations. The frequency of parental and nonparental ditype asci are expected to be unequal (they would be expected to be equal if the genes were not linked). The nonparental ditype asci, which are produced by four-strand double crossovers, should be much less numerous than the parental ditype asci.

SUMMARY OF KEY POINTS

In organisms amenable to tetrad analysis, linkage is indicated when parental ditype tetrads are more frequent than nonparental ditype tetrads. In *Drosophila*, genes can be assigned to chromosomes by analyzing crosses involving specially marked balancer chromosomes.

In organisms with ordered tetrads, the distance between a gene and the centromere is one-half the frequency of tetrads that show second division segregation. In Drosophila, genes can be localized on the polytene chromosome maps by testing recessive mutations against cytologically defined deletions and duplications. A deletion will reveal the phenotype of a recessive mutation located between its endpoints, whereas a duplication will conceal the mutant phenotype.

The lod score is a statistical method that tests genetic marker data in human families to determine whether two loci are linked. A gene can be located to a particular human chromosome by correlating its sequence or product with the human chromosomes that are retained in interspecific somatic cell hybrids. This location can be refined by using somatic cell hybrids that carry rearranged human chromosomes. The Human Genome Project is an international effort to map and sequence the entire human genome.

ANSWERS TO QUESTIONS AND PROBLEMS

1) synteny 2) tetratype 3) ditype 4) sex-linked genes 5) e 6) X-chromosome 7) chromosome 3 8) a. 50%, b. 5% 9 - 13) see *Approaches to Problem Solving* section below.

Ideas on thought challenging exercise:

Male *Drosophila* do not undergo recombination of linked genes because no crossing-over occurs in the males. They do undergo recombination due to independent assortment. Therefore, the fastest way to assign a gene to a chromosome is to cross a mutant female fly with a fly from a line that is homozygous for mutant genes marking chromosomes 2, 3, and 4. If your mutant is recessive and sex-linked, all the F_1 males will show the phenotype but the female will be normal. Select the F_1 dihybrid and use as a male parent in a testcross. If the gene is sex-linked you will observe all of the progeny with the mutant phenotype. The autosomally inherited genes will be assorting independently. If the gene of interest is autosomally inherited, it will segregate independently of two of the three genes of the tester line. But, the gene of interest will show no recombination with one of the genes. This gene is located on the same chromosome as the mutant and we therefore can designate the corresponding chromosome that houses the gene.

APPROACHES TO PROBLEM SOLVING

6. The gene must be located on the X-chromosome because the progeny of the yellow eye female to the $Cy/Pm; Tb/Sb$ tester stock consisted of yellow-eyed males and wild-type eye females. The pertinent part of the cross can be diagrammed as follows:

    Parents   ($y\,y$ female) X ($y^+/\,\grave{}\,$ male)

    Progeny y^+y = wild-type female
    $y\,/\,\grave{}\,$ = yellow-eyed males

7. When the F_1 curly, tubby flies were backcrossed to the blue female parent, the association of the mutant blue phenotype with curly but not tubby indicates that the blue eye gene is located on chromosome 3. The cross can be illustrated as follows:

P $\dfrac{Cy^+ \quad Pm^+ \quad b \quad Tb^+ \quad Sb^+}{Cy^+ \quad Pm^+ \quad b \quad Tb^+ \quad Sb^+}$ X $\dfrac{Cy \quad Pm^+ \quad b^+ \quad Tb \quad Sb^+}{Cy^+ \quad Pm \quad b^+ \quad Tb^+ \quad Sb}$

       blue eyes   curly, plum; tubby, stubble

F₁   Select curly, tubby males and backcross them to the blue-eyed female.

    $\dfrac{Cy^+ \quad Pm^+ \quad b \quad Tb^+ \quad Sb^+}{Cy \quad Pm^+ \quad b^+ \quad Tb \quad Sb^+}$ X $\dfrac{Cy^+ \quad Pm^+ \quad b \quad Tb^+ \quad Sb^+}{Cy^+ \quad Pm^+ \quad b \quad Tb^+ \quad Sb^+}$

Among the progeny the blue phenotype appears with curly, but not tubby. This means that the *b* allele is located on the same chromosome (#3) as is *Tb*.

    $\dfrac{Cy^+ \quad Pm^+ \quad b \quad Tb^+ \quad Sb^+}{Cy \quad Pm^+ \quad b \quad Tb^+ \quad Sb^+}$

8. (a) Since III-3 and III-4 already have one son with DMD, it means that III-3 is heterozygous for the DMD allele. Since the DMD gene is X-linked recessive, III-3 and III-4 have a 50% chance that if they parent another son he will have DMD.

(b) The father of III-3 did not have DMD, but III-3's mother had a brother with DMD. Therefore, III-3 inherited the chromosome carrying the DMD from her mother. The X-chromosome of III-3 carrying the recessive colorblind allele came from her father who was colorblind. Since the DMD and the colorblind loci are 10 cM apart, we would expect the one recombinant carrying both the DMD and colorblind alleles to occur with a frequency of 0.05. Therefore, if III-3 and III-4 have another son, the probability that he will be both colorblind and DMD is 5%.

9. Because the short legged phenotype shows up with plum eyes in half of the progeny of the cross between short-legged flies and plum eye flies, the deletion causing the plum phenotype must include the region occupied by short-legged. The short-legged allele was expressed in the hybrid because it is hemizygous. Therefore, the *sl* locus must be located somewhere in the region containing bands 87 - 95. When the pink eye flies are crossed to wingless flies, the wingless progeny have pink eyes. This indicates that the recessive *p* allele is expressed hemizygously and is located somewhere in the region (bands 22 - 30) that is deleted in the *Wl* mutant. The linkage group covering *eb* to *ov* is oriented with *ov* being closest to the end of the chromosome. It can also be concluded that the linkage group covers a minimal distance from bands 22 - 30 to bands 87 - 95.

10. The first step in solving this problem is to designate the genotypic patterns of the spore pairs as 1st division (no crossover between the gene and centromere) or 2nd division (crossover between the gene and the centromere) segregation.

Number	Genotype of spore pairs				Division of segregation
39	*m*	*m*	*m+*	*m+*	1st
40	*m+*	*m+*	*m*	*m*	1st
9	*m+*	*m*	*m+*	*m*	2nd
8	*m*	*m+*	*m*	*m+*	2nd
7	*m+*	*m*	*m*	*m+*	2nd
6	*m*	*m+*	*m+*	*m*	2nd

The total number of asci scored was 109, thirty of which resulted from crossing-over between the gene and the centromere. Since only two of the four chromatids are involved in a single crossover, The frequency of recombination is 1/2 the frequency of second division segregation:

Recombination = $\dfrac{0.5 \text{ (freq. of asci showing 2nd division segregation)}}{\text{total number of asci scored}} = \dfrac{0.5(30)}{109} = 0.1376 = 13.76$ cM.

11. This cross was between $p\ s$ and $+\ +$ strains. The recombinants will be $p\ +$ and $+\ s$. The tetrads are classified as parental ditype (PD), nonparental ditype (NPD), or tetratype (T). They can also be classified as 1st or 2nd division segregation patterns with respect to each marker.

Spore pairs				Number	Type of	segregation	segregation
1	2	3	4	of asci	ascus	for p	for s
$p\ +$	$p\ +$	$+\ s$	$+\ s$	15	NPD	1st division	1st division
$p\ s$	$p\ s$	$+\ +$	$+\ +$	99	PD	1st division	1st division
$p\ s$	$p\ +$	$+\ s$	$+\ +$	64	T	1st division	2nd division
$p\ s$	$+\ s$	$p\ +$	$+\ +$	3	T	2nd division	1st division
$p\ s$	$+\ +$	$p\ s$	$+\ +$	17	PD	2nd division	2nd division
$p\ +$	$+\ s$	$p\ +$	$+\ s$	2	NPD	2nd division	2nd division
$p\ s$	$+\ +$	$+\ s$	$p\ +$	4	T	2nd division	2nd division
				204			

(a) We analyze the data and conclude that the genes are linked because the NPD (15 + 2 = 17) are much less numerous than the PD tetrads (99 + 17 =116). The standard mapping formula can be used to estimate the distance between p and s.

$$\dfrac{(1/2)\ T + NPD}{\text{total}} = \dfrac{(1/2)\ (71) + 17)}{204} = .257 = 25.7 \text{ cM}$$

(b) The distance of p from the centromere is calculated as:

$$\dfrac{(1/2)\ (\text{second division segregation})}{\text{total asci scored}} = \dfrac{(1/2)(26)}{204} = 0.064 = 6.4 \text{ cM}$$

(c) The distance of s from the centromere is calculated as:

$$\dfrac{(1/2)\ (\text{second division segregation})}{\text{total asci scored}} = \dfrac{(1/2)(87)}{204} = 0.213 = 21.3 \text{ cM}$$

The genes are linked on opposite arms of the chromosome as follows:

$$p \text{ — } 6.4 \text{ — centromere —} 21.3 \text{ — } s$$

12. This type of problem is solved by simply correlating the presence and absence of a human gene product with the presence and absence of a specific human chromosome in the somatic cell hybrid clones.

ADK is detected in clones A, B, C, and E. Chromosome 10 is the only chromosome that is present in the four clones and absent in clone D. The gene coding for ADK is found on chromosome 10.

CAT is present only in clone A. Clone A has chromosome 10 and 11. This narrows the chromosomal location of CAT to either chromosome 10 or 11. Chromosome 11 is unique to clone A and must carry the gene for CAT.

ESD activity is detected only in clones C and D. Since chromosome 13 is unique to clones C and D, it must house the gene coding for ESD.

13. This problem is worked similar to problem 12. However, some of the somatic hybrid lines have only part of specific chromosomes (centromere with p arm or centromere with q arm). This allows some genes to be assigned to chromosome arms.

Hepatic lipase occurs in clones A, C, D, and F. Chromosome 15, or q arm, occurs in clones A, C, D, and F. Therefore, the gene coding for hepatic lipase must be present on the q arm of chromosome 15.

Esterase D activity is present in clone A and B. The q arm of chromosome 13 occurs in clone A and the p arm occurs in clone F. Chromosome 13 occurs in clone B. The gene coding for esterase D must be located in the q arm of chromosome 13.

Catalase activity is found in clones C, D, E, and F. Chromosome 11 occurs in clones C, D, and E. The q arm of chromosome 11 occurs in clone B and the p arm occurs in clone F. Chromosome 2 occurs in clones C, E , and F. The q arm of chromosome 2 occurs in clones A and B, and the p arm of chromosome 2 occurs in clone D. Therefore, the gene coding for catalase must reside in the p arm of either chromosome 2 or 11.

Alkaline phosphatase activity is found in all clones except D. Part or all of chromosome 2 occurs in all clones. Since the p arm of chromosome 2 by itself occurs only in clone D and no alkaline phosphatase activity is present in clone D, the gene coding for alkaline phosphatase must be found in the q arm of chromosome number 2.

9

DNA and the Molecular Structure of Chromosomes

FUNCTIONS OF THE GENETIC MATERIAL

PROOF THAT GENETIC INFORMATION IS STORED IN DNA
 Discovery of Transformation in Bacteria
 Proof That DNA Mediates Transformation
 Proof That DNA Carries the Genetic Information in Bacteriophage T2
 Proof that RNA Stores the Genetic Information in Some Viruses

THE STRUCTURE OF DNA AND RNA
 Nature of the Chemical Subunits in DNA and RNA
 DNA Structure: The Double Helix
 DNA Structure: Alternative Forms of the Double Helix
 DNA Structure: Negative Supercoils *In Vivo*

CHROMOSOME STRUCTURE IN PROKARYOTES

CHROMOSOME STRUCTURE IN EUKARYOTES
 Chemical Composition of Eukaryotic Chromosomes
 One Large DNA Molecule per Chromosome
 Three Levels of DNA Packaging of DNA in Eukaryotic Chromosomes
 Centromeres and Telomeres

EUKARYOTIC GENOMES: REPEATED DNA SEQUENCES
 Detection of Repeated Sequences: DNA Renaturation Kinetics
 Repeated DNA in the Human Genome

IMPORTANT CONCEPTS

   A. The genetic information of all living organisms, except certain viruses, is stored in DNA. RNA is the genetic material of some viruses.
   1. DNA is a polynucleotide that usually has a double helical structure in which the two strands are held together by hydrogen bonding between the two bases of each nucleotide pair and hydrophobic bonding in the core of stacked base pairs.
       a. The base pairing is specific: adenine always pairs with thymine, and guanine always pairs with cytosine.
       b. The two strands of a DNA double helix are complementary: once the sequence of bases in one strand is established, the sequence of bases in the other (complementary) strand is fixed.
       c. The two complementary strands of a DNA double helix have opposite chemical polarity, one 3' to 5' and the other 5' to 3'.
   B. Chromosomes of prokaryotic organisms are highly compacted *in vivo*.
   1. The DNA molecules are separated into many negatively supercoiled loops by RNA-containing cross-links.
   C. Each eukaryotic chromosome contains a single giant molecule of DNA that extends from one end of the chromosome through the centromere to the other end of the chromosome.
   1. The DNA of interphase chromatin is held in a negatively supercoiled configuration by association with histones.
       a. Interphase chromosomes are characterized by 10-nm-diameter fibers.
       b. Isolated interphase chromatin has a beads-on-a-string structure.

(1) Each bead or nucleosome is an ellipsoid like structure about 11 nm in diameter and 6 nm high.

(2) Each contains a highly conserved core consisting of a 146-nucleotide-pair-long segment of DNA wrapped around an octomer of histones.

(3) Each octomer contains two molecules each of the histones H2a, H2b, H3, and H4.

2. During mitosis and meiosis, the chromatin is further condensed, probably by a second level of coiling, into a 30-nm chromatin fiber.

   a. At metaphase, the coiled or folded 30-nm chromatin fiber is in turn coiled or folded into the highly condensed structure that can be seen with the light microscope; this third level of condensation involves a chromosomal scaffold that is composed of nonhistone chromosomal proteins.

3. The centromere (spindle fiber-attachment regions) and telomeres (ends) of the chromosomes have unique structures.

4. Eukaryotic genomes contain unique or single-copy sequences (sequences present once or a few times), moderately repetitive DNA sequences (those repeated 10 to 10^5 times), and highly repetitive DNA sequences (repeated more than 10^5 times).

   a. Repetitive DNA may be organized as tandem repeats or dispersed throughout the genome.

      (1) Highly repetitive sequences are often located in centromeric or telomeric regions of the chromosomes.

   b. In all eukaryotes studied to date, a major fraction of the genome contains single-copy sequences of DNA interspersed with middle-repetitive sequences.

   c. DNA sequences can move around in the genome as parts of transposable elements.

      (1) Two important types of moderately repetitive transposable elements are called LINEs (long interspersed nuclear elements) and SINEs (short interspersed nuclear elements).

      (2) The best known example of the human LINE family is the transposable element L1, which is present in 20,000 to 50,000 copies per genome.

      (3) The most extensively studied SINE sequences are the *Alu* sequences, which are present about every 5,000 nucleotide pairs in the human genome.

IMPORTANT TERMS

In the space allotted, concisely define each term.

replication:

deoxyribonucleic acid (DNA):

ribonucleic acid (RNA):

deoxyribonuclease (DNase):

ribonuclease (RNase):

protease:

nucleotide:

purines:

pyrimidines:

double helix:

complementarity:

hydrogen bonds:

antiparallel (opposite chemical polarity):

B-DNA:

A-DNA:

Z-DNA:

negative supercoiling:

topoisomerases:

DNA gyrase:

folded genome:

domains:

chromatin:

histones (H1, H2a, H2b, H3, and H4):

nonhistone chromosomal proteins:

protamines:

nucleosome:

multineme:

unineme:

lampbrush chromosomes:

autoradiography:

linkers:

nucleosome core:

chromatin fibers (10-nm and 30-nm diameter):

scaffold:

telomeres:

repetitive DNA:

satellite DNA:

denaturation:

renaturation:

unique (single-copy) DNA sequences:

moderately repetitive DNA sequences:

highly repetitive DNA sequences:

transposable genetic elements:

LINEs:

SINEs:

Alu sequences:

acholinesterasemia:

neurofibromatosis:

hypercholesterolemia:

thalessemia:

IMPORTANT NAMES

In the space allotted, concisely state the major contribution made by the individual or group.

Johann Freidrich Miescher :

Frederick Griffith:

Oswald Avery, Colin Macleod, and Maclyn McCarty:

Alfred Hershey and Martha Chase:

Erwin Chargaff:

James Watson:

Francis Crick:

Maurice Wilkins:

Rosalind Franklin:

Ruth Kavenoff, Lynn Klotz, and Bruno Zimm:

Barbara McClintock:

TESTING YOUR KNOWLEDGE

In this section, answer the questions, fill in the blanks, and solve the problems in the space allotted. Problems noted with an * are solved in the Approaches to Problem Solving section at the end of the chapter.

1. The basic building block of a nucleic acid is called a _____.

2. The proteins that are rich in basic amino acids and function in the initial packaging of DNA in eukaryotes are called _____.

3. The 11 nm by 6 nm particle that consists of an octomer of histones around which is coiled a 146-basepair-long sequence of DNA is called a _____.

4. If a double-stranded DNA molecule contains 15% guanine, it must contain _____ % thymine.

5. An enzyme that catalyzes the destruction of a DNA molecule by breaking phosphodiester bonds is called _____.

6. The two pyrimidines found in DNA are _____ and _____.

7. The purines found in DNA are _____ and _____.

8. If a double-stranded DNA molecule contains 12% cytosine, then it must contain _____ % guanine.

9. The individuals who first showed that DNA was the genetic material using transformation were _____, _____, and _____.

10. The pyrimidines found in RNA are _____ and _____.

11. The persons who provided additional evidence that DNA is the genetic material using bacteriophage T2 were _____ and _____.

*12. A segment of DNA has the base-pair sequence   A T G G C A
                                                     T A C C G T

How many hydrogen bonds occur in this six base-pair sequence?

*13. A nucleic acid with the base composition of 45% adenine, 45% cytosine, 5% thymine and 5% guanine is:

    (a) double-stranded DNA
    (b) double-stranded RNA
    (c) single stranded DNA
    (d) single stranded RNA

14. A nucleic acid with the base composition of 30% adenine, 20% cytosine, 30% thymine and 20% guanine is:

    (a) double-stranded DNA
    (b) double-stranded RNA
    (c) single stranded DNA
    (d) single stranded RNA

*15. A nucleic acid with a base composition of 22% adenine, 20% cytosine, 30% uracil and 28% guanine is:

    (a) double-stranded DNA
    (b) double-stranded RNA
    (c) single stranded DNA
    (d) single stranded RNA

16. The three scientists who shared a Nobel prize for discovering the structure of DNA were _____, _____, and _____.

17. Which one of the histones is <u>not</u> found in the nucleosome core particle?

18. Supercoils are introduced into DNA by enzymes called _____.

19. DNA that coils to the left and has 12 base-pairs per turn is called _____.

20. The interphase chromatin consists of:

    (a) supercoiled DNA molecules free of histones

(b) 30-nm fibers

(c) 10 nm-fibers and a scaffold of nonhistone proteins

(d) 10 nm-fibers consisting of DNA and histones

*21. One strand of a DNA molecule has the following sequence 5' - ACGTATGA - 3'. The complementary strand must be:

(a) 5'-TGCATACT-3'

(b) 5'-TCATACGT-3'

(c) 5'-UCAUACGT-3'

(d) 5'-ACGTATGA-3'

*22. An important crop plant has a genome size of about 1 pg (1 X 10^9 base pairs).

(a) How many nucleosomes would it have assuming each nucleosome and spacer DNA comprises 200 bp of DNA?

(b) How many turns of the double helix are present in the DNA of this plant?

(c) If this plant has its DNA equally partitioned among four chromosomes, how many cectimeters of DNA would be in each chromosome?

THOUGHT CHALLENGING EXERCISE

It has been observed in eukaryotes, that: (1) there appears to be a great excess in DNA, much more than that needed to code for proteins (< 1% of the DNA in some species has a protein-coding function); (2) DNA varies greatly among species (even among closely related species); and (3) DNA content does not correlate positively with evolutionary advancement or genetic complexity. These phenomena led to what has been called the DNA C-value paradox. What do you think the role is of the huge excess in DNA found in most eukaryotes, including humans?

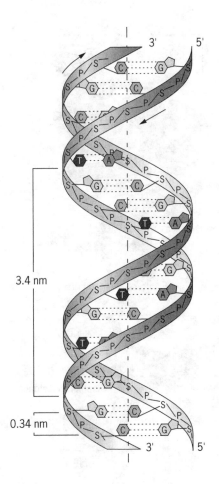

Figure 9.10   Diagram of the double-helix structure of DNA.

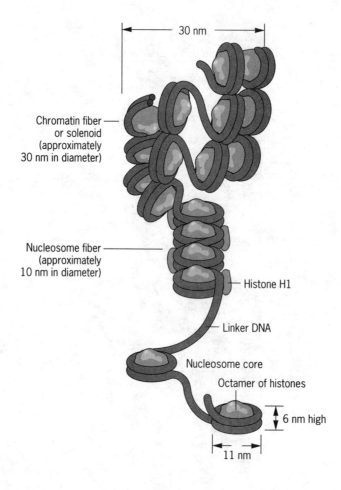

30 nm

Chromatin fiber
or solenoid
(approximately
30 nm in diameter)

Nucleosome fiber
(approximately
10 nm in diameter)

Histone H1

Linker DNA

Nucleosome core

Octamer of histones

6 nm high

11 nm

Figure 9.27 Diagram of the solenoid model of the 30-nm chromatin fiber. Histone H1 appears to stabilize the 10-nm nucleosome fiber and contribute to the formation of the 30-nm chromatin fiber.

SUMMARY OF KEY POINTS

The genetic material must provide the genotypic, phenotypic, and evolutionary functions of replication, gene expression, and mutation, respectively.

The genetic information of living organisms is stored in deoxyribonucleic acid (DNA). In some viruses ribonucleic acid (RNA) is the genetic material.

Nucleic acids are of two types: deoxyribonucleic acid and ribonucleic acid. DNA usually exists as a double helix, with the two strands held together by hydrogen bonds between the complementary bases: adenine paired with thymine and guanine paired with cytosine. The complementarity of the strands of a double helix makes DNA uniquely suited to store and transmit genetic information from generation to generation. The two strands of a DNA double helix have opposite polarity. RNA usually exists as a single-stranded molecule containing uracil instead of thymine. The DNA molecules present in chromosomes are negatively supercoiled.

The DNA molecules in prokaryotic chromosomes are partitioned into negatively supercoiled domains. Bacterial chromosomes contain circular molecules of DNA segregated into about 50 such domains.

Each eukaryotic chromosome contains one giant molecule of DNA packaged into 10-nm ellipsoidal beads called nucleosomes. In the condensed chromosome present during meiosis and mitosis, the 10-nm nucleosome fibers are further coiled into chromatin fibers about 30 nm in diameter. At metaphase, these 30-nm fibers are organized into domains by scaffolds composed of nonhistone chromosomal proteins. The spindle fiber-attachment regions (centromeres) and ends (telomeres) of chromosomes have unique structures that facilitate their respective functions.

Eukaryotic genomes contain repeated DNA sequences, with some sequences present a million times or more. Eukaryotic DNA sequences are commonly grouped into three classes: (1) unique or single-copy sequences present in one to a few copies per genome; (2) moderately repetitive sequences present in 10 to 10^5 copies; and (3) highly repetitive sequences present in over 10^5 copies per genome. Two families of moderately repetitive transposable elements make up ten percent of the human genome and play an important role in its evolution.

ANSWERS TO QUESTIONS AND PROBLEMS

1) nucleotide 2) histones 3) nucleosome 4) 35 5) nuclease 6) thymine and cytosine 7) adenine and guanine 8) 12 9) Avery, Macleod, and McCarty 10) uracil and cytosine 11) Hershey and Chase 12) 15 13) c 14) a 15) d 16) Watson, Crick, and Wilkins 17) H1 18) topoisomerases 19) Z-DNA 20) d 21) b 22) a. 5×10^6 b. 1×10^8 c. 4.5

Ideas on thought challenging exercise:

One of the great enigmas of modern biology is the apparently excessive amount of DNA beyond that needed for coding functions, and the abundant variation in DNA content among organisms. Some molecular biologists view this excessive DNA as junk or parasitic. In this theory, variation in DNA content is considered to be without phenotypic effects and accumulates simply because some sequences have the capability to differentially replicate themselves and spread throughout the genome. However, extensive studies over the last three decades have shown that DNA content variation does have measurable and somewhat predictable effects on the phenotype. Studies of plants show that DNA content positively correlates with nuclear volume, meristematic cell volume, and the duration of the mitotic cycle. A variety of correlations have also been observed between DNA content and the environment in which a plant grows. The effects of DNA content on the cell, independent of any coding function, have been called the nucleotype by Michael Bennett. Other investigators have postulated that heterochromatin, that often contains repetitive DNA sequences, may have multiple functions in the nucleus including gene regulation, positions of chiasmata, and chromosome pairing.

APPROACHES TO PROBLEM SOLVING

12. There are two hydrogen bonds between adenine and thymine, and three hydrogen bonds between guanine and cytosine. Therefore, fifteen hydrogen bonds occur in the six base-pair sequence containing three A:T base pairs and three G:C base pairs.

13. In a double stranded DNA molecule the number of purines (A + G) must equal the number of pyrimidines (C + T) and A = T and G = C. In this DNA there is 45% adenine and 5% thymine. Since A ≠ T, the DNA must be single-stranded.

15. This nucleic acid contains uracil instead of thymine; therefore, it must be RNA. Since A ≠ U, and C ≠ G, this RNA is single-stranded.

21. The strands of a DNA are held together by complementary base-pairs. The stands are also of opposite polarity. Therefore, the strand complementary to 5' -ACGTATGA -3' must be 3' - TGCATACT - 5'.

22. (a) Assume that each nucleosome and its spacer average 200 bp of DNA. To obtain the number of nucleosomes in 1 pg DNA, simply divide 1×10^9 base pairs by 2×10^2 bp per nucleosome = 5×10^6 nucleosomes.

    (b) The DNA helix makes a complete turn every 10 base-pairs. Therefore, 1×10^9 bp/10 bp per turn of the double helix = 1×10^8 turns.

    (c) The base-pairs of the DNA helix are spaced 3.4 Å or 3.4×10^{-10} meters apart. Multiply (1×10^9 bp)(3.4×10^{-10} meters per bp) to obtain the length of the DNA containing 1×10^9 bp. This calculates to be 0.34 meters or 34 cm. If the DNA is partitioned equally among four chromosomes, there must be 4.5 cm DNA per chromosome.

10

Replication of DNA and Chromosomes

BASIC FEATURES OF DNA REPLICATION *IN VIVO*
    Semiconservative Replication
    Visualization of Replication Forks by Autoradiography
    Unique Origins of Replication
    Bidirectional Replication

DNA POLYMERASES AND DNA SYNTHESIS *IN VITRO*
    Discovery of DNA Polymerase I in *Escherichia coli*
    Multiple DNA Polymerases
    DNA Polymerase III: The Replicase in *Escherichia coli*
    Proofreading Activities of DNA Polymerases

THE COMPLEX REPLICATION APPARATUS
    Continuous Synthesis of One Strand: Discontinuous Synthesis of the Other
    Covalent Closure of Nicks in DNA by DNA Ligase
    Initiation of DNA Chains with RNA Primer
    Unwinding DNA with Helicases, DNA-Binding Proteins, and Topoisomerases
    The Replication Apparatus: Prepriming Proteins, Primosomes, and Replisomes
    Rolling-Circle Replication

UNIQUE ASPECTS OF EUKARYOTIC CHROMOSOME REPLICATION
    The Cell Cycle
    Multiple Replicons per Chromosome
    Two DNA Polymerases at a Single Replication Fork
    Duplication of Nucleosomes at Replication Forks
    Telomerase: Replication of Chromosome Termini
    Telomere Length and Aging in Humans

IMPORTANT CONCEPTS

A. Replication of DNA is semiconservative in both prokaryotes and eukaryotes.
    1. During replication the two complementary strands unwind, and each single strand serves as a template directing the synthesis of a new complementary strand.
        a. The net result of replication is two progeny DNA molecules identical to the parental double helix.
    2. Replication often begins at unique origins and proceeds bidirectionally from these origins.
        a. Prokaryotic chromosomes usually contain a single primary origin of replication.
    3. DNA replication is catalyzed by enzymes called DNA polymerases.
        a. All known DNA polymerases, both prokaryotic and eukaryotic, have an absolute requirement for a 3'-hydroxyl primer.
        b. DNA polymerases catalyze covalent extension only in the 5' $\rightarrow$ 3' direction. Thus, synthesis of the strand growing in the 3' $\rightarrow$ 5' direction is discontinuous.
        c. Short segments called Okazaki fragments are synthesized in the 5' $\rightarrow$ 3' direction and then are covalently joined by DNA ligase.
        d. No known polymerases can catalyze the initiation of new DNA chains.
            (1) Nascent DNA chains are initiated by short RNA primers synthesized by DNA primase.
B. Chromosome replication in *E. coli* is accomplished by a complex replication apparatus or replisome composed of many different proteins and enzymes.
    1. DNA topoisomerases produce transient breaks in DNA that function as swivels during the unwinding of the complementary strands of a parental double-helix.

2. DNA helicase uses energy from ATP to unwind the complementary strands of a parental DNA molecule.
3. Single-strand DNA-binding protein coats the DNA single strands and keeps them in an extended configuration for replication.
4. The DNA polymerase III holoenzyme, which catalyzes semiconservative DNA replication in *E. coli*, is a multimeric protein containing at least 20 polypeptides.
   a. The 3' → 5' exonuclease activity built into prokaryotic DNA polymerases provides a proofreading function that is essential for accurate DNA replication.
5. Negative supercoils are introduced into and removed from *E. coli* DNA by DNA gyrase.

C. Eukaryotic DNA replication, with a few unique aspects, is basically the same as that in prokaryotes.
1. RNA primers and Okazaki fragments are shorter in eukaryotes than in prokaryotes.
2. Eukaryotic chromosomes contain multiple origins of replication.
3. Eukaryotic replisomes contain two different DNA polymerases.
   a. DNA polymerase δ catalyzes the replication of the leading strand.
   b. DNA polymerase α carries out discontinuous replication of the lagging strand.

D. DNA replication and histone synthesis are tightly coupled in eukaryotes, and the progeny DNA molecules are rapidly packaged into nucleosomes.

E. Telomerase, an enzyme with a built-in RNA template, adds the unique termini or telomere sequences to eukaryotic chromosomes.

IMPORTANT TERMS

In the space allotted, concisely define each term.

semiconservative replication:

replication fork:

origin of replication:

replication bubble:

autonomously replicating sequences (ARS elements):

DNA polymerase I:

primer DNA:

template DNA:

nuclease:

5' → 3' exonuclease activity:

3' → 5' exonuclease activity:

DNA polymerase II:

DNA polymerase III:

holoenzyme :

proofreading:

continuous synthesis:

discontinuous synthesis:

leading strand:

lagging strand:

DNA ligase:

RNA primer:

DNA primase:

DNA helicases:

single-stranded DNA-binding proteins (SSB proteins):

DNA topoisomerase I:

DNA topoisomerase II:

DNA gyrase:

OriC:

prepriming proteins:

DnaA protein:

DnaC protein:

Okazaki fragments:

primosome:

replisome:

rolling-circle replication:

G1 cyclins:

cyclin-dependent protein kinases (Cdks):

start kinases:

replicon:

DNA polymerase α:

DNA polymerase δ:

telomerase:

progerias:

IMPORTANT NAMES

In the space allotted, concisely state the major contribution made by the individual or group.

Matthew Meselson and Franklin Stahl:

J. Herbert Taylor, Philip Woods, and Walter Hughes:

John Cairns:

Arthur Kornberg:

Peter DeLucia:

TESTING YOUR KNOWLEDGE

In this section, answer the questions, fill in the blanks, and solve the problems in the space allotted. Problems noted with an * are solved in the Approaches to Problem Solving section at the end of the chapter.

1. The enzyme that catalyzes the formation of a covalent bond between the 5' and 3' ends of newly synthesized single-strands of DNA during DNA replication is called _____.

2. The enzyme that replicates most of the DNA in *E. coli* that also has a proofreading function is _____.

3. During DNA replication of *E. coli*, the enzyme that removes RNA primers and replaces them with DNA is called _____.

4. The relatively short, single-stranded DNA fragments that are synthesized during discontinuous DNA replication and joined together to form a continuous strand are called _____.

5. The enzyme that uncoils DNA into single strands prior to DNA replication is called _____.

6. The synthesis of the leading strand of DNA proceeds from _____ → _____ direction and the lagging strand is synthesized in the _____ → _____ direction.

7. Short RNA primers are synthesized during DNA replication by an enzyme called _____.

8. The enzyme that forms a phosphodiester bond between adjacent nucleotides, but does not extend the strand is called _____.

9. The sequence of DNA in *E. coli* that functions in the origin of DNA replication is called _____.

10. During DNA synthesis, the extended single strands of DNA are maintained by _____.

11. Enzymes that catalyze temporary single-strand breaks in DNA molecules during uncoiling, and use covalent linkages to themselves to hold onto the cleaved molecules are called _____.

12. What are the first three proteins necessary in *E. coli* for the initial steps in formation of the replication fork in the *OriC* region?

13. In *E. coli*, the initiation of DNA synthesis on the lagging strand is carried out by a protein complex containing DNA primase and DNA helicase called the _____.

14. The enzyme that removes supercoils from DNA, one at a time, by producing transient single-strand breaks is called _____.

15. The complete replication apparatus moving along the DNA molecule at the replication fork is called the _____.

16. The enzyme that removes or introduces supercoils from DNA, two at a time, by producing transient double-strand breaks is called _____.

17. An enzyme complex that plays a key role in triggering the onset of DNA synthesis in eukaryotes is called _____.

18. In eukaryotes, the DNA polymerase complex that also has primase activity, but is otherwise most similar to DNA polymerase III of prokaryotes is called _____.

19. In eukaryotes, the DNA polymerase that apparently catalyzes the replication of the leading strand is called _____.

20. The enzyme that has a built-in RNA strand template that initiates DNA synthesis at the end of chromosomes is called _____.

21. In humans, a type of disorder that is characterized by premature aging is called _____.

*22. A culture of bacteria has been grown for many generations in a medium containing N^{15}. Bacteria are removed and grown on a culture containing only normal nitrogen, i.e., N^{14}.

(a) Diagram the gradient density profiles that would be generated by the three possible types of replication after one and two generations of growth on the normal culture medium.

(b) After one generation of growth in the normal culture medium can you differentiate between a semiconservative, conservative, and dispersive form of DNA replication? Which mode of replication does the experimentally observed density gradient profile support and why?

THOUGHT CHALLENGING EXERCISE

Design an experiment to measure the length of the cell cycle in eukaryotes.

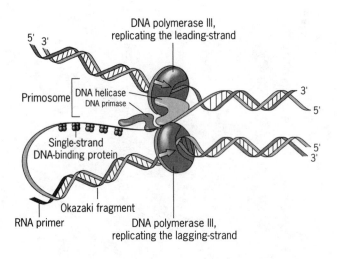

Figure 10.27   Diagram of the *E. coli* replisome, showing the two catalytic cores of the DNA polymerase III replicating the leading and the lagging strands and the primosome unwinding the parental double helix and initiating the synthesis of new chains with RNA primers. The entire replisome moves along the parental double helix, with each component performing its respective function in a concerted manner.

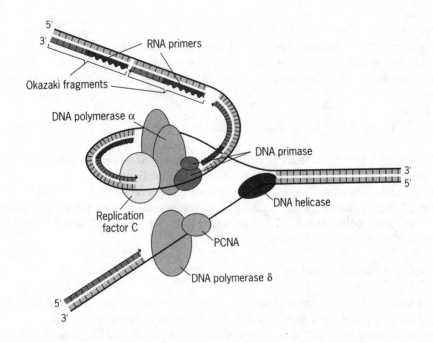

Figure 10.31 Two DNA polymerases, δ and α, function at the replication fork in eukaryotes. DNA polymerase δ replicates the leading strand, and DNA polymerase α replicates the lagging strand. PCNA = proliferating cell nuclear antigen.

SUMMARY OF KEY POINTS

DNA replication is semiconservative. As the two complementary strands of a parental double helix unwind and separate, each serves as a template for the synthesis of a new complementary strand. The hydrogen-bonding potentials of the bases in the template strands specify complementary base sequences in the newly synthesized strands. Replication is initiated at unique origins and usually proceeds bidirectionally from each origin.

DNA synthesis is catalyzed by enzymes called DNA polymerases. All DNA polymerases have an absolute requirement for a primer strand, which is extended, and a template strand, which is copied. All DNA polymerases have an absolute requirement for a free 3'-OH on the primer strand, and all DNA synthesis occurs in the 5' to 3' direction. The $3' \rightarrow 5'$ exonuclease activities of DNA polymerases proofread nascent strands as they are synthesized, removing any mispaired nucleotides at the 3' end of primer strands.

DNA replication is complex, requiring the participation of a large number of proteins. DNA synthesis is continuous on the progeny strand that is being extended in the overall $5' \rightarrow 3'$ direction, but is discontinuous on the strand that is being extended in the overall $3' \rightarrow 5'$ direction. New DNA chains are initiated by short RNA primers synthesized by DNA primase. The enzymes and DNA-binding proteins involved in replication assemble into a replisome at each replication fork and act in concert as the fork moves along the parental DNA molecule.

DNA replication is basically the same in both prokaryotes and eukaryotes, including humans. Replication of the giant DNA molecules in eukaryotic chromosomes occurs bidirectionally from multiple origins. Two DNA polymerases (α and δ) are present at each replication fork. Telomeres, the unique sequences at the ends of chromosomes, are added to chromosomes by a special RNA-containing enzyme called telomerase.

ANSWERS TO QUESTIONS AND PROBLEMS

1) DNA ligase 2) DNA polymerase III 3) DNA polymerase I 4) Okazaki fragments
5) DNA helicase 6) 5' → 3'; 5' → 3' 7) DNA primase 8) DNA ligase 9) *OriC*
10) single-stranded DNA-binding proteins 11) topoisomerases 12) DnaA; DnaB; DnaC
13) primosome 14) topoisomerase I 15) replisome 16) topoisomerase II 17) start kinases 18) DNA polymerase α 19) DNA polymerase δ 20) telomerase 21) progerias 22) see *Approaches to Problem Solving* section at the end of the chapter.

Ideas concerning thought challenging exercise:

Either root tips of a plant, or a cell culture of an animal could be treated with colchicine for a short period of time to inhibit spindle fiber formation that will result in tetraploid cells. After treatment, the tissue could be transferred to growth conditions lacking colchicine. Cells could be sampled every couple of hours over a period of two or three days. Slides prepared from the samples could be microscopically analyzed for diploid and tetraploid metaphases. The proportion of tetraploid metaphases plotted over time should yield cyclic peaks as the tetraploid nuclei pass through the cell cycles. The interval between the peaks is a measure of the length of the mitotic or cell cycle. A similar experiment could be carried out by pulse-labeling the cells with tritiated thymidine. The metaphases on microscope slides with incorporated radioactive thymidine could be detected by a procedure called autoradiography.

APPROACHES TO PROBLEM SOLVING

22. (a) The following density gradient profiles would be generated by the three possible types of replication following one generation of growth in normal medium after removing from the N^{15}-containing medium.

		semiconservative	conservative	dispersive
1st generation	light		1/2 _____	
	intermediate	all _____		all _____
	heavy		1/2 _____	

The following density gradient profiles would be generated by the three possible types of replication following two generations growth in normal medium after removing from the N^{15}-containing medium

		semiconservative	conservative	dispersive
2nd generation	light	1/2 _____	3/4 _____	
	intermediate	1/2 _____		all _____
	heavy		1/4 _____	

(b) Semiconservative and dispersive modes of replication would result in a single intermediate band after one generation. Therefore, you can't differentiate between these in the first generation. If a conservative mode of replication occurred there would be two bands, one heavy and one light, in equal proportions after the first generation. Therefore, if conservative replication occurred, it would be apparent in the first generation. A distinct centrifugation profile should be apparent for each mode of replication in the second generation. A dispersive mode would have one band halfway between the intermediate and the light band. The conservative mode would result in two bands; one at the heavy position and another about three times as intense at the light position. Semiconservative replication would result in two bands of equal intensity, one at the intermediate position and one at the light position. Data consistent with the semiconservative model of replication are experimentally observed.

11

Transcription and RNA Processing

IMPORTANT CONCEPTS

A. The pathway of information flow by which a gene exerts its effect on the phenotype of an organism is often very complex. The first two steps in this pathway are transcription and translation.

    1. Transcription involves the synthesis of a messenger RNA or pre-messenger RNA molecule, which functions as an intermediary in protein synthesis, using one of the two strands in the gene as a template.

    2. The messenger RNA produced is translated, according to the specifications of the genetic code, into the sequence of amino acids in the polypeptide gene-product.

    3. Transcription is catalyzed by a complex, multimeric enzyme, called RNA polymerase.

    4. Eukaryotes have three RNA polymerase complexes.

        a. RNA polymerase I is located in the nucleolus and catalyzes the synthesis of large ribosomal RNAs.

        b. RNA polymerase II transcribes the nuclear genes into mRNAs that encode protein products.

        c. RNA polymerase III catalyzes the synthesis of tRNAs, most of the snRNAs, and the 5S ribosomal RNA.

B. In eukaryotes, the primary transcripts usually undergo several types of processing.

1. Processing frequently includes the excision of segments of primary transcripts and the addition of 5' 7-methyl guanosine caps and 3' poly(A) tails.
2. In some cases, especially in mitochondria of trypanosomes and plants, the nucleotide sequences of primary transcripts are altered by RNA editing processes.

C. Many eukaryotic genes and a few prokaryotic genes contain noncoding sequences called introns that separate the coding sequences called exons.

1. The entire nucleotide-pair sequence of each split gene, including both intron and exon sequences, is transcribed to produce a pre-mRNA molecule.
2. The intron sequences are removed from the primary transcript or pre-mRNA by splicing processes to produce the mature, functional RNA molecule.

   a. Three distinct splicing processes are known.

   (1) Introns in many pre-tRNA molecules are excised by simple splicing nucleases and ligases.

   (2) Introns in certain rRNA precursors and some primary transcripts of genes in mitochondria and chloroplasts are excised autocatalytically (self-catalyzed with no protein involved).

   (3) Introns of nuclear pre-mRNAs are excised by complex ribonucleoproteins called spliceosomes.

3. The intron sequences of nuclear genes are highly divergent.

   a. In primary transcripts of nuclear genes, the only completely conserved sequences of different introns are the dinucleotide sequences at the ends of introns, i.e., 5' exon-GT-intron-AG-exon 3'.

   b. There is only one short, poorly conserved sequence, the TACTAAC box, located about 30 nucleotides upstream from the 3' splice site of introns.

   (1) The adenine base at position six in the TACTAAC box is completely conserved and plays a key role in the splicing reaction.

IMPORTANT TERMS

In the space allotted, concisely define each term.

messenger RNA (mRNA):

amino acids:

transcription:

translation:

genetic code:

codons:

ribosome:

primary transcript:

pre-mRNAs:

transfer RNAs (tRNAs):

ribosomal RNAs (rRNA):

small nuclear RNAs (snRNAs):

anticodon:

DNA template strand:

DNA nontemplate strand:

RNA polymerases:

promoters:

transcription bubble:

transcription unit:

RNA polymerase holoenzyme:

sigma (δ) factor:

recognition site:

elongation of RNA:

termination signal:

heterogeneous nuclear RNA (hnRNA):

RNA polymerase I, II, and III:

TATA box:

CAAT box:

GC box:

octomer box:

basal transcription factors :

poly(A) polymerase:

RNA editing:

guide RNAs:

introns:

exons:

R-loops:

TACTAAC box:

splicing endonuclease:

splicing ligase:

self splicing (autocatalytic):

spliceosomes:

snRNPs:

systemic lupus erythematosus:

IMPORTANT NAMES

In the space allotted, concisely state the major contribution made by each individual or group.

Phillip Leder:

Pierre Chambon:

Alan Jeffries:

Richard Flavell:

Tom Maniatus:

Sol Spiegelman:

Sydney Brenner, Francois Jacob, and Matthew Meselson:

Francois Jacob and Jacques Monod:

Oscar Miller and Barbara Hamkalo:

TESTING YOUR KNOWLEDGE

1. Noncoding intervening sequences in eukaryotic genes that interrupt the coding sequences are called
   _____.

2. A specific DNA sequence to which the RNA polymerase binds to start transcription is called a
   _____.

3. The molecule that transmits information, stored in the DNA of the nucleus, to the ribosomes in the cytoplasm where it is read is called _____.

4. The molecule that caps the 5' end of the pre-messenger RNA is _____.

5. The 3' end of a pre-mRNA is modified by the addition of a _____ tail.

6. The locally unwound segment of DNA which is produced by RNA polymerase is called
   a _____.

7. The protein in *E. coli* whose function is to recognize and bind RNA polymerase to the promoter site in DNA is called a _____.

8. In eukaryotes, the population of primary transcripts is called _____.

9. The conserved part of a promoter of eukaryotic genes that is centered at about thirty bases upstream of the transcription initiation site is called the _____.

10. The conserved part of the promoter of a eukaryotic gene that is located about eighty bases upstream of the transcription initiation site is called the _____.

11. The sequences of eukaryotic genes that code for amino acid sequences or RNA sequences or the processed gene product are called _____.

12. The enzyme that adds 3' poly(A) tails to pre-mRNAs is called _____.

13. The complex ribonucleoprotein structures that carry out pre-mRNA splicing are called _____.

14. The eukaryotic enzyme complex that catalyzes the transcription of protein-coding genes is called _____.

15. A disease that results from mutations that alter the splicing of human beta hemoglobin pre-mRNA is called _____.

16. The small ribonucleoproteins that are found in spliceosomes are called _____.

*17. The chicken ovalbumin gene has seven introns interrupting the coding sequences or exons. How many R-loops will be observed if DNA of the gene and the mRNA coded by it are mixed under conditions that allow the single-stranded mRNA to hybridize and displace one single strand of the DNA double helix?

*18. How many single-stranded loops will be observed if the ovalbumin mRNA is hybridized to single stranded DNA of the ovalbumin gene?

*19. The following is a partial sequence of gene in *E. coli* and sequences upstream from the coding sequence:

5'- A A G C T A T A A T C G T A C G T A C C T A A G G A G G T A A G C G -3'
3'- T T C G A T A T T A G C A T G C A T G G A T T C C T C C A T T C G C -5'

164   Chapter 11

Indicate the RNA that is transcribed by RNA polymerase for the portion of the gene shown.

*20. The following is a partial sequence of a gene in humans with some of the upstream sequences:

5'- T AT AAA ACG T GCAT CGACGT ACGAT AT CGCCGT ACCGCT GCAC -3'
3'- AT AT T T T GCACGT AGCT GCAT GCT AT AGCGGCAT GGCGACGT G -5'

Indicate the RNA that is transcribed by RNA polymerase II from the portion of the gene shown.

THOUGHT CHALLENGING EXERCISE

It has been observed that bacteria transformed with cloned eukaryotic genes often produce a nonfunctional eukaryotic gene product. What aspects of the flow of genetic information in prokaryotes compared to eukaryotes might explain this phenomenon? How might this problem be prevented so that a bacterial cell culture could be used to produce a functional eukaryotic protein?

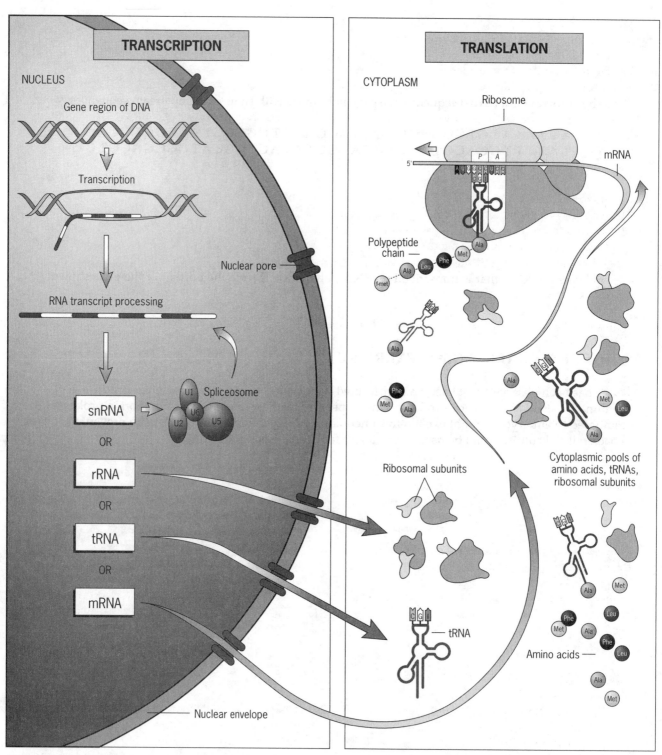

(a) Transcription and RNA processing occur in the nucleus.

(b) Translation occurs in the cytoplasm.

Figure 11.5   An overview of protein synthesis, emphasizing the transcriptional origin of snRNA, tRNA, rRNA, and mRNA, the splicing function of snRNA, and the translational roles of tRNA, rRNA, mRNA, and ribosomes.

SUMMARY OF KEY POINTS

Metabolism occurs by sequences of enzyme-catalyzed reactions, with each enzyme specified by one or more genes. Therefore, each step in a metabolic pathway is under genetic control.

The central dogma of molecular biology is that genetic information flows from DNA to DNA during chromosome duplication and from DNA to protein during gene expression. The flow of information from DNA to protein occurs in two steps: (1) transcription, DNA to RNA, and (2) translation, RNA to protein. Transcription involves the synthesis of an RNA transcript complementary to one strand of DNA of a gene(s). Translation is the conversion of information stored in the sequence of nucleotides in the RNA transcript into the sequence of amino acids in the polypeptide gene product, according to the specifications of the genetic code.

In eukaryotes, genes are present in the nucleus, whereas polypeptides are synthesized in the cytoplasm. Messenger RNA molecules function as intermediaries that carry genetic information from DNA to the ribosomes, where proteins are synthesized. RNA synthesis, catalyzed by RNA polymerases, is similar to DNA synthesis in many respects. However, RNA synthesis occurs within a localized region of strand separation, and only one strand of DNA functions as a template for RNA synthesis.

RNA synthesis occurs in three stages: (1) initiation, (2) elongation, and (3) termination. The RNA polymerases that catalyze transcription are complex, multimeric proteins. Covalent extension of RNA chains occurs within transcription bubbles, locally unwound segments of DNA. Chain elongation stops when RNA polymerase encounters transcription-termination signals in DNA. Transcription, translation, and degradation of a specific mRNA molecule often occur simultaneously in prokaryotes.

Three different RNA polymerases are present in eukaryotes, and each polymerase transcribes a distinct set of genes. Eukaryotic gene transcripts usually undergo three modifications: (1) the addition of 7-methyl guanosine caps to the 5' termini, (2) the addition of poly(A) tails to the 3' ends, and (3) the excision of noncoding intron sequences. The information content of some eukaryotic transcripts is altered by RNA editing, which changes the nucleotide sequences of transcripts prior to translation.

Most, but not all , eukaryotic genes are split into coding sequences called exons and noncoding sequences called introns. Some genes contain very large introns; others harbor large numbers of small introns. The biological significance of introns is still open to debate.

Noncoding intron sequences are excised from RNA transcripts in the nucleus prior to their transport to the cytoplasm. Introns in tRNA precursors are removed by the concerted action of a splicing endonuclease and ligase, whereas introns in some rRNA precursors are spliced out autocatalytically—with no protein involved. The introns in nuclear pre-mRNAs are excised on complex ribonucleoprotein structures called spliceosomes. The excision process must be precise, with accuracy to the nucleotide level, to assure that codons in exons distal to introns are read correctly during translation.

ANSWERS TO QUESTIONS AND PROBLEMS

1) introns 2) promoter 3) messenger RNA (mRNA) 4) 5', 7-methyl guanosine 5) poly (A)
6) transcription bubble 7) sigma factor 8) heterogeneous nuclear RNA (hnRNA) 9) *TATA box* (Hogness box) 10) *CAAT box* 11) exons 12) poly(A) polymerase 13) spliceosome 14) replisome
15) β -thalassemia 16) snRNPs 17 - 20) see *Approaches to Problem Solving* section at the end of the chapter

Some ideas concerning the thought challenging exercise:

Eukaryotic genes generally have introns interrupting the coding sequences, whereas prokaryotes do not. Therefore, if a eukaryotic gene containing introns was put into a bacterial cell, the mechanism to process out the introns would be lacking. The whole transcript including the sequences complementary to the introns would be translated. This would result in a nonfunctional protein. One way to avoid this problem is to isolate mRNA which has had the introns removed. RNA can be converted to DNA by an enzyme called reverse transcriptase. The resulting DNA sequences, called cDNAs, then can be combined with a bacterial promoter and inserted into a bacterial cell. The transcription and translation by bacterial systems produces a protein with the correct linear order of amino acids.

APPROACHES TO PROBLEM SOLVING

17. R-loops result when purified mRNA transcripts are annealed with double-stranded DNA of the corresponding gene under conditions that favor DNA-RNA duplex formation. The RNA strand displaces the homologous DNA strand. The single-stranded DNA loops are visualized using electron microscopy. The R-loops correspond to the nontranscribed strand of the segment of double-stranded DNA from which the ovalbumin pre-mRNA was transcribed. The introns occur as double-stranded and separate the R-loops. Therefore, there will always be one more R-loop than there are introns. The ovalbumin gene has seven introns and, therefore, would display eight R-loops.

18. If mRNA is hybridized to the DNA of the gene coding it, homologous pairing will occur. Since the mRNA does not have introns, there will be pairing between the mRNA and DNA, except for the introns. The DNA sequences comprising the introns will appear as single-stranded loops. Since the ovalbumin gene has seven introns, one will observe seven single-stranded DNA loops.

19. To approach this problem, one first needs to determine where transcription will initiate. In prokaryotes, transcription initiates at a site 10 base-pairs from the center of the *TATA box*. The sense strand of DNA is the strand opposite from which the *TATA box* is read. The location of the *TATA box*, the point of initiation of transcription, and the RNA transcribed by RNA polymerase for the portion of the gene shown are indicated in bold face below:

$$5'- \text{AAGC} \textbf{TATAAT} \text{CGTACGTACCTAAGGAGGTAAGCG} - 3'$$
$$3'- \text{TTCGATATTAGCATGCATGGATTCCTCCATTCGC} - 5'$$
$$5' - \textbf{AC} \text{CUAAGGAGGUAAGCG} - 3' \text{ Transcribed RNA}$$

20. The *TATA box* in eukaryotes is centered about 30 bp upstream of the point of transcription initiation. By analyzing the sequence of nucleotides in the portion of the gene indicated, the *TATA box* and the start point of transcription are found and indicated in bold face.

$$5'-\textbf{TATAAA} \text{ACGTGCATCGACGTACGATATCGCCGT} \textbf{A} \text{CCGCTGCAC} - 3'$$
$$3'-\text{ATATTTTGCACGTAGCTGCATGCTATAGCGGCATGGCGACGTG} - 5'$$
$$5' \text{ACCGGUGCAC} - 3'$$

The base sequence of the mRNA transcribed from the DNA is shown complementary to the sense strand of the DNA double-helix.

12

Translation and the Genetic Code

PROTEIN STRUCTURE
    Polypeptides: Twenty Different Amino Acid Subunits
    Proteins: Complex Three-Dimensional Structures

PROTEIN SYNTHESIS: TRANSLATION
    Overview of Protein Synthesis
    Components Required for Protein Synthesis: Ribosomes and Transfer RNAs
    Translation: The Synthesis of Polypeptides Using mRNA Templates

THE GENETIC CODE
    Properties of the Genetic Code: An Overview
    Three Nucleotides Per Codon
    Deciphering the Code
    Initiation and Termination Codons
    A Degenerate and Ordered Code
    A Nearly Universal Code

CODON tRNA INTERACTIONS
    Recognition of Codons by tRNAs: The Wobble Hypothesis
    Suppression Mutations That Produce tRNAs With Altered Codon Recognition

IN VIVO EVIDENCE CONFIRMS THE NATURE OF THE GENETIC CODE

IMPORTANT CONCEPTS

A. Genes control growth and development largely by acting through protein intermediaries.
1. Proteins play structural roles in cells and catalyze an enormous array of metabolic reactions.
   a. Proteins are composed of twenty different amino acids with highly varied chemical properties.
   b. Their three-dimensional structures provide proteins with enormous functional versatility.
   c. The primary structure, amino acid sequence, of proteins is determined by specific sequences of nucleotide-pairs in the genes that encode them.
2. The process of protein synthesis (translation) involves the translation of the genetic information stored in the nucleotide sequences of mRNA molecules into sequences of amino acids in polypeptide chains.
   a. Translation is complex and involves about one-third of the macromolecules in a cell.
   b. Translation takes place on ribosomes in the cytoplasm.
      (1) Each ribosome consists of two subunits, which together contain three to four different RNA molecules and 50 to 80 different proteins. The exact composition varies from species to species.
   c. Translation requires the participation of 40 to 60 small RNA molecules called transfer RNAs (tRNAs).
      (1) Each tRNA is activated for protein synthesis by the covalent attachment of a specific amino acid.
      (2) The attachment of an amino acid to a tRNA is catalyzed by an enzyme called aminoacyl-tRNA synthetase.
      (3) There is at least one tRNA and one aminoacyl-tRNA synthetase for each of the twenty amino acids that are commonly found in proteins.

d. Translation requires several soluble proteins called initiation, elongation, and termination factors.

e. Translation involves enzymatic activity by peptidyl transferase.

(1) Peptidyl transferase activity resides in the 23S rRNA molecule of the large subunit of the ribosome, rather than a ribosomal protein.

B. The genetic code has several important characteristics.

1. The genetic code is triplet, with a sequence of three nucleotides making up each codon.

2. All 64 possible triplet codons are used, 61 specify amino acids, two of which are used for polypeptide chain initiation, and three function in terminating translation.

3. The code is degenerate, i.e., several different codons often specify the same amino acid.

a. The base-pairing between the third (3') base of the codon and the 5' base of the anticodon does not follow strict base-pairing rules; instead, there is wobble at this site, permitting base-pairs to form other than the usual A:U and G:C. Therefore, the anticodon of a single tRNA may recognize one, two, or three different codons.

4. The genetic code is nonoverlapping. Each nucleotide in a mRNA belongs to just one codon.

5. The genetic code is nearly universal, i.e., the codons have the same meaning, with minor exceptions, in all species.

C. Mutations that affect nucleotides of a gene may affect the amino acid sequence of the encoded polypeptide.

1. Mutations that produce chain-termination triplets within genes are called nonsense mutations.

a. Some suppressor mutations alter the anticodons of tRNAs so that the mutant tRNAs recognize chain-termination codons and insert amino acids in response to their presence in mRNA molecules. The resulting polypeptide will be functional if the amino acid inserted by the suppressor mutation does not significantly alter the protein's chemical properties.

2. A single base change in a triplet codon so that it specifies a different amino acid is called a missense codon.

3. A gain or loss of one or a few base pairs in the protein coding region of a gene may alter the reading frame and the sequence of encoded amino acids.

IMPORTANT TERMS

In the space allotted, concisely define each term.

translation:

amino acid:

peptide:

primary structure:

secondary structure:

tertiary structure:

quaternary structure:

α helices:

β helices:

ionic bonds:

hydrogen bonds:

aminoacyl-tRNA synthetase:

A or aminoacyl site :

P or peptidyl site:

E or exit site:

initiation factors:

methionyl-tRNA^{fmet}:

Shine-Dalgarno sequence:

elongation factor Tu (EF-Tu):

EF-Tu·GTP:

elongation factor Ts (EF-Ts):

peptidyl transferase:

elongation factor G (EF-G):

chain termination codons:

release factors (Rfs):

triplet code:

suppressor mutations:

reading frame:

trinucleotides:

degenerate code:

nearly universal code:

wobble hypothesis:

nonsense mutations:

missense mutations:

reading frame (frameshift) mutation:

IMPORTANT NAMES

In the space allotted, concisely state the major contribution made by each individual or group.

James Herrick:

Vernon Ingram:

Robert Holley:

H. Ghobina Khorana:

Francois Chapeman and Gunter von Ehrenstein:

Francis Crick:

Marshall Nirenberg:

J. Heinrich Matthaei:

Severo Ochoa:

Harris Bernstein:

TESTING YOUR KNOWLEDGE

*In this section, answer the questions, fill in the blanks, and solve the problems in the space allotted. Problems noted with an * are solved in the Approaches to Problem Solving section at the end of the chapter.*

The following mRNA codons are useful in answering some of the questions and solving problems of this section and in subsequent chapters.

Triplet nucleotide sequence of mRNA codons

UUU	Phe	UCU	Ser	UAU	Tyr	UGU	Cys
UUC	Phe	UCC	Ser	UAC	Tyr	UGC	Cys
UUA	Leu	UCA	Ser	UAA	Stop	UGA	Stop
UUG	Leu	UCG	Ser	UAG	Stop	UGG	Trp
CUU	Leu	CCU	Pro	CAU	His	CGU	Arg
CUC	Leu	CCC	Pro	CAC	His	CGC	Arg
CUA	Leu	CCA	Pro	CAA	Gln	CGA	Arg
CUG	Leu	CCG	Pro	CAG	Gln	CGG	Arg
AUU	Ileu	ACU	Thr	AAU	Asn	AGU	Ser
AUC	Ileu	ACC	Thr	AAC	Asn	AGC	Ser
AUA	Ileu	ACA	Thr	AAA	Lys	AGA	Arg
AUG	Met	ACG	Thr	AAG	Lys	AGG	Arg
GUU	Val	GCU	Ala	GAU	Asp	GGU	Gly
GUC	Val	GCC	Ala	GAC	Asp	GGC	Gly
GUA	Val	GCA	Ala	GAA	Glu	GGA	Gly
GUG	Val	GCG	Ala	GAG	Glu	GGG	Gly

1. The energy for activating amino acids, i.e., attaching them to their respective tRNA, comes from _____.

2. A triplet of nucleotides in a tRNA that recognizes, by complementary base pairing, the triplets of nucleotides in the mRNA is called an _____.

3. The scientist most responsible for breaking the genetic code was _____.

4. A single base-pair change in a gene that results in a single amino acid substitution in the corresponding protein is called a _____ mutation.

5. The primary energy source for protein synthesis is _____.

6. A complex structure composed of protein and RNA molecules that functions as the site of amino acid polymerization is called a _____.

7. The person who was awarded a Nobel Prize for research on the structure of tRNA was _____.

8. A single base pair mutation that results in premature termination of translation of the corresponding mRNA is called a _____ mutation.

9. The phenomenon that one tRNA for a particular amino acid can recognize more than one codon for the amino acid can be explained by Crick's _____ hypothesis.

10. A base sequence in a mRNA, upstream from the point of initiation of translation, that hydrogen bonds to a sequence in the rRNA of the small ribosomal subunit is called the _____.

11. Which end (5′ or 3′) of a mRNA is synthesized first?

12. Which end of a polypeptide is the oldest (amino terminal or carboxyl terminal)?

13. If the anticodon in a tRNA is 3′ - CCU - 5′, what is the complementary sequence in the gene coding for this tRNA?

14. Which group of an amino acid covalently bonds to the tRNA?

*15. What is the minimum number of aa-tRNAs necessary to recognize all the codons for the amino acid arginine? Hint: Consider the base pairing between the 5′ base of the anticodon and the 3′ base of the codon as depicted in Table 12.3.

*16. Three of the genetic codons are nonsense and function in termination of translation. Like any other codons, these can change because of mutational base-pair substitutions in the DNA. Considering only one base pair change at a time, indicate which amino acids could be inserted in a protein by mutation of the nonsense codon UAG.

*17. A protein was sequenced in several mutant individuals. It was found that these mutants resulted in single amino acid substitutions at position number eight from the amino terminal end of the polypeptide. Arrows indicate the sequence in which the mutants were recovered. Assuming single nucleotide changes, indicate the most likely codon for each of the amino acids shown below.

Wild-type		mutant 1		mutant 2		mutant 3		mutant 4
Thr ____	→	Ala ____	→ Val ____		→ Met ____		→ Leu ____	

*18. A fragment of a polypeptide, Met-Leu-Ala-Gly, is encoded by the following sequence of DNA:

Strand A   -G T T A C A A C C G G C C A -
Strand B   -C A A T G T T G G C C G G T -

(a) Which strand is the template strand?

(b) Indicate the 3′ and 5′ ends of the sense strand of the DNA.

(c) Which polypeptide bond in the above polypeptide was formed first?

(d) Which amino acid is at the amino terminal end of the polypeptide?

*19. The amino acid sequence of a protein encoded by two mutant alleles was compared to the amino acid sequence of the protein encoded by the normal gene.

	1	2	3	4	5	6	7	8
Normal	Met	Lys	Tyr	Ser	Glu	Trp	Val	Val
Mutant I	Met	Lys	Tyr					
Mutant II	Met	Lys	Tyr	Arg	Ser	Gly	Trp	

(a) Indicate the type of change at the mRNA level that resulted from mutant number I and the specific codon affected.

(b) Indicate the type of change at the mRNA level that resulted from mutant number II and the initial codon affected, and which possible codons were utilized for each amino acid of the normal protein.

*20. The following is a sequence of nucleotides in a DNA double helix that codes for a short polypeptide. The messenger RNA encoded by this DNA has both translational initiation and termination codons.

Strand A -   A A A T C A A T A G T T A G A A C C C A T C T T G
Strand B -   T T T A G T T A T C A A T C T T G G G T A G A A C

(a) Which strand is the template strand?

(b) What is the polarity of the strands of the DNA double helix?

(c) What is the base sequence of the mRNA coded by this DNA?

(d) What is the amino acid sequence of the polypeptide encoded by this mRNA?

THOUGHT CHALLENGING EXERCISE

It was suggested about one-hundred years ago that nuclein (DNA) may be the genetic material. However, it took until the late 1950's before it was generally accepted that genes were sequences of polynucleotide in the DNA. Why do you think that it took over fifty years after the rediscovery of Mendel's laws in 1900 before much interest was shown by biochemists in the nucleic acid component of chromatin?

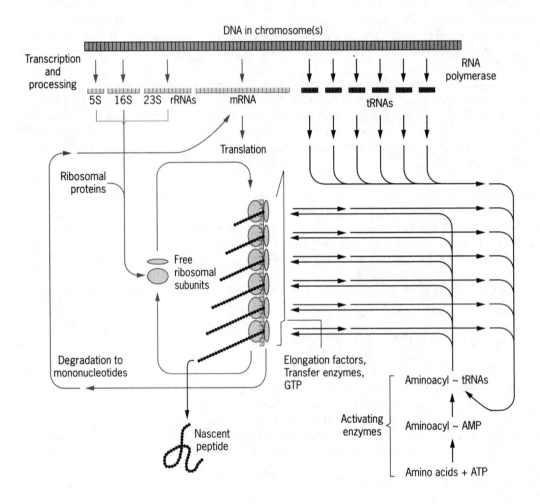

Figure 12.6 Overview of protein synthesis. The sizes of the rRNA molecules shown are correct for bacteria; larger rRNAs are present in eukaryotes. For simplicity, all RNA species have been transcribed from contiguous segments of a single DNA molecule. In reality, the various RNAs are transcripts of genes located at different positions on one to many chromosomes.

SUMMARY OF KEY POINTS

Most genes exert their effect(s) on the phenotype of an organism through proteins, which are large macromolecules composed of polypeptides. Each polypeptide is a chain-like polymer assembled from twenty different amino acids. The amino acid sequence of each polypeptide is specified by the nucleotide sequence of a gene. The vast functional diversity of proteins results in part from their complex three-dimensional structures.

Genetic information carried in the sequence of nucleotides in an mRNA molecule is translated into a sequence of amino acids in a polypeptide product by intricate macromolecular machines called ribosomes. The translation process is complex, requiring the participation of many different RNA and protein molecules. Transfer RNA molecules serve as adaptors, mediating the interaction between amino acids and codons in mRNA. The process of translation involves the initiation, elongation, and termination of polypeptide chains and is governed by the specifications of the genetic code.

Each of the twenty amino acids in proteins is specified by one or more nucleotide triplets in mRNA. Of the 64 possible triplets, given the four bases in mRNA, 61 specify amino acids and three signal chain termination. The code is nonoverlapping, with each nucleotide part of a single codon, degenerate, with most amino acids specified by two or four codons, and ordered, with similar amino acids specified by related codons. The genetic code is nearly universal; with minor exceptions, the 64 triplets have the same meaning in all organisms.

The wobble hypothesis explains how a single tRNA can respond to two or more degenerate codons. According to this hypothesis, the pairing between the third base of a codon and the first base of an anticodon is less stringent than normal, permitting wobble at this site. Some suppressor mutations alter the anticodons of tRNAs so that the mutant tRNAs recognize chain-termination codons and insert amino acids in response to their presence in mRNA molecules.

Comparisons of the nucleotide sequences of genes with amino acid sequences of their polypeptide products verify the assignments deduced from *in vitro* studies.

ANSWERS TO QUESTIONS AND PROBLEMS

   1) ATP **2)** anticodon **3)** Nirenberg **4)** missense **5)** GTP **6)** ribosome **7)** Holley **8)** nonsense **9)** wobble **10)** Shine-Dalgarno sequence **11)** 5′ **12)** amino terminal **13)** 5′ GGA 3′ **14)** carboxyl **15)** 3 **16)** tyrosine, serine, leucine, trytophan, lysine, glutamic acid, glutamine **17)** ACG, GCG, GUG, AUG, UUG or CUG **18) a.** strand A; **b.** 3′- GTTACAACCGGCCA - 5′; **c.** Met-Leu; **d.** Met **19)** see Approaches to Problem Solving section **20. a.** strand A; **b.** Strand A, 5′ → 3′; Strand B, 3′ → 5′; **c.** 5′-AUGGGUUCUAACUAUUGAUUU-3′; **d.** fMet-Gly-Ser-Asn-Tyr.

Ideas concerning the thought challenging exercise:

Advancements in science often come very rapidly and many top scientists often do their research in areas that are most exciting and receiving attention at a given time. During the first half of the 20th century, exciting advances were being made in the discipline of biochemistry. Much was being learned about proteins and their functions as enzymes. This was an era where much of the attention was centered around proteins. The chemical components of DNA, i.e., nucleotides, had been characterized in the 1920s and it was know that nucleic acids were polymers of nucleotides. However, it was thought that nucleic acids were simply a polymer consisting of a repeated tetrameric nucleotide unit, e.g., A - T - G - C, and hence were too simple to be the genetic material. The fallacious idea that protein was likely the genetic material did not die easily. This view could still be found in the literature of the late 1950s and early 1960s, long after conclusive evidence had been presented that DNA was the genetic material.

APPROACHES TO PROBLEM SOLVING

15. To solve this problem, first list the six codons for arginine.

CGU
CGC
CGA
CGG
AGA
AGG

Next, recall that, due to wobble, there does not have to be a different tRNA for each codon of an amino acid. Therefore, a tRNA with 3' - UCU - 5' anticodon can recognize both the 5' - AGA - 3' and 5'- AGG - 3'. The tRNA with the anticodon GCI can recognize the codons CGA, CGU, and CGC. It requires a third tRNA with the anticodon GCU or GCC to recognize the codon CGG.

16. The approach to working this problem is straight-forward. Simply list all the codons that can result from single base changes of UAG and then check their coding properties from the table of codons.

UAG

UAC	tyrosine
UAU	tyrosine
UAA	nonsense
UCG	serine
UUG	leucine
UGG	tryptophan
AAG	lysine
CAG	glutamine
GAG	glutamic acid

17. Since all mutations are assumed to have occurred by single base-pair changes in the DNA, first look at the possible codons for each of the amino acids substituted at position number eight. All of these can be accounted for by changes at the first or second position of the codon. Any problem with the degeneracy is minimized because methionine has only one codon, AUG. Therefore, we know that the third base in the codons must be G.

Thr <u>ACG</u> --> Ala <u>GCG</u> --> Val <u>GUG</u> --> Met <u>AUG</u> --. Leu <u>UUG</u> or <u>CUG</u>

18. (a) The template strand can be determined by looking for the strand of DNA that is complementary for the codons for the Met-Leu-Ala-Gly. The codon for Met is 5' - AUG - 3'. The complementary strand of DNA must contain a 3'-TAC-5' sequence. TAC starts with the third base from the left on strand A. The transcription of strand A, 3' - GTTACAACCGGCCA - 5', results in a mRNA with the reading frame sequence of 5' - CAAUGUUGGCCGGU- 3' which translates as fMet.Leu.Ala.Gly.

(b) Since the messenger RNA is translated 5' → 3' and has reversed polarity of the sense strand of DNA, the sense strand of the DNA must be

3' - GTTACAACCGGCCA - 5'.

(c) Translation of the mRNA proceeds from 5′ → 3′. Therefore, the peptide bond connecting methionine and leucine was the first formed.

(d) Since polymerization of amino acids occurs by formation of a peptide bond between the carboxyl group of the growing polypeptide chain and the amino group of the incoming amino acid, methionine must occupy the amino terminal end of the above polypeptide.

19. (a) When you look at the product of mutant I, it is noticed that it is only a partial polypeptide. This suggests that a nonsense mutation may account for it. A single base substitution in the DNA could change a codon for serine to a nonsense codon, i.e., UCA → UGA, or UCG → UAG.

(b) Mutant II has the first three amino acids in its polypeptide identical to that of the normal. The rest of the amino acids are different, suggesting that a single addition or deletion of a base has affected the reading frame. To determine what has happened and to deduce, as much as possible about the codons utilized to produce the normal, start by inserting possible codons for the amino acids of the normal and mutant protein. When this is done we don't have to designate any changes in the first three codons. The mutant and normal differ, starting at the fourth position. A deletion of the **U** base of the serine codon UCG or the **U/C** base of the tyrosine codon shifts the reading frame one base to the right, so that the next codon is CGG for Arg.

Normal	Met	Lys	Tyr	Ser	Glu	Trp	Val	Val
	AUG	AAG	UAU	UCA	GAA	UGG	GUG	GU_
		A	C	G	G			
	AUG	AAG	UAU	CGG	AGU	GGG	UGG	
		A	C					
Mutant II	Met	Lys	Tyr	Arg	Ser	Gly	Trp	

The next codon is either AAU or AGU. Since the next amino acid in the mutant protein is serine, the codon must be AGU. The next codon is GGG for glycine, followed by UGG for tryptophan. The assignments of the frameshifted codons in many cases dictates the codons used for the normal protein. The codon assignments for the normal mRNA are Met (AUG) - Lys (AAG or AAA) - Tyr (UAU or UAC) - Ser (UCG) - Glu (GAA) - Trp (UGG) - Val (GUG) - Val (GUU, GUC, GUA, or GUC).

20. The key to solving this problem is that both translational initiation (AUG) and termination sequences (UAA, UAG, or UGA) are present in the mRNA. Generally, you can determine the sense strand of the DNA by looking for an open reading frame flanked by sequences complementary to initiation (3′-TAC-5′) and termination (3′-ATT-5′, 3′-ATC-5′ or 3′-ACT-5′) codons. These are found in Strand A of the DNA:

Strand A- A A A **T C A** A T A G T T A G A A C C **C A T** C T T G-
Strand B- T T T A G T T A T C A A T C T T G G G T A G A A C-

(a) Strand A is the template strand.

(b) The polarity of the strands is as follows:

Strand A 5′ - A A A **T C A** A T A G T T A G A A C C **C A T** C T T G - 3′
Strand B 3′ - T T T A G T T A T C A A T C T T G G G T A G A A C - 5′

(c) The translated mRNA sequence coded by the sense strand has the following nucleotide sequence:

5'- AUGGGUUCUAACUAUUGAUUU-3'

(d) By reading the triplet codons starting with the initiation codon AUG and terminating with the nonsense codon UGA, we get the following polypeptide:

fMet-Gly-Ser-Asn-Tyr-

13

Mutation, DNA Repair, and Recombination

MUTATION: SOURCE OF THE GENETIC VARIABILITY REQUIRED FOR EVOLUTION

MUTATION: BASIC FEATURES OF THE PROCESS
    Mutation: Somatic or Germinal
    Mutation: Spontaneous or Induced
    Mutation: Usually a Random, Nonadaptive Process
    Mutation: A Reversible Process

MUTATION: PHENOTYPIC EFFECTS
    Mutations with Phenotypic Effects: Usually Deleterious and Recessive
    Effects of Mutations in Human Hemoblobin Genes
    Mutations in Humans: Blocks in Metabolic Pathways
    Conditional Lethal Mutations: Powerful Tools for Genetic Studies

MULLER'S DEMONSTRATION THAT X-RAYS ARE MUTAGENIC

THE MOLECULAR BASIS OF MUTATION
    Mutations Induced by Chemicals
    Mutations Induced by Radiation
    Mutations Induced by Transposable Genetic Elements
    Expanding Trinucleotide Repeats and Inherited Human Disease

SCREENING CHEMICALS FOR MUTAGENICITY: THE AMES TEST

DNA REPAIR MECHANISMS
    Light-Dependent Repair
    Excision Repair
    Mismatch Repair
    Postreplication Repair
    The Error-Prone Repair system

INHERITED HUMAN DISEASES WITH DEFECTS IN DNA REPAIR

DNA RECOMBINATION MECHANISMS
    Recombination: Cleavage and Rejoining of DNA molecules
    Gene Conversion: DNA Repair Synthesis Associated with Recombination

---

IMPORTANT CONCEPTS

  A. Mutations are heritable changes in the genetic material.
    1. Mutations that involve changes at specific sites of a gene are called point mutations.
    2. Forward mutations are those that change a wild-type allele to an allele that results in an abnormal phenotype. A mutation of a mutant allele to the wild-type allele is a back mutation.
    3. Mutations may occur spontaneously or may be induced by agents that interact with DNA and RNA.
      a. Irradiation can induce mutation.
        (1) Examples of mutagenic ionizing radiation are X-rays, gamma rays, and neutrons.
        (2) Ultraviolet light is nonionizing. It is maximally absorbed by DNA at a wavelength of 254 nm, the wavelength at which it is most mutagenic. Thymine dimerization is the major mutagenic effect of UV.

b. Many chemicals that react with DNA and RNA are potent mutagenic agents.
   (1) Base analogs have structures similar to normal bases, are incorporated into DNA during replication, and cause transition mutations (purine substituting for purine, and pyrimidine substituting for pyrimidine) through base mispairing due to tautomeric shifts.
   (2) Nitrous acid deaminates adenine to hypoxanthine, which pairs with cytosine during replication, and cytosine to uracil, which pairs with adenine during replication. It causes transition mutations.
   (3) Acridine dyes such as proflavin and acridine orange cause frameshift mutations because they intercalate between stacked base pairs in DNA and cause addition or deletion of one or a few base pairs during replication.
   (4) Alkylating agents, exemplified by ethyl methanesulfonate (EMS), add alkyl groups to bases in the DNA. Alkylating agents cause transversions, transitions, frameshifts, and chromosomal breaks which may lead to aberrations.
   (5) Hydroxylamine (NH_2OH) reacts specifically to hydroxylate cytosine to hydroxylaminocytosine, which pairs with adenine during DNA replication. Therefore it causes only GC -> AT transitions.

4. Mutations may also be induced by transposable genetic elements.
5. New mutations provide the raw material for evolution. Thus, some level of mutation is needed to allow organisms to adapt to changes in the environment.
6. Most mutations are detrimental to the organism in which they reside. Mutation rates that are too high would be disadvantageous to a species, except possibly in a rapidly changing environment.
7. The potential benefits of the use of X-rays, nuclear reactors, and other types of ionizing radiation must be carefully weighed against the known and estimated potential risks. Similar precautions must be taken to prevent the continued pollution of our environment with mutagenic and carcinogenic chemicals. These risk estimates and precautions must take into account the potential harm to future generations of living organisms, including humans, keeping in mind the increased frequencies of deleterious mutations that may result.

B. The integrity of the genetic information of living organisms is protected by a whole battery of repair mechanisms, each functioning to correct a specific type of defect in DNA.
   1. E. coli has at least five distinct mechanisms for the repair of DNA defects.
      a. Photoreactivation is catalyzed by photolyase and repairs thymine dimers by removing cross-linkages.
      b. Excision repair involves recognition of a damaged region by an enzyme complex that excises bases of the defective single strand, fills in the gap with DNA polymerase activity, and seals the break left by DNA polymerase by DNA ligase.
         (1) Base excision repair systems remove abnormal or chemically modified bases from DNA.
         (2) Nucleotide excision repair pathways remove larger defects like thymine dimers.
      c. The mismatch repair pathway provides a backup to replicative proofreading by correcting mismatched bases. This system uses a difference in methylation state to excise mismatched nucleotides in the newly synthesized strand. This requires a GATC-specific endonuclease, exonucleases, DNA polymerase, and DNA ligase.
      d. Postreplicative repair is a recombination-dependent repair process.
      e. The error-prone repair system operates when the DNA has been heavily damaged. During the SOS response, a whole battery of DNA repair, recombination, and repair proteins are synthesized. This allows DNA replication to proceed across damaged segments in template strands and subsequently results in an increased frequency of replication errors and mutations

C. Individuals with inherited disorders in DNA repair pathways suffer from developmental abnormalities and health problems, often including susceptibility to certain types of cancer.

1. An example of a human inherited disease with defects in DNA repair is *Xeroderma pigmentosum*.
   a. Individuals with *Xeroderma pigmentosum* are extremely sensitive to sunlight which results in a high frequency of skin cancer.
D. Recombination of genes located on homologous chromosomes is produced by breakage and rejoining of the resident DNA molecules.
   1. Recombination involves the breakage of DNA molecules in individual chromatids and the exchange of parts.
      a. The breakage and reunion process usually is associated with a small amount of DNA repair synthesis.
   2. When crosses are performed using genetic markers that are far apart, recombination is usually reciprocal. However, when markers are closely linked, a type of nonreciprocal recombination called gene conversion is often observed.
      a. Gene conversion results from the repair of nucleotide-pair mismatches that are produced during the recombination process.

IMPORTANT TERMS

In the space allotted, concisely define each term.

mutation:

point mutations:

germinal mutations:

somatic mutations:

spontaneous mutation:

induced mutation:

mutagen:

replicate plating:

forward mutation:

back mutation:

neutral mutation:

auxotrophs:

conditional lethal mutation:

CIB method:

tautomeric shift:

transition:

transversion:

alkylating agents:

base analogs:

nitrous acid:

acridine dyes:

hydroxylating agents:

ionizing radiation:

ultraviolet radiation:

transposons:

expanding trinucleotide repeats:

light-dependent repair (photoreactivation):

DNA photolyase:

base excision repair:

excinuclease:

mismatch repair:

postreplication repair:

SOS response:

RecA protein:

single-strand assimilation:

gene conversion:

heteroduplexes:

Tay-Sach's disease:

phenylketonuria:

alkaptonuria:

albinism:

sickle cell anemia:

IMPORTANT NAMES

In the space allotted, concisely state the contribution made by each individual or group.

Herman J. Muller:

Bruce Ames:

Robin Holliday:

Joshua and Esther Lederberg:

Jean Lamarck:

Trofid Lysenko:

Robert Edgar, Jonathan King and William Wood:

TESTING YOUR KNOWLEDGE

*In this section, answer the questions, fill in the blank, and solve the problems in the space allotted. Problems noted with an * are solved in the Approaches to Problem Solving section at the end of the chapter.*

1. A single base pair change in a gene that results in a single amino acid substitution in the corresponding protein is called a _____ mutation.

2. Nitrous acid is mutagenic because it deaminates _____ to _____, and _____ to _____.

3. If a human has a genetic block between phenyalanine and tyrosine, a disease called _____ results.

4. What type of mutation may result if a single base pair is inserted or deleted in the DNA double helix?

5. If a human cannot metabolize the brain ganglioside GM2, a disease results that is called _____ .

6. Humans who lack homogentisic acid oxidase activity have a disease called _____.

7. If a human cannot repair thymine dimers, which disease will occur?

8. A point mutation that results in premature termination of translation of the corresponding mRNA is called a _____ mutation.

9. Sickle cell hemoglobin differs from hemoglobin A in that it has _____ instead of _____ at the sixth position in the beta polypeptide.

10. A mutation of a DNA strand in which a purine replaces a pyrimidine and vice versus is called a _____ .

11. The chemical mutagen, hydroxylamine (H_2NOH), reacts with cytosine to form _____.

12. An enzyme catalyzed, light-independent, process of repair of thymine dimers in DNA that involves removal of a dimer and some flanking nucleotides, synthesis of a new segment of DNA complementary to the undamaged strand, and DNA ligase is called _____.

13. An example of a mutagenic base analog is _____.

14. An example of an intercalating agent, that inserts itself between adjacent base pairs and causes base pair addition or deletions during DNA replication is _____.

15. The enzyme required for light-dependent repair of thymine dimers is called _____.

16. A mutation of a DNA strand in which a different purine replaces a purine, and a pyrimidine replaces a pyrimidine is called a _____.

17. Sickle cell hemoglobin differs from hemoglobin C only in that it has _____ instead of _____ at the sixth position in the beta polypeptide.

*18. A culture of *Bacillus* was treated with nitrous acid and several mutants were recovered. From one of these mutants, wild type mutations revertants (back mutations) could be obtained by treating with 5-bromouracil, but not with hydroxylamine.

(a) Illustrate, using a DNA base pair, the pathway by which the original mutant was generated.

(b) Indicate the steps by which bromouracil may have produced the back mutation.

*19. The base-pair sequence of a gene coding for a very small polypeptide is:

Strand A   3' -T T T A C C A C C C C T A C C A G A A C A C T T -5'
Strand B   5' - A A A T G G T G G G G A T G G T C T T G T G A A -3'

(a) If this gene codes for a polypeptide that contains three tryptophans, which strand is transcribed?

(b) What is the nucleotide sequence of the mRNA transcribed from this DNA?

(c) What is the polypeptide sequence of the protein translated from this mRNA?

(d) Hydroxylamine affects the bold-faced base pair and causes a mutation during DNA replication. What is the amino acid sequence of the polypeptide coded by the mutant gene?

*20. The following is an eight amino acid sequence from the amino terminal end of a wild-type enzyme of *E. coli*.

fMet - Trp - Cys - Trp - Gln - Leu - Asn - Trp - (aa)$_n$

After treatment of the bacteria culture with an acridine, a mutant is isolated that lacks the enzyme activity. Only a partial polypeptide consisting of five amino acids is detected as the gene product in the mutant strain. The amino acid sequence is:

fMet - Gly - Ala - Gly - Asn

A culture of the mutant bacteria is grown in the presence of an acridine and a revertant with partial enzyme activity is recovered. The mutant had a normal sized polypeptide chain, differing from the wild type in that it had different amino acids at positions two, three, four and five. The amino acid sequence of the partial revertent was:

fMet - Gly - Ala - Gly - Asn - Leu - Asn - Trp - (aa)$_n$

Determine, as much as possible, the nucleotide sequence of the mRNA coding for the polypeptide of the wild-type, mutant and partial revertent strains.

KEY FIGURES

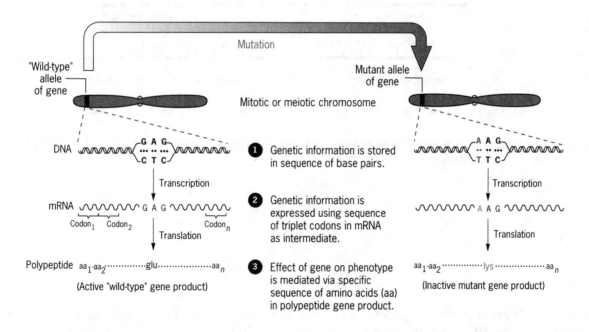

Figure 13.7 Overview of the mutation process and the expression of wild-type and mutant alleles. Mutations alter the sequence of nucleotide pairs in genes, which, in turn cause changes in the amino acid sequences of the polypeptides encoded by these genes. A G:C base pair (top, left) has mutated to an A:T base pair (top, right). This mutation changes one mRNA codon from GAG to AAG and one amino acid in the polypeptide from glutamic acid (glu) to lysine (lys). Such changes often yield nonfunctional gene products.

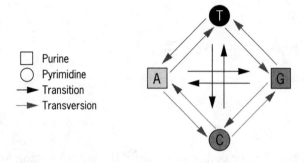

(a) Twelve different base substitutions can occur in DNA.

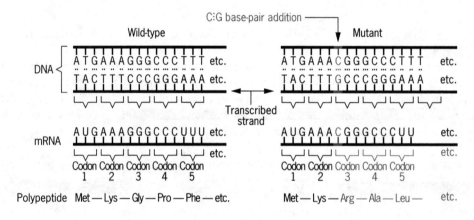

(b) Insertions or deletions of one or two base pairs alter the reading frame of the gene distal to the site of the mutation.

Figure 13.15 Types of point mutations that may occur in DNA: (a) base substitutions and (b) frameshift mutations. (a) The base substitutions include four transitions (purine for purine and pyrimidine for pyrimidine; solid arrows) and eight transversions (purine for pyrimidine and pyrimidine for purine; dashed arrows). (b) A mutant gene (top, right) was produced by the insertion of a C: G base pair between the sixth and seventh base pairs of the wild-type gene (top, left). This insertion alters the reading frame of that portion of the gene distal, relative to the direction of transcription and translation (left to right, as diagrammed) to the mutation. The shift in reading frame, in turn, changes all of the codons in the mRNA and all of the amino acids in the polypeptide specified by the base-pair triplets distal to the mutation.

THOUGHT CHALLENGING EXERCISE

Mutation can affect the genome in many different ways. For now, lets consider mutations that (1) result in changes of the coding sequences of genes, and (2) result in changes of nucleotide sequences of introns. Which type of mutation do you feel would have the most evolutionary potential, and why?

SUMMARY OF KEY POINTS

Mutations are heritable changes in the genetic material that provide the raw material for evolution.

Mutations occur in both germ-line and somatic cells, but only germ-line mutations are transmitted to progeny. Mutations may occur spontaneously or be induced by mutagenic agents in the environment. Mutation usually is a nonadaptive process, with an environmental stress simply selecting organisms with preexisting, randomly occurring mutations. Restoration of the wild-type phenotype in a mutant organism may occur by either back mutation or the occurrence of a suppressor mutation.

The effects of mutations on the phenotype of living organisms range from minor to lethal changes. Most mutations exert their effects on phenotype by altering the amino acid sequences of polypeptides, the primary gene products. The mutant polypeptides, in turn, cause blocks in metabolic pathways. Conditional lethal mutations provide powerful tools with which to dissect biological processes such as morphogenesis.

Herman J. Muller first demonstrated that an external agent can be mutagenic when, in 1927, he showed that X rays induce recessive lethals on the X chromosome of *Drosophila*.

Mutations are induced by chemicals, ionizing irradiation, ultraviolet light, and endogenous transposable genetic elements that jump from one position in the genome to new sights. Point mutations are of three types: (1) transitions, purine for purine and pyrimidine for pyrimidine substitutions, (2) transversions, purine for pyrimidine and pyrimidine for purine substitutions, and (3) frameshift mutations, additions or deletions of one or two nucleotide pairs, which alter the reading frame of the gene distal to the site of the mutation. Several inherited human diseases are caused by expanded trinucleotide repeats.

By using *Salmonella* tester strains that carry various types of mutations in genes encoding histidine biosynthetic enzymes, Bruce Ames and co-workers developed an inexpensive and sensitive method for detecting the mutagenicity of chemicals.

Multiple DNA repair systems have evolved to safeguard the integrity of genetic information in living organisms. Each repair pathway corrects a certain type of damage in DNA.

The importance of DNA repair pathways is documented convincingly by inherited human disorders that result from defects in DNA repair. Recent evidence indicates that the onset of certain types of cancer may be associated with defects in specific DNA repair pathways.

Crossing over involves breakage of homologous DNA molecules and rejoining of parts in new combinations. When genetic markers are closely linked, the four products of meiosis often contain three copies of one marker and one copy of the other marker. This gene conversion results from repair synthesis that occurs during the recombination process.

ANSWERS TO QUESTIONS AND PROBLEMS

1) missense **2)** adenine to hypoxanthine and cytosine to uracil **3)** phenylketonuria (PKU) **4)** frameshift **5)** Tay-Sach **6)** alkaptonuria **7)** *Xeroderma pigmentosum* **8)** nonsense **9)** valine, glutamic acid **10)** transversion **11)** hydroxylaminocytosine **12)** excision repair **13)** 5-bromouracil **14)** acridine orange, proflavin **15)** photolyase **16)** transition **17)** valine, lysine **18-20)** see *Approaches to Problem Solving* section

Some ideas concerning the thought challenging exercise:

The amino acid sequence of proteins is the result of millions of years of evolution. Therefore mutations that affect the coding regions of genes are likely to be detrimental or neutral in their effects, rather than improving upon the function of the protein. There is a lot more to learn about the nature of introns. However, we do have examples of mutations of introns that affect the rate of processing of pre-mRNAs and reduce the amount of mRNA and protein produced. Mutations that affect the activity, or timing of the activity, of genes are considered to be much more likely to be of evolutionary significance. It is likely that some mutations of introns may affect splicing choices and result in the production of a different but related protein. A mutation resulting in the production of a novel protein would more likely to be of evolutionary significance than a mutation that effects the amino acid sequence of an existing protein.

APPROACHES TO PROBLEM SOLVING

18. A key to working this problem is to know the action of the chemical mutagens. The transition mutation induced by nitrous acid could be reverted by treatment with 5-bromouracil, but not with hydroxylamine. Therefore, the original mutant involved a **CG -> AT** transition. Hydroxylamine reacts specifically with cytosine and can not cause transitions of A-T base pairs.

(a) The original mutation occured as follows:

CG ---> nitrous acid → UG - DNA replication → UA   DNA replication → **TA**
deaminates   uracil pairs with   adenine pairs
cytosine to   adenine   with thymine
uracil

(b) 5-bromouracil led to the back mutation as follows:

TA---> 5-BU (B) present during → BA --5-BU shifts to its → BG --DNA replication → **CG**
DNA replications gets   enol tautomer during   guanine pairs
incorporated in its keto   the next replication   with cytosine.
tautomer in place   and pairs with guanine.
of thymine.

19. (a) To solve this problem, first pick the DNA strand that has three triplets, ACC, that are complementary to the tryptophan codon, UGG. That is strand A.

(b) Next transcribe strand A of the DNA to obtain the mRNA sequence:

5′- A A A U G G U G G G G A U G G U C U U G U G A A -3′

(c) Translate the mRNA starting with the initiation codon AUG and terminating with a nonsense codon, UGA in this example.

5′- A A AUG GUG GGG AUG GUC UUG UGA A -3′

fMet Val Gly Met Val Leu

(d) Hydroxylamine reacts specifically with cytosine to form hydroxlyaminocytosine, which in turn pairs with A during DNA replication to cause CG → TA transitions. Therefore, if the bold-faced base pair is affected, the **C** will be substituted with **T** in the mutant.
G **A**

This will result in the following nucleotide sequence of the affected mRNA with just a single base change indicated by bold-facing:

5′- A A AUG GUG G**A**G AUG GUC UUG UGA A -3′

The new codon GAG, resulting from the single base change, codes for Glu. Therefore, the amino acid sequence coded by the mutant gene is Met Val **Glu** Met Val Leu.

20. The key to solving this problem is knowing that acridines cause single base-pair additions or deletions in the DNA double helix. When these occur in the coding region of a gene they result in frameshift mutations.

The next step is to indicate all possible codons for the amino acids of the wild type protein as follows:

fMet -	Trp -	Cys -	Trp -	Gln -	Leu -	Asn -	Trp - $(aa)_n$
AUG	UGG	UGU	UGG	CAA	UUA	AAU	UGG
		UGC		CAG	UUG	AAC	
					CU_		

Compare the amino acid sequence of the mutant with that of the wild-type. Find the first amino acid at which they vary and see whether it is a single base addition or deletion that has caused the change in amino acid sequence. The deletion of the **U** in the first Trp codon in the mRNA changes the reading frame to generate the codon GGU for Gly.

fMet - Gly - Ala - Gly - Asn

AUG**U**GGUGC

The codons for the amino acids of the mutant are determined by comparing to the possible wild type codons and realizing that the reading frame has been shifted one base to the right. The nucleotide sequence of the mutant mRNA is therefore determined to be:

```
fMet -  Gly -   Ala -   Gly -   Asn

AUGUGGU     GCU    GGC    AAU   UGA   AUUGG
                   C      A           C
```

The nonsense codon generated downstream from the initial point of the frameshift causes premature termination of translation of the mutant mRNA and only a polypeptide that is five amino acids long.

The nucleotide sequence of the wild-type mRNA is:

```
fMet -  Trp -  Cys -  Trp -  Gln -  Leu -  Asn -  Trp - (aa)n

AUG   UGG   UGC   UGG   CAA   UUA   AAU   UGG
                        C     G     C
```

Only amino acids two through five differ between the partial reverent and the wild type protein. The mutation that partially restored enzymatic activity was induced by an acridine. An addition of a base pair (AT in this case) in the gene and a **U** in the mRNA restored the correct reading frame.

```
fMet - Gly -  Ala -  Gly -  Asn -  Leu -  Asn -  Trp - (aa)n

AUG   GGU   GCU   GGC   AAU   UUA   AAU   UGG
                        C     G     C
```

14

Definitions of the Gene

IMPORTANT CONCEPTS

A. The definition of a gene has changed as more genetic knowledge accumulated.

1. The gene was first defined as the unit of genetic material controlling the interaction of one phenotypic characteristic or one trait.
2. Subsequently, the gene was more precisely defined as the unit of genetic material encoding one polypeptide (or one RNA molecule in the case of genes not coding for proteins).
3. Prior to 1941, the gene was also believed not to be subdivisible by mutation or recombination.
   a. Many mutable sites separable by recombination exist within each gene.
      (1) The unit of genetic material not subdivisible by mutation or recombination is now known to be the single nucleotide pair.

B. The gene is operationally defined by the complementation test.

1. A *cis-trans* test is performed by constructing *cis* heterozygotes and *trans* heterozygotes and determining whether they have mutant or wild-type phenotypes.
   a. *Cis* heterozygotes carry the two mutations on one chromosome and the corresponding wild-type alleles on the homologous chromosome.
      (1) The *cis* heterozygotes must exhibit the wild-type phenotype or the results of the *trans* test will not be informative.
   b. *Trans* heterozygotes carry the two mutations on different homologous chromosomes.
      (1) The complementation test, the *trans* part of the *cis-trans* test, is used to determine whether two mutations are in the same gene or in different genes. If a *trans* heterozygote has the mutant phenotype, the two mutations are in the same gene. If a *trans* heterozygote has the wild-type phenotype, the two mutations are in different genes.

2. The utility of the test is limited in some cases by intragenic complementation, polar effects of chain-termination mutations in operons, and intergenic noncomplementation resulting from the interaction between gene products.
   a. Intragenic complementation frequently occurs when the active form of the protein gene product is a multimer consisting of two or more polypeptide products of a given gene.
   b. Delimiting genes by complementation tests should be done whenever possible using null mutations (mutations resulting in no gene product, a partial gene product, or an otherwise totally nonfunctional gene product) to minimize the possibility of confounding effects of intragenic complementation.
C. Eukaryotic genes frequently are mosaics of introns and exons.
   1. In some cases, the exons of a gene are spliced together in different combinations so that a family of protein isoforms results. In these cases, the gene can be defined as a DNA sequence that is a single unit of transcription and encodes a set of closely related polypeptides.
D. In the case of antibody synthesis in vertebrates, functional genes are assembled from gene segments during the development of the antibody-producing cells of the immune system. Thus, our definition of the gene must remain flexible enough to encompass several variations on the one gene-one polypeptide concept.

IMPORTANT TERMS

In the space allotted, concisely define each term.

gene:

complementation test:

alkaptonuria:

one mutant gene-one metabolic block:

one gene-one enzyme:

one gene-one polypeptide:

colinear:

position effect:

cis-trans position effect:

coupling (*cis*) configuration:

repulsion (*trans*) configuration:

complementation test (*trans* test):

cis-trans test:

cistron:

homoalleles:

intragenic complementation:

polar mutation:

protein isoforms:

gene segments:

antibodies:

IMPORTANT NAMES

In the space allotted, concisely state the major contribution made by the individual or group.

Sir Archibald Garrod:

George W. Beadle and Edward L. Tatum:

Wilhelm Johannsen:

Clarence Oliver:

Seymour Benzer:

E. B. Lewis:

Melvin M. Green:

Charles Yanofsky:

Sydney Brenner:

George Beadle and Boris Ephrussi:

TESTING YOUR KNOWLEDGE

*In this section, fill in the blanks , answer the questions, and solve the problems in the space allotted. Problems noted with an * are solved in the Approaches to Problem Solving section at the end of the chapter.*

1. The *one gene-one enzyme* hypothesis was proposed by _____ and _____.

2. What are the instances in which the results of complementation tests **cannot** be used to unambiguously delimit genes?

3. The interaction of *trans* coded gene products is called _____.

4. If a *trans*-heterozygote has the mutant phenotype, are two mutations of the same or different genes?

5. If a *trans*-heterozygote has the wild-type phenotype, are the two mutants in the same or different genes?

6. The term *cistron* is synonymous with the term _____.

7. Who first proposed the concept of *one mutant gene-one metabolic block*?

8. The three basic tools of genetics are _____, _____, and _____.

9. Mutations of the same site that do not undergo recombination are called _____ alleles.

10. Two mutations that do not complement are of the same gene and are called _____ alleles.

11. Mutations that are both structurally and functionally allelic are called _____.

12. A mutation that not only results in a defective product of the gene in which it is located, but also interferes with the expression of one or more adjacent alleles is called a _____ mutation.

*13. In barley, single gene mutations occur that result in hood-like structures on the awns. Two recessive mutants with similar phenotypes are *subjacent-hooded* and *calcaroides*. A cross between these two mutants produced hybrids with hood-like structures on the awn. Are these mutants of the same or different genes? Justify your answer.

*14. Bacteriophage T4 that are mutant for the rapid lysis gene *rII* will enter but not lyse *E. coli* strain K12(λ). Wild-type T4 will infect and lyse strain K12(λ) bacteria. Benzer performed complementation tests between numerous pairs of *rII* mutants. The results of several complementation tests, done by simultaneously infecting K12(λ) bacteria with pairs of *rII* mutants are presented in the following table ("+" represents complementation and "o" no complementation):

Mutant:	r1	r2	r3	r4	r5	r6	r7	r8	r9
r9	o	+	+	o	+	+	o	o	o
r8	o	+	+	o	+	+	o	o	
r7	o	+	+	o	+	+	o		
r6	+	o	o	+	o	o			
r5	+	o	o	+	o				
r4	o	+	+	o					
r3	+	o	o						
r2	+	o							
r1	o								

(a) From these data, the nine *rII* mutants can be assigned to how many genes?

(b) Which mutants are assignable to the same gene?

*15. Seven different mutants of *Salmonella typhimurium* were isolated which were unable to grow in the absence of tryptophan (trp). These were examined in all possible *cis* and *trans* heterozygotes (partial diploids). All of the *cis* heterozygotes were able to grow in the absence of tryptophan. Some of the *trans* heterozygotes grew in the absence of tryptophan; some did not. The experimental results, using (+) to indicate growth and (o) to indicate no growth, are given in the following table. How many genes do these seven mutations represent? Which mutations are in the same gene(s)?

Growth of *Trans* heterozygotes (minus tryptophan)

Mutant	1	2	3	4	5	6	7
7	+	+	+	+	+	+	o
6	+	+	+	+	o	o	
5	+	+	+	+	o		
4	+	+	o	o			
3	+	+	o				
2	o	o					
1	o						

*16. Srb and Horowitz isolated several *Neurospora* mutants that were auxotrophic for arginine, i.e., they could not produce their own arginine and to grow required the addition of arginine to the culture media. The mutants were individually tested for their ability to grow in the presence of compounds suspected to be intermediate in the biochemical pathway leading to arginine. Mutant 1 grew if ornithine, citrulline or arginine was added to the media. Mutant 2 grew if citrulline or arginine, but not ornithine was added to the media. Mutant 3 grew if arginine was added to the media, but did not grow if either citrulline or ornithine was added.

(a) From analysis of the above data propose a metabolic pathway leading to arginine and indicate which step is controlled by each gene.

(b) Which precursor would be expected to accumulate in mutant 2?

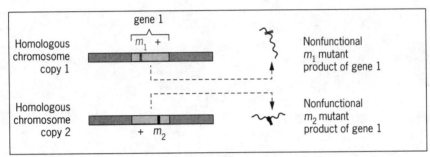

No functional gene 1 product is synthesized in the *trans* heterozygote; therefore, it will have a mutant phenotype.

(a) *trans* heterozygote: mutations in one gene.

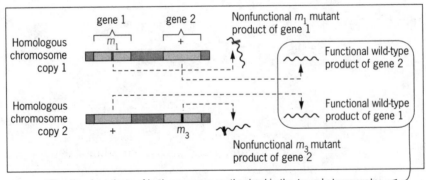

Functional products of both genes are synthesized in the *trans* heterozygote; therefore, it will have the wild-type phenotype.

(b) *trans* heterozygote: mutations in two different genes.

Figure 14.16. The *trans* test. The *trans* heterozygote should have (a) the mutant phenotype if the two mutations are in the same gene, and (b) the wild-type phenotype if the mutations are in two different genes.

THOUGHT CHALLENGING EXERCISE

Albinism in humans is a recessive genetic disorder resulting from the lack of the melanin pigment which is derived from the amino acid tyrosine. Generally, all children of parents who are both albino are also albino. However, some families where both parents are albino have only normally pigmented children. What does this tell us about the genetic nature of albinism and about the derivation of melanin?

SUMMARY OF KEY POINTS

The concept of the gene has undergone many refinements since its discovery by Mendel in 1866. Most genes encode one polypeptide and can be operationally defined by the complementation test.

The existence of a basic genetic element, the gene, that controlled a specific phenotypic trait was established by Mendel's work in 1866. Since the discovery of Mendel's results in 1900, the concept of the gene has evolved from the unit that can mutate to cause a specific block in metabolism, to the unit specifying one enzyme, to the sequence of nucleotide pairs in DNA encoding one polypeptide chain.

The concept of the gene has evolved from a bead on a string, not divisible by recombination or mutation, to a sequence of nucleotide pairs in DNA encoding one polypeptide chain. The unit of genetic material not divisible by recombination or mutation is the single nucleotide pair.

The complementation or *trans* test provides an operational definition of the gene; it is used to determine whether mutations are in the same gene or different genes. Intragenic complementation may occur when a protein is a multimer containing at least two copies of one gene product.

The transcripts of some genes undergo alternate pathways of splicing to produce mRNAs with different exons joined together. Translation of these mRNAs produce closely related polypeptides called protein isoforms. Other genes, such as those encoding antibody chains, are assembled from gene segments during development by regulated processes of somatic recombination.

ANSWERS TO QUESTIONS AND PROBLEMS

1) Beadle and Tatum **2)** dominant and codominant mutations, mutations that exhibit intragenic complementation, polar mutations, and *cis* acting genes that don't code for diffusible products **3)** complementation **4)** same **5)** different **6)** gene **7)** Garrod **8)** mutation, recombination, complementation **9)** structural **10)** functional **11)** homoalleles **12)** polar **13-16)** see *Approaches to Problem Solving* section at the end of the chapter.

Some ideas concerning thought challenging exercise:

The albino phenotype is caused by a recessive mutant. If there was just one gene that could mutate to cause the albino phenotype, then we would expect all progeny from parents who were both albino to have only albino children. The normal children born to albino parents indicates complementation. Therefore, two different genes must be involved, with one parent homozygous at one locus and the other parent homozygous at a second locus. The fact that there are two different recessive mutant genes that can result in albinism indicates that there must be at least two steps in the biochemical pathway leading from tyrosine to melanin.

APPROACHES TO PROBLEM SOLVING

13. This is an example of using a complementation test in a plant. The hybrid from the cross between *subjacent-hooded* and *calcaroides* displayed the hood-like structure on the awns. Therefore the two independently derived mutants must be of the same gene, i.e., they are allelic.

14. (a) To approach this problem remember that mutants of the same gene do not generally complement, whereas mutants of different genes do. The *r* mutants 1, 4, 7, 8, and 9 don't complement each other, and therefore are of the same gene. The r mutants 2, 3, 5, and 6 don't complement each other, but do complement 1, 4, 7, 8, and 9. These mutant can be assigned to two genes. (b) Mutants *r1*, *r4*, *r7*, *r8*, and *r9* are assigned to one gene. Mutants *r2*, *r3*, *r5*, and *r6* are assigned to a second gene.

15. The approach to this problem is the same as the previous problem. Simply see which mutants complement (mutants of different genes) and which ones do not (mutants of the same gene). To double check your assignments, make sure that all mutants that you have assigned to one gene don't complement, but that they do complement all mutants assigned to different genes. Mutant 7 complements all other mutants. Therefore, 7 is in one gene. Mutants 5 and 6 don't complement each other, but complement all other mutants listed in the table. This tells us that mutants 5 and 6 are of a second gene. Mutant 3 and 4 don't complement but complement all the other mutants. Mutants 3 and 4 must belong to a third gene. Mutants 1 and 2 don't complement, but complement all other mutant genes. Mutants 1 and 2 must be of a fourth gene.

16. The data presented allows a biochemical pathway and the step affected by each gene to be deduced.

   (a) Since all mutants were screened for the inability to grow without the addition of arginine, they must be unable to carry out one or more steps in the production of arginine. Mutant 3 grows only if arginine is added. Therefore, it cannot convert a precursor to arginine.

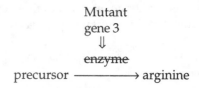

The wild-type allele of mutant 3 must code for the enzyme that catalyzes this step.

Mutants 2 grows if citrulline or arginine is added. Therefore, citrulline must be the immediate precursor to arginine, and mutant 2 blocks the pathway at the step that in normal strains converts a precursor to citrulline.

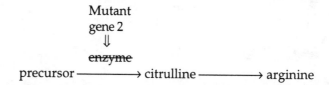

Mutant 1 grows if ornithine, citrulline, or arginine is added to the medium. It must affect a step in the pathway before these three compounds are produced.

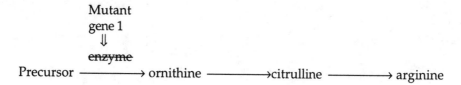

The wild-type allele of mutant gene 1 must code for the enzyme that converts a precursor to ornithine.

The biochemical pathway and the step controlled by each gene are indicated below:

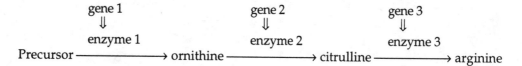

(b) In mutant 2, ornithine is not converted to citrulline. Therefore, ornithine would accumulate.

15

The Genetics of Viruses

IMPORTANT CONCEPTS

A. Bacterial viruses (bacteriophages) were discovered and recognized as distinct entities in the early part of the twentieth century.
    1. Bacteriophages were first used as research tools in genetic studies in the 1930s.

B. The bacteriophage T4 has a life cycle of about 22 minutes.
    1. The phage attaches to the host cell and injects its DNA into the cell.
        a. The host cell's metabolic machinery is used for production of progeny phage which are released by lysis of the cell.
        b. The phage DNA contains hydroxymethylcytosine instead of cytosine and this serves to protect it from the host cell's DNA degrading enzymes.
        c. Phage genes are mapped by mixedly infecting bacterial cells with genetically different phage and then analyzing the progeny for recombination events.
        d. The mechanism of recombination in phage is different from that in sexually reproducing organisms, but the strategy of constructing a phage genetic map is similar.

C. The rII region in T4 contains two genes, A and B.
    1. Over 2000 mutants in these two genes were analyzed by Benzer using pairwise crosses and deletions.
        a. The mutants mapped to about 300 different sites.
        b. This was one of the most detailed genetic analyses of a gene ever undertaken.
    2. The phage T4 has a circular genetic map but a physically linear chromosome.
        a. This paradox is explained by the terminally redundant, circularly permutated T4 chromosome.

D. The bacteriophage φ174 codes for more proteins than its DNA could linearly encode.

a. This paradox was resolved when it was discovered that φ174 has genes contained within genes. Sequences of nucleotides are read in different frames so that a single nucleotide may be part of more than one gene.

E. HIV is the retrovirus that causes AIDS.
1. This virus infects and destroys a type of white blood cell called the helper T-cell which directs the other cells of the immune system.
   a. When the helper T-cells are destroyed, the entire immune system fails and opportunistic infections occur.
2. The HIV life cycle begins with the binding of the gp120 region of a glycoprotein to the CD4 receptor of a helper T-cell and the membrane of the HIV virus fuses with the host cell membrane. The viral capsid is taken up by the host cell.
   a. The RNA chromosome of the HIV virus is converted to double-stranded DNA by the enzyme reverse transcriptase.
   b. The HIV DNA migrates to the nucleus and is integrated into the host nuclear genome by the action of a virally encoded enzyme called integrase.
   c. The integrated HIV genome may be inactive or it may be transcriptionally active, producing progeny viral genomes.
      (1) An active integrated HIV genome is transcribed into either genomic or mRNA.
      (2) Following transcription, HIV mRNA is translated into viral proteins.
      (3) The viral core, composed of core proteins surrounding genomic RNA and enzymes, is formed at the cellular membrane.
      (4) Budding of the progeny viruses occur through the host cell membrane, where the core acquires its external envelope.
      (5) The mature HIV virus externally looks like a twenty-sided soccer ball, the shape of which is determined by the proteins that make up the outer protein coat or capsid. Overlying the capsid is a lipid membrane derived from the host.
3. Currently, a tremendous amount of energy and research dollars are being invested in the understanding and control of the HIV virus.

IMPORTANT TERMS

In the space allotted, concisely define each term.

bacteriophage:

early phage proteins:

late phage proteins:

lysozyme:

restriction enzymes:

plaque morphology:

rapid lysis:

phage complementation analysis:

phage deletion mapping:

concatomers:

headful mechanism:

overlapping genes:

mottled plaques:

DNA heteroduplexes:

acquired immune deficiency syndrome (AIDS):

human immunodeficiency virus (HIV):

reverse transcriptase:

CD4 receptor:

helper T cell:

macrophage:

retrovirus:

integrase:

long terminal repeats (LTRs):

IMPORTANT NAMES

In the space allotted, concisely state the major contribution made by the individual or group.

Frederick Twort:

Felix d'Herelle:

Max Delbruck:

Alfred Hershey:

Seymour Benzer:

A. H. Doermann, George Streisinger, and Franklin Stahl:

TESTING YOUR KNOWLEDGE

*In this section, fill in the blanks, answer the questions and solve the problems in the space allotted. Problems noted with an * are solved in the Approaches to Problem Solving section at the end of the chapter.*

1. What is the smallest unit of genetic material that can be changed by mutation?

2. A virus that infects a bacterial cell is called a _____.

3. What are the two types of bacteriophage mutations that are easiest to study genetically?

4. The smallest unit of recombination is the _____ .

5. Name two examples of conditional mutations of bacteriophages.

6. A clear area in a confluent lawn of bacterial cells that results from lysis or killing of contiguous cells by several cycles of bacteriophage growth is called a _____.

7. What are two unusual features about the DNA of the bacteriophage φX174?

8. Why do occassional plaques occur when *rII* mutants are plated on a lawn of *E. coli* K12 (λ)?

9. Why do plaques never occur when *rII* deletion mutants are plated on a lawn of *E. coli* K12 (λ)?

10. What are the two human blood cell types infected by the HIV virus?

11. The HIV virus belongs to a class of viruses called _____.

12. The cellular surface molecule to which the HIV virus binds is called_____ .

13. Which class of enzymes is used by the bacterium to protect it from invading viruses?

14. The DNA of bacteriophage T4 may be protected from destruction by enzymes of the bacteria because it has _____ instead of cytosine in its DNA.

*15. Strain B *E. coli* cells were simultaneously infected with two *rII* mutants, *r1* and *r2*. Identically diluted aliquots of the progeny phage were plated on (1) *E. coli* strain B to detect the total number of phage progeny, and on (2) *E. coli* strain K12(λ) to detect the number of wild-type phage. The plaques on strain B numbered 10,000 and those on K12(λ) numbered 10. What is the frequency of recombination between these mutants?

*16. A series of *rII* mutants were crossed and the following frequencies of wild-type recombinants were recovered:

$r1 \times r2 \rightarrow$ .011

$r1 \times r3 \rightarrow$ .005

$r1 \times r4 \rightarrow$ .003

$r2 \times r3 \rightarrow$ .006

$r2 \times r4 \rightarrow$ .008

$r3 \times r4 \rightarrow$ .002

Indicate the order of the mutants and the map distances between them.

*17. Strain B *E. coli* cells were coinfected with *rII* mutants *x* and *y*. Progeny phage plated in a dilution series on both *E. coli* strains B and K12(λ) gave the following results:

Bacterial lawn	Dilution	Number of Plaques
Strain B	10^{-8}	6
Strain K12(λ)	10^{-4}	4

(a) What is the recombination frequency between the sites of mutants *x* and *y*?

(b) If one map unit is 1 kb pairs, how many nucleotides apart are these mutants?

*18. The following illustrates a set of Benzer A cistron deletion mutants with the boundary of each mutant numbered:

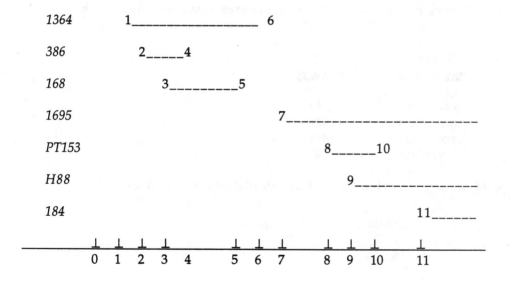

Four *rII* point mutations of the bacteriophage T4 were crossed with the seven deletion mutants where (+) indicates the appearance of r^+ recombinants and (-) indicates the lack of recombinants among the progeny.

Mutant #	1364	386	168	1695	PT153	H88	184
			Deletion Mutant				
r1	+	+	+	-	+	+	+
r2	-	+	+	+	+	+	+
r3	+	+	+	-	-	+	+
r4	-	+	-	+	+	+	+

Indicate the region , e.g., 0 - 1, 1 - 2, 2 - 3, etc., of the A cistron where each of the mutants occur.

*19. Hershey and coworkers demonstrated that two-point mapping in bacteriophage T2 could be done using rapid lysis mutants (r^+ = small plaque, r = large plaque) and host range mutants (h^+ infects only *E. coli* strain B, h infects strains B and B/2). Phage of the genotype hr^+ = and h^+r were used to coinfect *E. coli* strain B. The progeny phages were collected and plated on a mixed culture of *E. coli* B and B/2. From the following results calculate the frequency of recombination between the h and r genes.

Plaque morphology	Frequency
small, turbid	49
small, clear	425
large, turbid	413
large, clear	45

The turbid plaques result from only the lysis of strain B bacteria of the mixture of B and B/2.

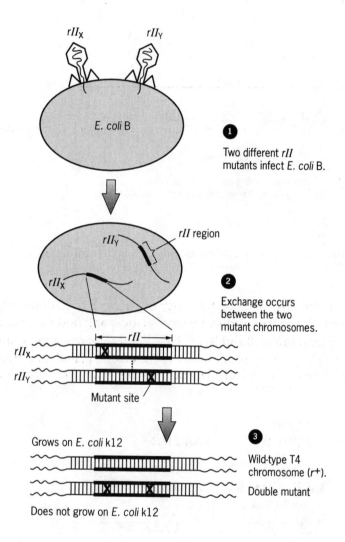

Figure 15.8   A genetic cross between two different *rII* mutants on *E. coli* B cells. The recombinant progeny are wild type (*r*⁺) and double mutants.

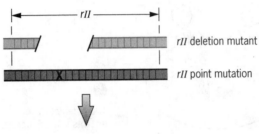

No wild type (r+) possible when point
mutation lies within area defined by the deletion.

Figure 15.9 A cross between an *rII* point mutation and an *rII* deletion mutation.

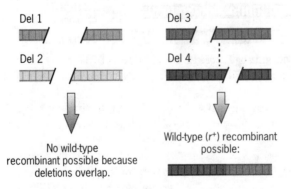

No wild-type
recombinant possible because
deletions overlap.

Wild-type (r+) recombinant
possible:

Figure 15.10 Crosses between pairs of deletion mutants. If the deletions overlap, no wild-type recombinants are produced; if they do not overlap, *r+* progeny are produced.

THOUGHT CHALLENGING EXERCISE

One way scientists are trying control the HIV virus and treat AIDS is to target genes that regulate the HIV infection process and look for ways to disrupt the viral life cycle. Another way is to try to obtain a vaccine that would induce antibodies to destroy key proteins of the HIV infection process. Why has this been so difficult with the HIV virus? You may want to do some additional reading on the replication properties of reverse transcriptase.

SUMMARY OF KEY POINTS

Initially, viruses were thought to be a type of bacteria, but key observations by Twort and d'Herelle suggested that viruses were distinct organisms that infected and then destroyed bacteria. Delbruck and Ellis recognized the genetic significance of viral simplicity and launched studies into their life cycle.

Bacteriophage require a bacterial host to complete their life cycle. The phage attach to the host cell and inject their DNA. The phage then takes over the cellular protein-synthesizing machinery. Phage DNA directs the synthesis of progeny phage that are released from the cell and infect other cells. T4 viruses have glucosylated hydroxymethylcytosine instead of cytosine in their DNA. This protects them against degradation by bacterial restriction enzymes.

Phage phenotypes, such as plaque morphology and host range, were utilized in early mapping studies. The mechanism of recombination in phage is different than it is in sexually reproducing organisms. Benzer mapped over 2400 mutations in the rII region of T4 and demonstrated that the base pair was the smallest unit of mutation and that recombination probably occurred between adjacent base pairs. His results solidified the concept of the gene as a unit of function that is divisible by mutation and recombination.

Phage T4 has a circular genetic map, but the chromosome is physically linear. To resolve the paradox, experiments showed that the chromosome was terminally redundant and circularly permutated.

The bacteriophage ϕX174 has enough genetic information in its single-stranded DNA chromosome to code for about four or five proteins. However, because its genes overlap, ϕX174 codes for 11 proteins.

DNA molecules may break in such a way that they are left with single-stranded ends. If DNA fragments with complementary single-stranded ends come together, they may form a heteroduplex molecule: One DNA strand in a specific region carries genetic information for one allele, and the other DNA strand carries genetic information for a different allele. This discovery provided insight into the molecular mechanism of recombination.

ANSWERS TO QUESTIONS AND PROBLEMS

1) the individual base pair **2)** bacteriophage **3)** plaque morphology; conditional **4)** the individual base pair **5)** host range; temperature sensitive **6)** plaque **7)** single-stranded DNA; overlapping genes **8)** $r \rightarrow r^+$ back mutations occurred **9)** once part of a gene is lost, it generally can't be regained **10)** helper T-cells, macrophages **11)** retrovirus **12)** CD4 **13)** restriction enzymes **14)** glucosylated hydroxymethylcytosine **15 - 19)** see *Approaches to Problem Solving* section that follows.

Comments on thought challenging exercise:

A major difficulty in controlling the HIV virus is its very rapid rate of evolution and the occurrence of multiple strains. This is due, in part, to the fact that reverse transcriptase does not have proofreading capabilities. Therefore, mistakes made during the synthesis of DNA from RNA are not corrected and appear as mutations. The occurrence of multiple strains complicates the development of vaccines. Another problem in the development of vaccines is that in the intact HIV virus, the targeted key antigens (proteins) may not be exposed where they can be recognized by antibodies.

APPROACHES TO PROBLEM SOLVING

15. Since the aliquots were of the same dilution they should contain equal numbers of progeny phage. Therefore, we can calculate the number of recombinants [number of plaques on strain K12(λ)] divided by the total number of phage in the aliquot (number of plaques on strain B). Remember, both r mutants and r^+ phage infect and lyse strain B, whereas only wild-type ($r1^+ r2^+$) will infect and lyse strain K12(λ). Also note that only the $r1^+ r2^+$ recombinants are detected. The double mutant $r1\ r2$ does not infect and lyse strain K12(λ). Since only half of the recombinants are detected we use the formula:

$$\text{Recombination (R)} = \frac{2[\text{number plaques on strain K12}(\lambda)]}{\text{number of plaques on strain B}} = \frac{2(10)}{10,000} = 0.002$$

16. The wild type phage that infect *E. coli* strain K12(λ) represent only one-half of the recombinants. Therefore, double the detected recombination frequencies to estimate the total frequency of recombination. Since one map unit equals 1% recombination, multiply the recombination frequencies by 100.

$r1 \times r2 \rightarrow$	$.011 \times 2 = .022$	map distance = 2.2
$r1 \times r3 \rightarrow$	$.005 \times 2 = .010$	map distance = 1.0
$r1 \times r4 \rightarrow$	$.003 \times 2 = .006$	map distance = 0.6
$r2 \times r3 \rightarrow$	$.006 \times 2 = .012$	map distance = 1.2
$r2 \times r4 \rightarrow$	$.008 \times 2 = .016$	map distance = 1.6
$r3 \times r4 \rightarrow$	$.002 \times 2 = .004$	map distance = 0.4

A map is constructed by assigning mutants a sequence that is additive and compatible with the individual map distances. The two mutants displaying the largest map distance are often at the opposite ends of the map, in this case $r1$ and $r2$.

$$r1 \underline{\hspace{1.5cm} 2.2 \hspace{1.5cm}} r2$$

Next find the compatible order of the other mutants.

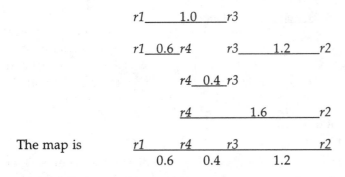

The map is

r1   r4   r3   r2
   0.6   0.4   1.2

17. The problem is approached the same as problem 15, except that the dilution of the aliquot is different for the phage plated on strain B compared with that plated on *E. coli* strain K12(λ). Since the dilution factor is 10,000-fold, the number of plaques on strain B must be multiplied by 10,000 (10,000 X 6 = 60,000). Four recombinants were detected out of 60,000 plaques. Remember, only one-half of the recombinants are detectable.

(a) Recombination (R) = $\dfrac{2[\text{number plaques on strain K12}(\lambda)]}{\text{number of plaques on strain B}} = \dfrac{2(4)}{60,000} = 0.00013$ (map distance = 0.013).

(b) If one map unit is 1 kb pairs, then (.013 X 1000 = 13) is the number of nucleotides that separate these mutants.

18. To assign each mutant to a region requires an analysis of the recovery or lack of recovery of wild type recombinants when the mutants are crossed with different deletion mutants. Wild type recombination indicates that the mutant is not in the area of the gene missing in the deletion mutant. Lack of recombinants indicates that the deletion covers the site corresponding to the position of the point mutation.

Mutant *r1* shows lack of recombination only with deletion mutant *1675*. This limits the location of the point mutation to the area covered by *1675*. The fact that it shows recombination with deletion mutants *PT153*, *H88*, and *184*, limits the location of mutant *r1* to region 7 - 8.

Mutant *r2* does not show recombination with deletion mutant *1364*, but does with all others. This limits the location of the point mutation *r2* to the area covered by *1364*. It must reside in either region 1 - 2 or region 5 - 6, because regions 1 - 2 and 5 - 6 are the only areas covered by *1364* that do not overlap the other deletion mutations.

Mutant *r3* shows recombination with all deletion mutants except *1695* and *PT153*. This indicates that *r3* is located within the area covered by deletions *1695* and *PT153*, or 8 - 10. Because the *r3* mutant shows recombination with *H88*, its can be more narrowly defined to region 8 - 9.

Mutant *r4* does not recombine with deletion mutants *1364* and *168*, but recombines with all others. This restricts it somewhere to the overlapping region of deletion mutants *1364* and *168* or region 3 - 5. Because it recombines with deletion mutant *368*, it must be located outside *368* or in region 4 - 5.

19. To approach this problem note that all the genotypes, including both recombinants, can be detected on the mixed lawn of *E. coli* strain B and strain B/2. The genotype of each plaque is indicated as follows:

Plaque morphology	Frequency	Genotype
small, turbid	49	$r^+ \ h^+$
small, clear	425	$r^+ \ h$
large, turbid	413	$r \ h^+$
large, clear	<u>45</u>	$r \ h$
	total = 932	

The recombinants r^+h^+ and $r\,h$ comprise 94 of the total of 932 plaques. The frequency of recombination is therefore 0.101.

16

The Genetics of Bacteria

IMPORTANT CONCEPTS

A. Bacteria do not reproduce sexually, but rather have developed parasexual ways to generate new gene combinations. There are three main processes of parasexual recombination: transformation, conjugation, and transduction.

    1. Transformation is the genetic alteration of bacteria brought about by the incorporation of foreign DNA into the bacterial cells.

        a. Linkage relationships can be determined in transformation experiments by determining the frequencies of co-transformation.

    2. Conjugation is a unidirectional transfer of genetic information from a donor cell to a recipient cell via a conjugation tube.

        a. A plasmid called the F factor mediates this transfer.

        b. Cells that carry the F factor are F$^+$ and those lacking it are F$^-$.

        c. If the F factor is integrated into the chromosome, the cell is Hfr (high frequency of recombination).

            (1) Hfr cells transfer chromosomal DNA from the donor to F$^-$ recipients.

            (2) Sometimes, when the F factor exits the chromosome, it carries with it a piece of the host chromosome. This F factor is called an F' factor.

        d. Conjugation can be used to map genes.

            (1) Conjugation mapping is based on the time of transfer of donor genes to recipient cells.

    3. Transduction is the transfer of DNA and recombination in bacteria mediated by bacterial viruses called bacteriophages.

a. In generalized transduction, any segment of the bacterial genome may be transferred by a phage from one bacterium to another.
    (1) Mapping experiments can be done by analyzing which genes are cotransformed on the same DNA fragment.
b. In specialized transduction, only the specific region of the bacterial chromosome bracketing the prophage insertion site is transferred from one bacterium to another.
    (1) In abortive transduction, bacterial DNA is injected into a bacterium but it does not replicate.
  B. Distinguishing between transduction, transformation, and conjugation involves determining whether cell contact is required and whether the process is sensitive to DNase.
    1. This is experimentally tested by employing a U-tube experiment.

IMPORTANT TERMS

In the space allotted, concisely define each term.

plasmid:

R-plasmid:

parasexual:

transformation:

transduction:

conjugation:

prototropic:

auxotropic:

U-tube experiment:

competent cells:

competent factor:

cotransformed:

F factor:

insertion sequences:

episome:

Hfr cell:

F′ factor:

sexduction:

interrupted mating experiment:

generalized transduction:

specialized transduction:

lambda phage:

temperate phage:

lysogenic:

prophage:

low-frequency transduction lysates (LFT lysates):

helper phage:

unstable transductant:

high-frequency transduction lysate (HFT lysate):

stable transductant:

IMPORTANT NAMES

In the space allotted, concisely state the major contribution made by the individual or group.

Frederick Griffith:

Avery, MacLeod, and McCarty:

A. D. Hershey and M. Chase:

Joshua Lederberg and Edward L. Tatum:

Bernard Davis:

William Hayes:

Norton Zinder:

TESTING YOUR KNOWLEDGE

In this section, answer the questions, fill in the blanks, and solve the problems in the space allotted. Problems noted with an * are solved in the Approaches to Problem Solving section at the end of the chapter.

1. When a bacteriophage genome has integrated into the host bacterium's DNA, it is called a (an) _____.

2. The uptake of exogenous DNA from the culture medium and its incorporation by recombination into the DNA of the bacterium is called _____.

3. A bacteriophage that lyses the host bacterium is called a _____.

4. The incorporation of bacterial genes into an F′ factor and their subsequent transfer by conjugation to an F⁻ cell is called _____.

5. A male bacterial cell, with the F factor integrated into its DNA is called a (an) _____.

6. An organism such as a bacterium that requires for growth a specific compound added to the minimal medium is called _____.

7. The ability of a bacterial cell to incorporate exogenous DNA and become genetically transformed is called _____.

8. A partial diploid in bacteria produced by the process of partial genetic exchange, such as sexduction, is called a (an) _____.

9. The bacterial plasmid that confers the ability to function as a genetic donor in conjugation is called a (an) _____.

10. A phage that invades but does not destroy (lyse) the host bacterium is called a _____.

11. The chromosomal fragment, homologous to a part of the bacterial genome, that is donated to a merozygote is called a (an) _____.

12. The life cycle of a bacteriophage which leads to the production of progeny phages, and the lysis of the host bacterial cell is called the _____ pathway.

13. The phenomenon when a bacteriophage particle carries a random or near random segment of bacterial DNA from one bacterium to another is called _____.

14. The phenomenon of a partially diploid bacterium containing an unintegrated, transduced DNA fragment is called _____.

15. The life-style of a bacteriophage which leads to integration via site-specific recombination into the bacterial chromosome is called the _____ pathway.

16. The phenomenon when a bacteriophage particle transduces only genetic markers located in one small region of the bacterial chromosome is called _____.

*17. In interrupted mating experiments involving *E. coli,* five different Hfr strains donate the following markers in the order shown.

<u>Order of entry of markers</u>

Hfr strain	1st	2nd	3rd	4th	5th
A	arg	xyl	metC	lysA	aroD
B	purE	metD	arg	xyl	metC
C	trpA	purE	metD	arg	xyl
D	xyl	arg	metD	purE	trpA
E	metC	lysA	aroD	trpA	purE

On a generalized map, indicate the order of these markers on the circular chromosome, the position where the sex factor integrated in each Hfr strain, and the direction of transfer mediated by each integrated F factor.

*18. In *E. coli*, the genes *ton*A (resistance to phages) and *pan* (auxotropic for pantothenate) are closely linked and map to position 3 on the time map. Another gene *azi*C (azide resistance) maps one time unit from these genes. Hfr bacteria of the genotype *azi*⁺ *tonA*⁺ *pan*⁺ were mated with an F⁻ strain of the genotype *azi* *ton*A *pan* for a time interval sufficient for the entry of the genes. The number of colonies scored of the different recombinant classes follows:

Genotype	Number
aziC⁺ *ton*⁺ *pan*⁺	833
aziC⁺ *ton* *pan*	200
aziC⁺ *ton* *pan*⁺	45
aziC⁺ *ton*⁺ *pan*	2

What is the linkage order and recombination frequency between the genes?

*19. In *E. coli*, a series of three-point transduction tests were performed to map the order of mutant sites in the *leu* gene. The unselected, linked marker was the *thr* gene. In each cross, *leu*⁺ recombinants were selected and then scored for *thr*⁺ or *thr*⁻. From the following table of results, deduce the linear order of the *leu*⁻ mutant.

Cross		*thr* allele in wild-type recombinants	% *thr*⁺
Donor markers	Recipient markers		
thr⁺ *leu*¹	*thr*⁻ *leu*²	60 *thr*⁺:300 *thr*⁻	17
thr⁺ *leu*³	*thr*⁻ *leu*²	185 *thr*⁺:175 *thr*⁻	51
thr⁺ *leu*²	*thr*⁻ *leu*¹	350 *thr*⁺:349 *thr*⁻	50
thr⁺ *leu*²	*thr*⁻ *leu*³	50 *thr*⁺:250 *thr*⁻	17

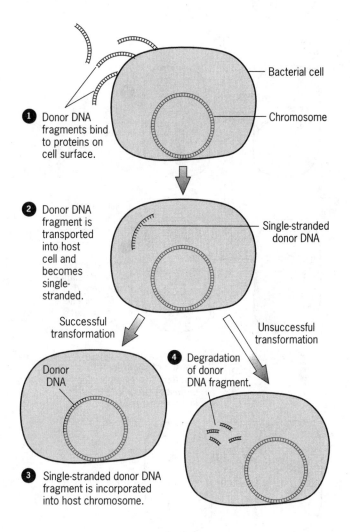

① Donor DNA fragments bind to proteins on cell surface.

Bacterial cell

Chromosome

② Donor DNA fragment is transported into host cell and becomes single-stranded.

Single-stranded donor DNA

Successful transformation

Unsuccessful transformation

④ Degradation of donor DNA fragment.

Donor DNA

③ Single-stranded donor DNA fragment is incorporated into host chromosome.

Figure 16.6   Bacterial transformation.

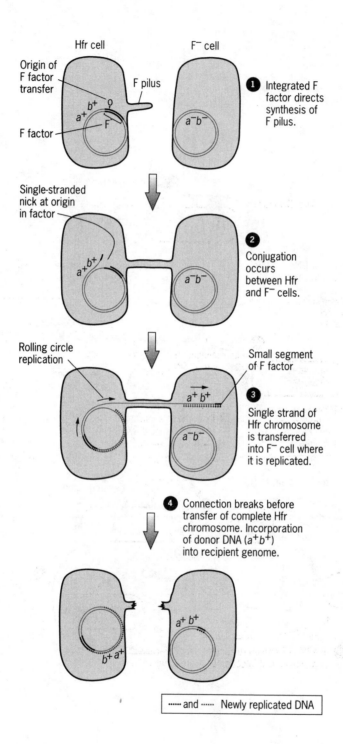

Hfr cell

Origin of F factor transfer

F pilus

a^+ b^+

F factor

F⁻ cell

a^-b^-

❶ Integrated F factor directs synthesis of F pilus.

Single-stranded nick at origin in factor

a^+b^+

a^-b^-

❷ Conjugation occurs between Hfr and F⁻ cells.

Rolling circle replication

Small segment of F factor

a^+ b^+

a^-b^-

❸ Single strand of Hfr chromosome is transferred into F⁻ cell where it is replicated.

❹ Connection breaks before transfer of complete Hfr chromosome. Incorporation of donor DNA (a^+b^+) into recipient genome.

a^+ b^+

b^+a^+

┈┈ and ┈┈ Newly replicated DNA

Figure 16.11 Hfr × F⁻ conjugation.

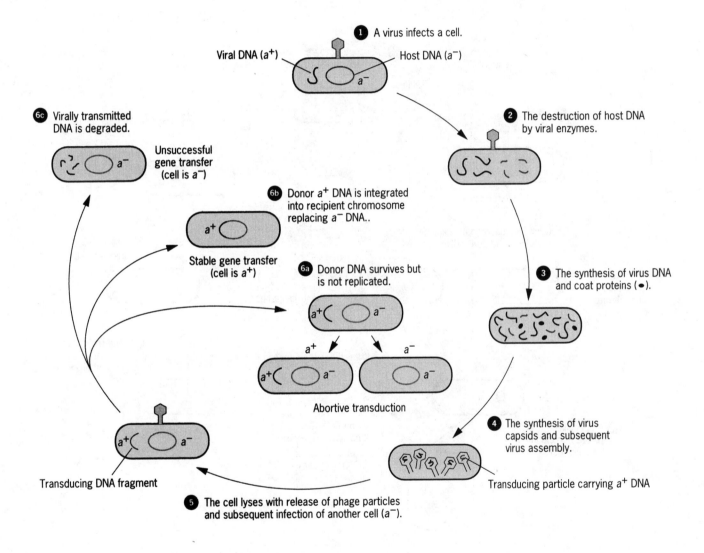

1 A virus infects a cell.

Viral DNA (a^+)

Host DNA (a^-)

a^-

6c Virally transmitted DNA is degraded.

Unsuccessful gene transfer (cell is a^-)

a^-

2 The destruction of host DNA by viral enzymes.

6b Donor a^+ DNA is integrated into recipient chromosome replacing a^- DNA..

a^+

Stable gene transfer (cell is a^+)

6a Donor DNA survives but is not replicated.

a^+ a^-

a^+ a^-

a^+ a^- a^- a^-

Abortive transduction

3 The synthesis of virus DNA and coat proteins ($\bullet$).

4 The synthesis of virus capsids and subsequent virus assembly.

Transducing particle carrying a^+ DNA

a^+ a^-

Transducing DNA fragment

5 The cell lyses with release of phage particles and subsequent infection of another cell (a^-).

Figure 16.19 **A model for the generalized transducing viruses.**

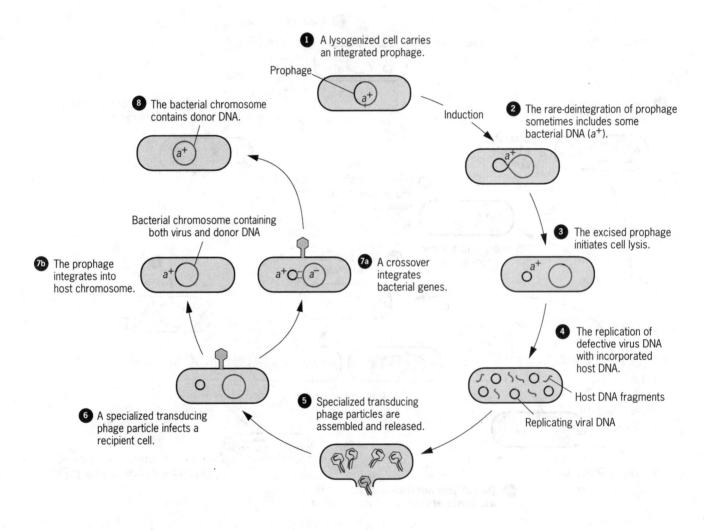

1 A lysogenized cell carries an integrated prophage.

Prophage

2 The rare-deintegration of prophage sometimes includes some bacterial DNA (a⁺).

Induction

8 The bacterial chromosome contains donor DNA.

3 The excised prophage initiates cell lysis.

Bacterial chromosome containing both virus and donor DNA

7b The prophage integrates into host chromosome.

7a A crossover integrates bacterial genes.

4 The replication of defective virus DNA with incorporated host DNA.

Host DNA fragments

Replicating viral DNA

6 A specialized transducing phage particle infects a recipient cell.

5 Specialized transducing phage particles are assembled and released.

Figure 16.21 Specialized transduction.

THOUGHT CHALLENGING EXERCISE

In the beef production business, antibiotics are often used as a supplement in cattle feed, even though no cattle are sick with bacterial infections. Also, well-meaning physicians tend to readily provide broad spectrum antibiotics to relieve potential discomfort or suffering due to bacterial or suspected bacterial infections. In light of your newly acquired knowledge of bacterial reproduction, comment on the positive and potentially negative aspects of the use of antibiotics for disease prevention.

SUMMARY OF KEY POINTS

Three main mechanisms of genetic exchange in bacteria are transformation, conjugation, and transduction.

Three main types of bacterial mutants commonly used for genetic analysis are antibiotic-resistant mutants, nutritional mutants, and carbon source mutants.

To test for which of the three parasexual processes may be operating in a particular bacterial species, it is necessary to determine if cell-cell contact is required and whether the process can be disrupted by the presence of DNase in the medium.

In transformation of some bacterial species, a fragment of double-stranded DNA is transported across the cell membrane of competent cells, is converted into a single-stranded DNA, and then is physically integrated into the recipient cell's genome, replacing its homologous sequence. Linkage in transformation experiments means that the two genes are frequently found on the same DNA fragment.. For a cell to be doubly transformed by genes far apart, two fragment insertions must occur. If the two genes are close together on the same DNA fragment, only one fragment insertion is necessary for double transformation.

Conjugation is a process in which bacteria make contact with each other and genetic material is transferred unidirectionally from donor to recipient cells. The F factor directs conjugation. Cells carrying an F factor are F+ and those without it are F-. A cell with an F factor integrated into the chromosome is called an Hfr. An F factor that has incorporated some genes of the main chromosome is called an F' factor. Interrupted conjugation experiments showed that the Hfr chromosome was transferred unidirectionally and that the first genes to be transferred were located close to the point of F-factor integration. The *E. coli* chromosome is partitioned into minutes.

Transduction is the transport of bacterial genetic material from one bacterium to another by a phage. A generalized transducing phage may carry any segment of the bacterial genome. If the transduced DNA is not incorporated into the recipient genome, abortive transduction may result. Specialized transducing phage insert into a specific site on the bacterial chromosome. They are able to incorporate only those bacterial genes located around the insertion site. Generalized transducing particles are useful for mapping bacterial genes. The closer two genes are, the greater the likelihood that they will be cotransduced. Specialized transducing particles can be used to map phage attachment sites and to analyze the genes closely linked to these sites.

ANSWERS TO QUESTIONS AND PROBLEMS

1) prophage 2) transformation 3) virulent 4) sexduction 5) Hfr 6) auxotrophic 7) competence 8) merozygote 9) F factor 10) temperate 11) exogenote 12) lytic 13) generalized transduction 14) abortive transduction 15) lysogenic 16) specialized transduction 17 to 19) see *Approaches to Problem Solving* section at the end of the chapter.

Ideas concerning thought stimulating exercise:

Cattle fed with antibiotic-supplemented feed generally grow faster. However, the prolonged use of antibiotics may create an environment that selects for those bacteria, nonpathogenic and pathogenic, that carry antibiotic-resistance genes. The use of antibiotics in treatment of human bacterial diseases prevents much suffering and in many cases is essential for complete healing or even survival. The extensive use of antibiotics in medicine does create problems in that it results in selection for antibiotic-resistant bacteria. If the resistance genes end up in pathogenic bacteria, their control by antibiotics is more difficult.

APPROACHES TO PROBLEM SOLVING

17. To approach this problem, the data on the order of the genes can be assembled either directly on a circle or on an overlapping linear map that subsequently can be presented as a circle. Using the latter approach we obtain the following overlapping linkages. The integrated F factors are indicated in bold face and the arrow indicates the direction of transfer during conjugation.

$$\leftarrow \textbf{F}^\text{E} - metC - lysA - aroD - tryA - purE -$$
$$\leftarrow \textbf{F}^\text{A} - arg - xyl - metC - lysA - aroD-$$
$$\leftarrow \textbf{F}^\text{B} - purE - metD - arg - xyl - metC -$$
$$\leftarrow \textbf{F}^\text{C} - trpA - purE - metD - arg - xyl -$$
$$- tryA - purE - metD - arg - xyl - \textbf{F}^\text{D} \rightarrow$$

The linear linkage maps above can be presented in circular form as follows.

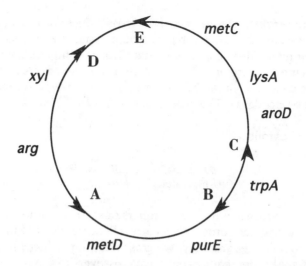

18. In this problem it is observed that all the progeny scored are recombinants for $aziC^+$, which is a marker in this Hfr strain that enters the F⁻ cell after ton^+ and pan^+. The amount of recombination is measured relative to the total number of $aziC^+$ colonies. All viable recombinants require even numbers of crossovers.

The $aziC^+\ ton^+\ pan^+$ recombinants resulted from crossovers flanking these three genes. These do not represent crossing over between $aziC$ and ton or between ton and pan.

The $aziC^+\ ton\ pan$ recombinants resulted from two crossovers, one before $aziC^+$ and one in the region between $aziC^+$ and the $ton\ pan$; $\underline{aziC^+\ ton^+\ pan^+}$.
$$aziC\ \ ton\ \ pan$$

The $aziC^+\ ton\ pan^+$ recombinants allow the order of the genes to be determined. This is done by determining which order requires the least number of crossover to give $aziC^+\ ton\ pan^+$. If the order is $aziC\ ton\ pan$, four crossovers, $\underline{aziC^+\ ton^+\ pan^+}$, would be needed to generate the $aziC^+\ ton\ pan^+$
$$aziC\ \ ton\ \ pan$$
recombinant. However, if the order was $aziC\ pan\ ton$, only two crossovers, $\underline{aziC^+\ pan^+\ ton^+}$, would
$$aziC\ \ pan\ \ ton$$
be required to generate the recombinant. The gene order is, therefore, $aziC\ pan\ ton$.

The $aziC^+\ ton^+\ pan$ recombinant requires four crossovers to occur, $\underline{aziC^+\ \boldsymbol{pan^+}\ ton^+}$. This is the
$$aziC\ \ \boldsymbol{pan}\ \ ton$$
rarest recombinant class which is compatible with our determined gene order.

The total recombination between $aziC$ and pan is $200 + 2/833 + 200 + 45 + 2 = 202/1080 = 0.1870 = 18.70$ map units.

The total recombination between pan and ton is $45 + 2/1080 = 0.04352 = 4.35$ map units.

19. To determine the order of the mutants in the leu gene, compare the number of crossovers it takes to produce the wild-type recombinants relative to thr^+ and thr for the two possible gene orders. The first cross is as follows.

$$\frac{thr^+\quad leu_1^-\quad leu_2^+}{thr\quad leu_1^+\quad leu_2}\qquad \text{or}\qquad \frac{thr^+\quad leu_2^+\quad leu_1}{thr\quad leu_2\quad leu_1^+}$$

If the first order was the correct one, it would take four crossovers to produce wild-type recombinants for thr^+ and two for thr^-. Therefore, we would expect the frequencies of thr^+ to be much less than thr^- among the wild-type recombinants. This is compatible with the observed 17% of the wild-type recombinants that are thr^+. If the second gene order is correct, then it would take two crossovers to produce either thr^+ or thr^- alleles among the wild-type recombinants and they would occur in more equal proportions. The gene order is thr - leu_1 - leu_2 .

The second cross is depicted as follows:

$$\begin{array}{ccc} thr^+ & leu_3 & leu_2^+ \\ thr^- & leu_3^+ & leu_2 \end{array} \quad \text{or} \quad \begin{array}{ccc} thr^+ & leu_2^+ & leu_3 \\ thr^- & leu_2 & leu_3^+ \end{array}$$

The correct gene order will produce about equal numbers of thr^+ (51%) and thr^- alleles among the wild-type recombinants. The first gene order is not correct because it would take four crossovers to produce thr^+ wild-type recombinants, and only two crossovers to produce thr^- wild type recombinants. The second gene order requires only two crossovers to produce either thr^+ or thr^- wild-type recombinants. We would then expect thr^- and thr^+ recombinants to occur in about equal frequency among the wild-type recombinants, as observed. Therefore, the correct gene order is thr - leu_2 - leu_3.

The third cross is as follows:

$$\begin{array}{ccc} thr^+ & leu_2 & leu_1^+ \\ thr^- & leu_2^+ & leu_1 \end{array} \quad \text{or} \quad \begin{array}{ccc} thr^+ & leu_1^+ & leu_2 \\ thr^- & leu_1 & leu_2^+ \end{array}$$

The second gene order requires only two crossovers to produce either thr^+ or thr^- among the wild-type recombinants. This is compatible with the observed frequency of thr^+ (50%) among the wild-type recombinants. The gene order is thr -leu_1 - leu_2.

The fourth cross is as follows:

$$\begin{array}{ccc} thr^+ & leu_2 & leu_3^+ \\ thr^- & leu_2^+ & leu_3 \end{array} \quad \text{or} \quad \begin{array}{ccc} thr^+ & leu_3^+ & leu_2 \\ thr^- & leu_3 & leu_2^+ \end{array}$$

If the first order was correct we would expect to see far fewer thr^+ (require four crossovers) than thr^- (require two crossovers) among the wild-type recombinants. Since only 17% of the recombinants are thr^+, this gene order appears to be correct. If the second gene order was correct, we would expect to see about equal numbers of thr^+ and thr^- among the wild-type recombinants, since each type would require only two crossovers. The correct gene order is thr - leu_2 - leu_3.

Combining the results from the four crosses we conclude that the order of the mutants of the leu gene is thr - leu_1 - leu_2 - leu_3.

17

Transposable Genetic Elements

TRANSPOSABLE ELEMENTS IN BACTERIA
   IS Elements
   Composite Transposons
   Tn3 Elements
   The Medical Significance of Bacterial Transposons

TRANSPOSABLE ELEMENTS IN EUKARYOTES
   Ac and Ds Elements in Maize
   P Elements and Hybrid Dysgenesis in *Drosophila*
   mariner, an Ancient and Widespread Transposon

RETROTRANSPOSONS
   Retrovirus-like Elements
   Retroposons

THE GENETIC AND EVOLUTIONARY SIGNIFICANCE OF TRANSPOSABLE ELEMENTS
   Transposons and Genome Organization
   Transposons and Mutation
   Evolutionary Issues Concerning Transposable Elements

IMPORTANT CONCEPTS

  A. Transposable elements, or transposons, are present in the genomes of many different organisms. These move from one position to another in the genome and are responsible for inducing mutations and breaking chromosomes.

  B. In bacteria, there are three main types of transposons: the Insertion Sequences (IS elements), the composite transposons, which are formed when two IS elements capture unrelated DNA sequences between them, and the Tn3 elements.

   1. All three types of transposons contain genes for proteins involved in transposition. In addition, the composite transposons and the Tn3 elements carry genes for antibiotic resistance.

  C. In eukaryotes, there are many different families of transposons, including the Ac/Ds elements of maize and the P elements of *Drosophila*.

   1. Like bacterial transposons, these types of eukaryotic transposons have inverted DNA sequences at their termini and create target site duplications when they insert into DNA molecules.

   2. Some members of eukaryotic transposon families encode transposase that catalyzes the movement of elements from one position to another.

  D. Eukaryotes possess different types of retrotransposons in their genomes.

   1. The movement of these elements depends on the reverse transcription of RNA into DNA by an enzyme encoded by the elements themselves.

   2. The retrovirus-like elements resemble the integrated forms of retroviruses.

   3. The retroposons have a tract of A:T base pairs at one end.

  E. In nature, transposons are responsible for a large fraction of all spontaneous mutations.

  F. Telomere-associated retroposons, which regenerate the ends of chromosomes in *Drosophila*, may perform an important function in their hosts; others may be genetic parasites.

IMPORTANT TERMS

In the space allotted, concisely define each term.

Insertion Sequences (IS elements):

inverted terminal repeats:

transposase:

target site duplication:

composite transposons:

Tn3 elements:

cointegrate:

conjugate R plasmids:

Ds (Dissociation):

Ac (Activator):

controlling elements:

hybrid dysgenesis:

P elements:

transposon tagging:

retrotransposons:

retrovirus-like elements:

long terminal repeats (LTRs):

Ty transposon:

retroposons:

long interspersed nuclear elements (LINES):

TART:

ectotopic exchanges:

retroelements:

IMPORTANT NAMES

In the space allotted, concisely state the major contribution made by the individual or group.

Barbara McClintock:

Margaret and James Kidwell:

John Sved:

Michael Simmons and Johng Lim:

Paul Bingham and Gerald Rubin:

A. J. Kingsman and S. M. Kingsman:

TESTING YOUR KNOWLEDGE:

*In this section, answer the questions, fill in the blanks, and solve the problems in the space allotted. Problems noted with an * are solved in the Approaches to Problem Solving section at the end of the chapter.*

1. In *Drosophila*, a phenomenon called _____ occurs in progeny of M females mated to P males.

2. The simplest bacterial transposons, carrying inverted terminal repeats (9 to 40 nucleotide pairs), that can insert at many different sites and mutate genes are called _____.

3. The IS elements encode a protein called _____ that is necessary for excising the element from the chromosome or plasmid.

4. The staggered cleavage of the double-stranded DNA molecule during the insertion of a IS element results in a short direct repeat called _____.

5. A _____ results when two IS elements insert near each other and capture a DNA sequence between them.

6. The enzyme encoded by retrotransposons and retroviruses that catalyzes the formation of DNA from RNA is called _____.

7. The transposition of *Ty* elements occurs through an _____ intermediate.

8. The *Ds* element of maize cannot transpose unless a second element called _____ is present.

9. In maize, two structurally related transposons that can create mutation by inserting in or near a gene were named _____ and _____ by McClintock.

10. The procedure of cloning a transposon-mutated gene is called _____ .

11. The two major types of retrotransposons are _____ and _____.

12. Transposable elements that are characterized by long terminal repeats which are oriented in the same direction are called _____.

13. The only retroposon known to be active in the human genome is known as _____.

14. The best-studied retrovirus-like elements in yeast is called _____.

15. A widely distributed class of retrotransposons that do not have inverted or direct repeats as integral parts of their termini, but instead have a homogeneous sequence of A:T base pairs at one end are called _____.

16. How do the *Ds* elements differ from the *Ac* elements of maize?

*17. A cross is made in maize between a female parent (lacking Ac elements) of the genotype *CC* with a male homozygous for a C^i allele (dominant inhibitor of color formation). The male parent contains a *Ds* element (but no *Ac* elements) proximal to the *C* locus on chromosome 9. What is the genotype and phenotype of the aleurone layer of the endosperm?

*18. A second cross is made between a female parent of the genotype *CC* (carrying *Ac* elements) with the same male as in the previous problem. What is the genotype and phenotype of the aleurone layer of the endosperm?

THOUGHT CHALLENGING EXERCISE

In light of the concepts presented in this chapter of the text, reevaluate and expand upon your discussion of the *Thought Challenging Exercise* of Chapter 16.

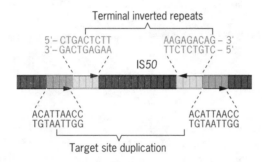

Figure 17.1   Structure of an inserted IS50 element showing its terminal inverted repeats and target site duplication.   The terminal inverted repeats are imperfect because the fourth nucleotide pair from each end is different.

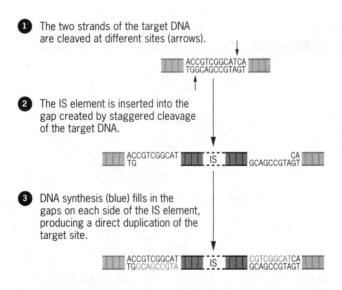

Figure 17.2   Production of target site duplication by the insertion of an IS element.

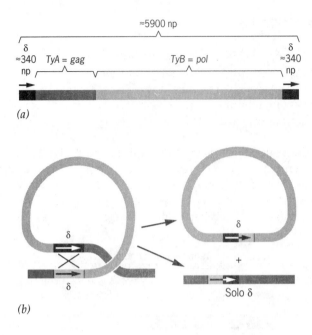

Figure 17.13 (a) Genetic organization of yeast Ty element, showing the long terminal repeat sequences (LTRs, denoted by the Greek letter delta) and the two genes (*Ty*A and *Ty*B). Lengths of the sequences are in nucleotide pairs (np). (b) Formation of a solo delta sequence by homologous recombination between the delta sequences at the ends of the element.

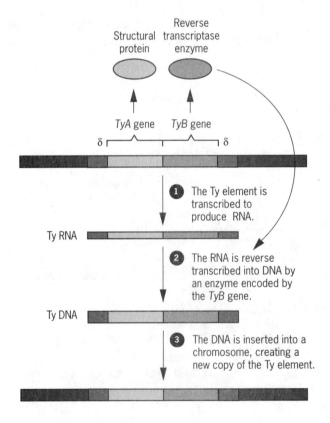

Figure 17.14 Transposition of the yeast Ty element.

SUMMARY OF KEY POINTS

Bacterial transposons, including the IS elements, composite transposons made from them, and Tn3, have short inverted repeats at their termini. Upon insertion into a chromosome or plasmid, they create a short, direct duplication of DNA at the insertion site. Transposition is catalyzed by a protein, the transposase, produced by a gene within the transposon. Other transposon-encoded proteins may repress transposition. Composite transposons and Tn3 elements carry genes that confer resistance to antibiotics.

The maize transposable element Ds, discovered through its ability to break chromosomes, is activated by another transposable element, Ac, which encodes a transposase. In *Drosophila*, transposable P elements are responsible for hybrid dysgenesis, a syndrome of germ line abnormalities that occurs in the offspring of crosses between P and M strains. The ancient *mariner* transposable elements are present in many distantly related organisms and probably have spread horizontally between species during the course of evolution.

The movement of retrovirus-like elements and retroposons depends on reverse transcription of RNA into DNA. Retrovirus-like elements have Long Terminal Repeat (LTR) sequences at both ends and resemble the integrated chromosomes of retroviruses. Retroposons have a sequence of A:T base pairs at one end. The retroposon *L1* is the only transposable element known to be active in the human genome. The HeT-A and TART retroposons are associated with the ends of *Drosophila* chromosomes.

Transposons constitute a significant fraction of the DNA in the genomes of some organisms. In nature, they may be a major cause of mutations and chromosome rearrangements. Some transposons may confer a selective advantage on their carriers; others may simply be genetic parasites.

ANSWERS TO QUESTIONS AND PROBLEMS

1) hybrid dysgenesis 2) IS elements 3) transposase 4) target site duplication 5) composite transposon 6) reverse transcriptase 7) RNA 8) Ac 9) activator, dissociation 10) transposon tagging 11) retro-like elements and retroposons 12) retro-like elements 13) Line-1 retrotransposon 14) Ty transposon 15) retroposons 16) The Ds elements lack internal sequences that are present in Ac. Since Ds also lacks the ability to code for transposase, it requires Ac for transposition 17, 18) see Approaches to Problem Solving section.

Ideas concerning the thought stimulating exercise:

The evolution of antibiotic resistance is much more dynamic than simply selecting for variability within the bacterial population. Since bacterial plasmids can carry genes for antibiotic resistance, they can move from chromosome to plasmid, and can be spread horizontally as well as vertically in bacterial populations. Conjugate R plasmids carrying resistance genes can be spread from one species to another. Antibiotic resistance that may have evolved in nonpathogenic lines of a species, or even different species of bacteria, can easily spread to pathogenic bacteria. Therefore, the indiscriminate use of antibiotics should be discouraged.

APPROACHES TO PROBLEM SOLVING

17. The cross CC (female) X $C^i C^i$ male will produce triploid endosperm with the genotype $CC\ C^i$, because the endosperm is formed from the fusion of the two polar nuclei of the female gametophyte with a sperm nucleus. There is a Ds element proximal to C^i on chromosome 9, but no Ac element is present to activate it. Therefore the endosperm will be colorless because the C^i allele is dominant to the C allele.

18. In this cross, the genotype of the endosperm is the same as above, i.e., $CC\ C^i$. However, the Ac element enters with the female chromosomes. This Ac element can allow Ds to leave its site on chromosome 9 and transpose to a new site. When this occurs, the chromosome may break at the Ds site, losing the chromosomal region distal to the break. In this case, the lost chromatin contains the C^i allele. When the C^i allele is lost the color allele, C, will be expressed. The endosperm will be primarily colorless, but will have colored spots or stripes.

18

The Genetics of Mitochondria and Chloroplasts

IMPORTANT CONCEPTS

A. The transmission of mitochondria and chloroplasts from cell to cell is through the cytoplasm.
  1. During sexual reproduction, these organelles are preferentially inherited through the cytoplasm of the gametes of one sex, which in higher eukaryotes is usually the female. This results in inheritance patterns of mitochondria and chloroplasts characterized by unequal contributions of the two parents and by an irregular segregation of alleles.
    a. Examples of these non-Mendelian patterns are the inheritance of leaf variegation in ornamental plants, cytoplasmic male-sterility in maize, antibiotic resistance in *Chlamydomonas* and metabolic defects in yeast.
      (1) All these traits appear to be controlled by genes located in either mitochondrial (mt) or chloroplast (cp) DNA.
B. The RNAs and proteins of mitochondria are encoded by genes found either in nuclear or mitochondrial DNA.
  1. Mitochondrial DNA molecules vary in size and structure.
    a. The circular mtDNA molecules encode some of the RNAs and polypeptides used within the mitochondrion, including proteins involved in aerobic metabolism.
    b. Most mitochondrial proteins are encoded by nuclear genes.
  2. In some organisms, the transcripts of mtDNA are spliced to remove introns and edited to alter nucleotide sequences.
  3. The translational machinery within mitochondria employs a genetic code that is slightly different from that used in the cytosol.
  4. Mutations in mtDNA are responsible for at least two human diseases, Leber's hereditary optic neuropathy and Pearson marrow-pancreas syndrome.
C. Chloroplast DNA molecules also vary in size and structure, and most are circular.

1. Chloroplast DNA molecules encode some of the RNAs and proteins used within the chloroplast, including some proteins that are involved in photosynthesis.
   D. Both mitochondria and chloroplasts apparently originated when prokaryotic organisms were internalized by primitive eukaryotic cells.
      1. In the course of evolution, these endosymbionts have exchanged genes with the nucleus.
      2. Mitochondria and chloroplasts cannot exist by themselves and are dependent on nuclear genes.

IMPORTANT TERMS

In the space allotted, concisely define each term.

chloroplasts:

mitochondria:

endosymbiosis:

heteroplasmy:

homoplasmy:

maternal inheritance:

non-Mendelian, biparental inheritance:

petite mutants:

mitochondrial DNA (mtDNA):

guide RNAs (gRNAs):

trans-splicing:

Leber's hereditary optic neuropathy (LHON):

Pearson marrow-pancreas syndrome:

chloroplast DNA (cpDNA):

biogenesis:

phytochromes:

endosymbiosis:

IMPORTANT NAMES

In the space allotted, concisely state the major contribution made by the individual or group.

Carl Correns and Erwin Baur:

Marcus Rhoades:

Ruth Sager:

Boris Ephrussi:

J. E. Wallin:

Allan Wilson:

TESTING YOUR KNOWLEDGE

*In this section, answer the questions, fill in the blanks, and solve the problems in the space allotted. Problems noted with an * are solved in the Approaches to Problem Solving section at the end of the chapter.*

1. The presence of a single genotype of an organelle within a cell is called _____.

2. The exception given in the text to the typical inheritance of chloroplasts only through the female gametophyte in plants is _____.

3. Cytoplasmic male sterility in maize is caused by a mutation in the DNA of _____.

4. The presence of two organelle genotypes within a cell is called _____.

5. Mitochondrial mutations in yeast that result in tiny colonies when grown on rich glucose-containing medium are called _____ mutants.

6. In humans, a condition characterized by the sudden onset of blindness in adults due to mitochondrial mutations that reduce the efficiency of oxidative phosphorylation is called _____.

7. A frequently lethal human disorder caused by a large deletion of the mtDNA and loss of bone marrow cells during childhood is called _____.

8. A CO_2 fixing enzyme that has one subunit coded by a nuclear gene and the other subunit coded by a chloroplast gene is called_____.

*9. Name, with examples, the types of genes encoded by mtDNA.

*10. Name, with examples, the types of genes encoded by cpDNA.

*11. What is the molecular basis of suppressive petite mutants and neutral petite mutants of yeast?

*12. Name three peculiarities of plant mitochondrial gene expression.

*13. Distinguish between a maternal effect and maternal inheritance.

*14. Discuss observations suggesting that genes may have moved from chloroplasts to the nucleus and from chloroplast to mitochondria in higher plants?

*15. Why is the genetic code not completely universal?

*16. How would you determine whether a variegated plant is due to mutants of chloroplasts or mutant nuclear genes?

*17. In the snail, *Limnaeae peregra*, shell coiling may be to the left (dextral) or to the right (sinistral). When snails from a sinistral line (female parent) are crossed with snails from a dextral line (male parent), the F_1 progeny all had sinistral shells. The F_2 progeny produced by self-fertilization of the F_1s were all sinistral. When self-fertilized, about 3/4 of the F_2 snails produced progeny with sinistral shells and 1/4 of the F_2 snails produced progeny with dextral shells. In the reciprocal cross, dextral female X sinistral male, all the F_1 progeny were dextral. All the F_2 progeny were sinistral. When self-fertilized, about 3/4 of the F_2 snails produced progeny with sinistral shells and 1/4 of the F_2 snails produced progeny with dextral shells. Is the direction of shell coiling in snails due to cytoplasmic inheritance? Why or why not?

THOUGHT CHALLENGING EXERCISE

A geneticist crossed two independently derived variegated *Pelargonium zonale* plants. In one of the variegated progeny, a dark green branch occurred from a pure white region of the plant. Seed obtained from flowers of this dark green branch produce dark green plants that in turn bred true for dark green. The geneticist then isolated chloroplast DNA from dark green regions and white regions of the plants and digested it with a variety of restriction endonucleases. It was established that the cpDNA of dark green regions of the parental variegated plants had similar restriction patterns, but the white regions of the two variegated plants each had a distinct restriction pattern. The DNA isolated from the white sectors of the hybrid plants possessed a restriction pattern that was additive of the white regions of the parental plants. The dark green plants derived from the white sector of the hybrid variegated plant had a DNA restriction pattern identical to the dark green regions of the parental plants. Explain these observations and the origin of the chloroplast DNA of derived dark green plants.

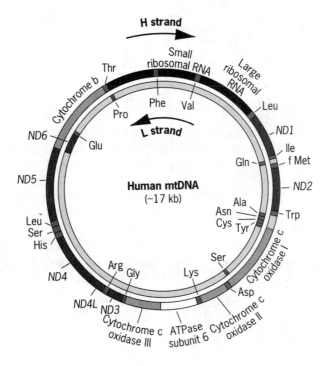

Figure 18.14 Map of human mtDNA showing the pattern of transcription. Genes on the inner circle are transcribed from the L strand of the DNA, whereas genes on the outer circle are transcribed from the H strand of DNA. Arrows show the direction of transcription. ND1-6 are genes encoding subunits of the enzyme NADH reductase; the tRNA genes in the mtDNA are indicated by abbreviations for the amino acids.

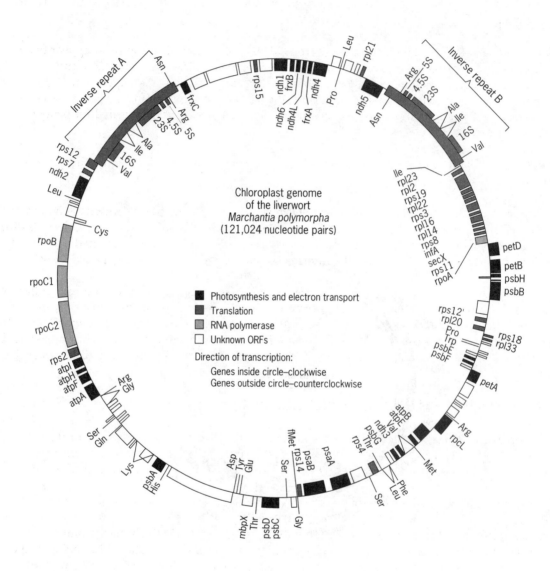

Figure 18.17 Genetic organization of the chloroplast DNA in the liverwort *Marchantia polymorpha*. Symbols: rpo, RNA polymerase; rps, ribosomal proteins of the small subunit; rpl and secX, ribosomal proteins of the large subunit; 4.5S, 5S, 16, 23S, rRNAs of the indicated size; rbs, ribulose bisphosphate carboxylase; psa, photosystem I; psb, photosystem II; pet, cytochrome b/f complex; atp, ATP synthesis; infA, initiation factor A; frx, iron-sulphur proteins; ndh, NADH reductase; mpd, chloroplast permease (?); tRNA genes are indicated by the abbreviations for the amino acids.

SUMMARY OF KEY POINTS

Organelle heredity is characterized by unequal contributions of the two parents and by irregular segregation of phenotypes. These non-Mendelian phenomena are due to the preferential transmission of chloroplasts or mitochondria through the gametes of one sex, which in higher eukaryotes is usually the female. Traits due to mitochondrial or chloroplast genes are therefore often inherited in a strictly maternal fashion.

Mitochondrial DNA (mtDNA) molecules range from 16 kb to 2500 kb in size, and most appear to be circular. These molecules contain genes for some of the ribosomal RNAs, transfer RNAs, and polypeptides used within the mitochondrion. The structure, organization, and expression of these genes varies among species. In some groups of organisms, the transcripts of mitochondrial genes are edited after they are synthesized. Both mitochondrial and nuclear gene products are needed for proper mitochondrial function.

Mutations in mtDNA can cause human diseases such as Leber's hereditary optic neuropathy and Pearson marrow-pancreas syndrome.

Circular chloroplast DNA (cpDNA) molecules are typically 120 to 292 kb in size and may contain 150 or more genes. The organization of these genes varies among species of plants and algae. Other plastids also contain cpDNA molecules. Light induces chloroplasts to develop from unpigmented proplastids. This process, called biogenesis, involves the interplay of chloroplast and nuclear gene products.

Mitochondria and chloroplasts probably originated as bacteria that were incorporated into eukaryotic cells about a billion years ago. The mutually beneficial relationship that has existed between these cells and their internalized bacteria has evolved into a relationship of complete interdependence. In the course of this endosymbiosis, genes have been shuffled among the mitochondria, chloroplast, and nuclear DNAs. For unknown reasons, mitochondrial genes seem to be evolving 5 to 10 times faster than nuclear genes.

ANSWERS TO QUESTIONS AND PROBLEMS

1) homoplasmy 2) leaf variegation in *Pelargonium zonale* 3) mitochondria 4) heteroplasmy 5) petite 6) Leber's hereditary optic neuroplasmy 7) Pearson marrow-pancreas syndrome 8) ribulose 1, 5-bisphosphate carboxylase 9 -17) see *Approaches to Problem Solving* section.

Ideas concerning thought challenging exercise:

The results suggest that the white region of the hybrid variegated plant likely had mutant chloroplasts derived from each parent. The dark green genotypes formed in the white region of the variegated plant may have resulted from genetic recombination between the DNA of two mutant chloroplasts.

APPROACHES TO PROBLEM SOLVING

9. Mitochondrial genes encode some of the RNAs and polypeptides that are used in the mitochondria. These include two to three types of ribosomal RNA, up to 25 different tRNAs and several polypeptides, e.g., some of the ribosomal structural proteins and several enzymes involved in aerobic metabolism, e.g., subunits of cytochrome c oxidase.

10. All cpDNA molecules carry the same basic set of genes including those for rRNA, some ribosomal proteins, various polypeptide components of the photosystems that are involved in capturing solar

energy, the catalytically active subunit of 1, 5-bisphosphate carboxylase, and four subunits of a chloroplast-specific RNA polymerase.

11. Neutral petite mutants lack mitochondrial DNA. Suppressive petite mutants have a smaller DNA molecule that is relatively A:T-rich.

12. Mitochondrial DNA is organized into many separate transcriptional units, some containing information for more than one gene. RNA editing alters the composition of codons in some plant mitochondrial transcripts; some mitochondrial mRNAs are formed by the process of trans-splicing.

13. A maternal effect is when the genes of the mother are determining a phenotype of the progeny. Maternal inheritance is when genes (generally in organelle DNA) are inherited through the cytoplasm of the egg.

14. Several lines of evidence suggest that genes have transposed among the nucleus, chloroplasts, and mitochondria. This evidence includes (1) mtDNA in some higher plants contains a tRNA gene from cpDNA, (2) in one plant species, the mtDNA contains part of the chloroplast gene for the large subunit of 1, 5-bisphosphate carboxylase, and (3) in most plant species, the small subunit of 1, 5-bisphosphate carboxylase is encoded by a nuclear gene, but in some algae, it is encoded by a chloroplast gene.

15. In mammalian mitochondrial mRNA translation, AGA and AGG are termination codons, whereas in the cytosol they specify the incorporation of arginine into a polypeptide. The termination codon (UGA) in cytosol translation functions as a tryptophan codon in the mitochondria. The codon, AUA, codes for isoleucine in the cytosol, but is the methionine initiation codon in the mitochondria.

16. One would determine this by performing reciprocal crosses between variegated and green plants. If the variegation is due to mutants of the chloroplast, the trait will be maternally inherited. If the trait is determined by a nuclear gene, then some form of dominance will be displayed in the F_1 progeny. Segregation of the trait will be apparent in the F_2 generation, regardless of the direction in which the cross was made.

17. The data in this example on the direction of shell coiling do not adhere to a pattern of true cytoplasmic inheritance, because segregation of the traits is apparent. The results can be interpreted as conforming to what is called a maternal effect, i.e., the genotype of the mother determines a phenotype of the progeny. The crosses can be genetically diagrammed as follows:

P	DD (female) X dd (male)			P	dd (female) X DD (male)		
	sinistral dextral				dextral sinistral		
F_1	Dd			F_1	Dd		
	sinistral				dextral		
F_2	all sinistral			F_2	all sinistral		
	1/4 DD	1/2 Dd	1/4 dd		1/4 DD	1/2 Dd	1/4 dd
	↓	↓	↓		↓	↓	↓
F_3	DD	3/4 $D\_$	dd	F_3	DD	3/4 $D\_$	dd
		1/4 dd				1/4 dd	
	sinistral	sinistral	dextral		sinistral	sinistral	dextral

The sinistral (D) allele is dominant to the dextral (d) allele. The phenotypic ratio lags the genotypic ratio by one generation due to the maternal effect.

19

The Techniques of Molecular Genetics

BASIC TECHNIQUES USED TO CLONE GENES
    The Discovery of Restriction Endonucleases
    The Production of Recombinant DNA Molecules *In Vitro*
    Amplification of Recombinant DNA Molecules in Cloning Vectors

CONSTRUCTION AND SCREENING OF DNA LIBRARIES
    Construction of Genomic Libraries
    Construction of cDNA Libraries
    Screening DNA Libraries for Genes of Interest
    Biological and Physical Containment of Recombinant DNA Molecules

THE MANIPULATION OF CLONED DNA SEQUENCES *IN VITRO*
    Phagemids: The Biological Purification of DNA Single-Strands
    Transcription Vectors: The Synthesis of RNA Transcripts *In Vitro*
    Joining DNAs with Linker and Adapter Molecules
    In Vitro Site-Specific Mutagenesis

THE MOLECULAR ANALYSIS OF DNA, RNA, AND PROTEIN
    Analysis of DNAs by Southern Blot Hybridizations
    Analysis of RNAs by Northern Blot Hybridizations
    Analysis of Proteins by Western Blot Techniques

THE MOLECULAR ANALYSIS OF GENES AND CHROMOSOMES
    Amplification of DNAs by the Polymerase Chain Reaction (PCR)
    Physical Maps of DNA Molecules Based on Restriction Enzyme Cleavage Sites
    Nucleotide Sequences: The Ultimate Fine Structure Maps

IMPORTANT CONCEPTS

A. Molecular technologies have been developed that allow the construction of recombinant DNA molecules that contain sequences from unrelated species, and the cloning of these molecules in appropriate host cells.

1. The ability to clone and sequence any gene or DNA sequence of interest depends on a class of enzymes called restriction endonucleases.

    a. Restriction endonucleases, discovered by Hamilton Smith and David Nathans in 1970, cut DNA molecules at specific sequences called restriction sites.

        (1) Restriction sites are commonly palindromes, i.e., nucleotide-pair sequences that read the same forward and backward ($5' \rightarrow 3'$) from a center of symmetry.

    b. Many restriction endonucleases produce DNA fragments (restriction fragments) with complementary single-stranded ends, but some produce DNA fragments with blunt ends.

    c. Two restriction fragments, regardless of source, with the same single-stranded ends can be covalently fused to produce a recombinant DNA molecule.

        (1) The first recombinant DNA molecules were produced in Paul Berg's laboratory at Stanford University.

2. Various applications of recombinant DNA technology require not only the construction of recombinant DNA molecules, but also the amplification or cloning of these molecules in cloning vectors.

    a. A cloning vector has three essential components: an origin of replication; a dominant selectable marker gene, usually a gene that confers drug resistance to the host cell; and at least one unique restriction endonuclease cleavage site.

(1) Stanley Cohen and his colleagues at Stanford University inserted an *EcoRI* restriction fragment from one DNA molecule into a cleaved, unique *EcoRI* restriction site of a circular, self-replicating DNA molecule. This recombinant DNA molecule, when transformed into *E. coli* cells exhibited normal autonomous replication, thus being the first cloning vector.

(2) More versatile plasmids containing several unique restriction enzyme cleavage sites and genes conferring resistance to more than one antibiotic were soon developed.

(3) Many of the cloning vectors used today are second- and third-generation derivatives of the plasmid pBR322, the first widely used cloning vector.

b. Cloning vectors have been developed to clone larger DNA fragments.

(1) The chromosome of bacteriophage lambda, with the central one-third of its DNA deleted, can be used as a vector to accommodate inserts of 10 kb to 15 kb.

(2) Cosmids are recombinant DNA molecules of plasmid and lambda chromosomes that combine the plasmid's ability to replicate autonomously in *E. coli* cells and the *in vitro* packaging capability (35 to 45 kb) of the λ chromosome.

(3) Shuttle vectors have been developed that contain both *E. coli* and yeast *S. cerevisiae* origins of replication and, therefore, can replicate in both cells of *E. coli* and yeast.

(4) Yeast artificial chromosomes (YACs) and bacterial artificial chromosomes (BACs) are cloning vectors that were developed to clone DNA fragments several hundred kb in length.

B. The first step in cloning a gene or DNA sequence of a given organism usually involves the construction of a library for screening cloned DNA sequences.

1. Genomic DNA libraries are usually prepared by isolating total genomic DNA of an organism, digesting the DNA with a restriction endonuclease, and ligating the restriction fragments into an appropriate cloning vector.

2. Libraries can also be produced from complementary DNAs (cDNAs).

a. Complementary DNAs are double-stranded DNA molecules produced from mRNAs by an enzyme called reverse transcriptase.

3. Screening eukaryotic DNA libraries for a specific gene or other DNA sequence of interest requires the identification of a single DNA sequence among thousands to millions of other such sequences.

a. The most powerful screening procedure is genetic selection: searching for a DNA sequence in the library that can restore the wild-type phenotype to a mutant organism.

b. The first eukaryotic DNA sequences to be cloned were genes that are highly expressed in specialized cells, which facilitate the purification of mRNA and the production of cDNA.

(1) Radioactive ovalbumin and hemoglobin cDNAs were used to screen genomic libraries by *in situ* colony or plaque hybridization.

C. Various techniques are available to prepare cloned DNA and RNA sequences for *in vitro* manipulation.

1. Phasmid cloning vectors (e.g., pUC118 and pUC119), contain components from both phage chromosomes and plasmids. After addition of a helper phage, they produce single-stranded DNA replicates and package them into phage coats.

a. Single-stranded DNA molecules are used in DNA sequencing and *in vitro* mutagenesis protocols.

2. Transcription vectors synthesize RNA transcripts of their cloned sequence *in vitro*.

a. Transcription vectors are used to prepare radioactive hybridization probes that are used in genomic and cDNA library screening, in various analyses of genome structure, and in studies of gene expression.

3. DNA fragments with different termini can be joined *in vitro* with synthetic olignucleotides called linkers and adaptors.

4. Oligonucleotide-directed site-specific mutagenesis allows scientists to change nucleotide sequences in a specific manner.

D. The development of recombinant DNA techniques has spawned a whole array of new approaches to analyze genes and gene products.

1. Gel electrophoresis provides a powerful tool for the separation of macromolecules with different sizes and charges.
2. In 1975, E. M. Southern reported an important new procedure subsequently nicknamed "Southern blotting", that allowed investigators to identify the location of genes and other DNA sequences on restriction fragments separated by electrophoresis.
   a. The essential feature of Southern blotting is the transfer of the DNA molecules, separated by gel electrophoresis, to nitrocellulose or nylon membranes followed by the hybridization of a radioactive DNA probe of interest. The probe will hybridize only with the immobilized DNA on the membrane that contains a nucleotide sequence complementary to the sequence of the probe.
3. RNA molecules separated by gel electrophoresis also can be transferred to nitrocellulose or nylon membranes and hybridized with radioactive probes. This is called northern blotting.
4. Western blotting involves transfer of proteins, separated on polyacrylamide electrophoresis, to a nitrocellulose membrane followed by the detection of individual proteins by using specific antibodies.

E. The structure of genes and chromosomes can be analyzed by an array of molecular techniques.
   1. The polymerase chain reaction (PCR), devised by Kerry Mullis, is an extremely powerful procedure that allows the amplification of a selected DNA sequence in the genome a millionfold or more without the use of living cells during the process.
      a. PCR technologies provide shortcuts for many cloning and sequencing applications.
         (1) Amplification of small available amounts of DNA by PCR allows for structural data on genes and DNA sequences to be obtained.
         (2) PCR is used for prenatal diagnosis of inherited human diseases where limited amounts of fetal DNA are available.
         (3) PCR DNA finger printing is used in forensic cases involving the identification of individuals from very small tissue samples.
   2. Physical maps of DNA molecules can be made that are based on restriction enzyme cleavage sites.
   3. The nucleotide pair sequence of a gene or DNA sequence can be determined by either of two sequencing procedures developed in the period from 1974 to 1977.
      a. The Maxam and Gilbert procedure uses four different chemical reactions to cleave DNA chains at As, Gs, Cs, or Ts.
      b. The second approach, developed by Fred Sanger and colleagues, uses an enzymatic procedure and specific chain-terminators to generate four populations of fragments that terminate at As, Gs, Cs, and Ts, respectively.

IMPORTANT TERMS

In the space allotted, concisely define each term.

restriction endonuclease:

palindrome:

restriction fragments:

polylinker site:

cosmids:

shuttle vectors:

yeast artificial chromosome (YAC):

bacterial artificial chromosome (BAC):

genomic DNA library:

complementary DNA (cDNA):

terminal transferase:

colony hybridization:

plaque hybridization:

phagemids:

helper phage:

forced cloning:

fusion proteins:

linkers:

adapters:

site-specific mutagenesis:

oligonucleotide-directed site-specific mutagenesis:

nick-translation:

Southern blots:

northern blots:

western blotting:

polymerase chain reaction (PCR):

Taq polymerase:

PCR DNA finger printing:

restriction enzyme cleavage site maps (restriction maps):

2′, 3′-dideoxyribonucleoside triphosphates:

IMPORTANT NAMES

In the space allotted, concisely state the major contribution of the individual or group.

Hamilton Smith and Daniel Nathans:

Werner Arber:

Paul Berg:

Stanley Cohen:

E. M. Southern:

Kerry Mullis:

Allan Maxam and Walter Gilbert:

Fred Sanger:

TESTING YOUR KNOWLEDGE:

*In this section, answer the questions, fill in the blanks, and solve the problems in the space allotted. Problems noted with an * are solved in the Approaches to Problem Solving section at the end of the chapter.*

1. An enzyme that cleaves DNA at a specific sequence in which the bases read the same on complementary strands ($5' \rightarrow 3'$ and $3' \rightarrow 5'$) from a point of symmetry is called a (an) _____ _____.

2. A double-stranded DNA complementary to a mRNA that is produced by reverse transcriptase is called a (an) _____.

3. Which cloning vector would be the best for producing a genomic DNA library of mean insert size of about 1000 bp?

4. Which cloning vectors are best suited for producing a genomic DNA library of average insert size of about 200 Kbp?

5. Which cloning vector is best suited for cloning cDNAs?

6. The ability to insert a foreign DNA in a predetermined orientation into the polycloning region of pUC118-pUC119 vector system is called _____.

7. Three individuals who shared the 1986 Nobel Prize in Physiology and Medicine for experiments that lead to, and the discovery of, restriction endonucleases were _____, _____, and _____.

8. A cloning vector that can replicate in both bacteria and yeast is called a (an) _____.

9. The individual who was a co-recipient of the 1980 Nobel Prize in Chemistry for his pioneering research on recombinant DNA molecules was _____.

10. The analysis of the functioning of a gene after changing one nucleotide at a time *in vitro* is called _____.

11. The individual who received a Nobel Prize in 1993 for inventing the polymerase chain reaction was _____.

12. The procedure by which proteins are transferred from electrophoretic gels to membranes and detected with antibodies is called _____.

13. A hybrid cloning vector used for the biological purification of single-stranded DNA from the double-stranded DNA insert are called _____.

14. The procedure by which DNA fragments, separated by electrophoresis, are transferred from the gel to nitrocellulose or nylon membranes, is called _____.

*15. Which of the following RNA sequences is complementary to a palindrome sequence in the encoding DNA?

   (a) AUCUUCUA
   (b) CCUAUAGG
   (c) AAAAGGGG
   (d) AUUACGCC
   (e) none of the above

*16. A geneticist is producing a genomic BAC library of a plant that has a genome size of 10^9 bp. The average insert size in the library is 100 Kbp. How many recombinant clones will have to be constructed in order to be 99% sure that the library will have a specific DNA sequence represented?

*17. A molecular geneticist determined the order of nucleotides in a ten base-pair region of a gene by the Sanger dideoxyribonucleoside method. The fragments in the four syntheses were electrophoresed, Southern blotted, and visualized by autoradiography. From an analysis of the autoradiogram diagrammed below, determine the 5′ → 3′ sequence of nucleotides in the strand sequenced.

lane 1	lane 2	lane 3	lane 4	Direction of
ddA	ddT	ddG	ddC	migration
				↓
___				
		___		
			___	
	___			
___				
___				
		___		
			___	
			___	
_____				

*18. A geneticist was developing a restriction map of a 10,000 bp linear DNA sequence. The DNA fragment was treated individually and in combination with the restriction enzymes *Eco*RI, *Bam*HI, and *Hind*III. The enzyme reactions were allowed to go to completion and the DNA fragments were separated and sized by agarose gel electrophoresis. Produce a restriction map from the following fragment sizes detected from the various digestions.

	*Eco*RI	*Bam*HI	*Hind*III	*Eco*RI + *Bam*HI	*Eco*RI + *Hind*III	*Bam*HI + *Hind*III	*Eco*RI + *Bam*HI + *Hind*III
	500	2000	1500	500	500	500	500
	3500	8000	3500	1500	1000	1500	1000
	6000		5000	2000	2500	3000	2000
				6000	5000	5000	5000
total length	10000	10000	10000	10000	9000	10000	9000

THOUGHT CHALLENGING EXERCISE

Some of the greatest discoveries in science that later had tremendous practical applications have come from purely basic research. Using examples, discuss scientific discoveries that fall into this category.

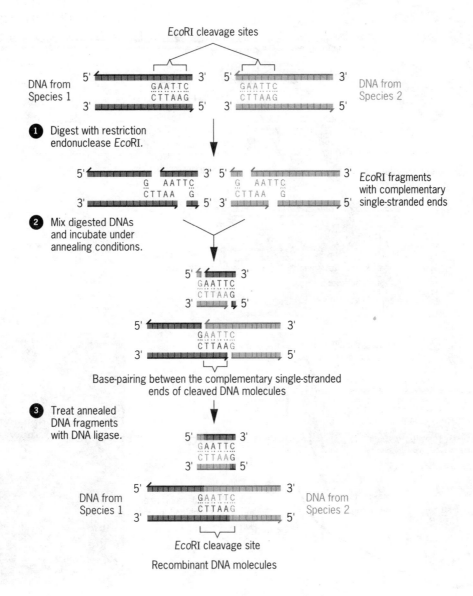

Figure 19.1   The construction of recombinant DNA molecules *in vitro*. DNA molecules isolated from two different species are cleaved with a restriction enzyme, mixed under annealing conditions, and covalently joined by treatment with DNA ligase. The DNA molecules can be obtained from any species — animal, plant, or microbe. The digestion of DNA with the restriction enzyme *Eco*RI produces the same complementary single-stranded AATT ends regardless of the source of the DNA.

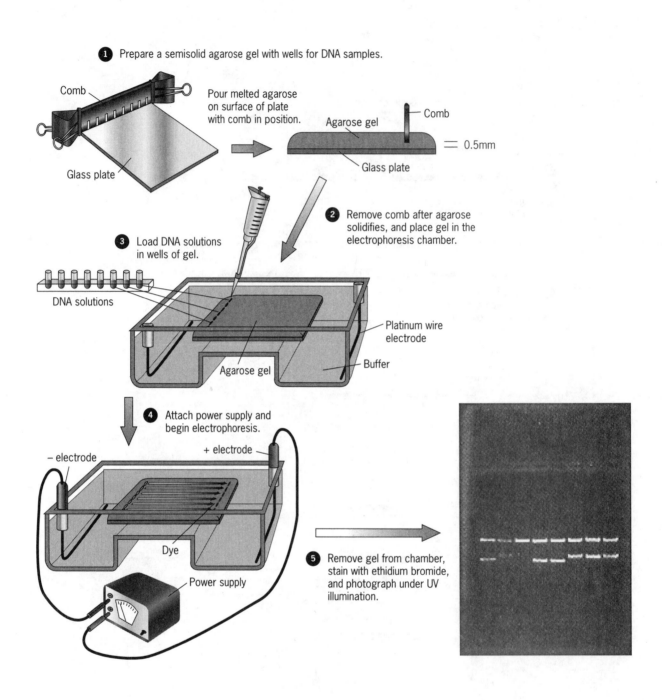

Figure 19.20 The separation of DNA molecules by agarose gel electrophoresis. The DNAs are dissolved in loading buffer with density greater than that of the electrophoresis buffer so that DNA samples settle to the bottom of the wells, rather than diffusing into the electrophoresis buffer. The loading buffer also contains a dye to monitor the rate of migration of molecules through the gel. Ethidium bromide binds to DNA and fluoresces when illuminated with ultraviolet light.

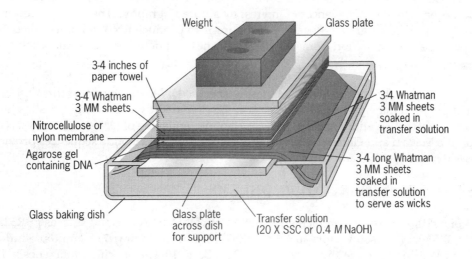

Figure 19.21 Procedure used to transfer DNAs separated by gel electrophoresis to nitrocellulose or nylon membranes. The transfer solution carries the DNA from the gel to the membrane as the dry paper towels on top draw the salt solution from the reservoir through the gel to the towels. The DNA binds to the membrane on contact. The membrane with the DNA bound to it is dried and baked under vacuum to firmly affix the DNA prior to hybridization.

SUMMARY OF KEY POINTS

Recombinant DNA and gene-cloning techniques allow scientists to isolate and characterize virtually any gene or other DNA sequence of interest from any organism. These techniques became possible with the discovery of restriction endonucleases, enzymes that recognize and cleave DNA in a sequence-specific manner. DNA sequences of interest are inserted into small, self-replicating DNA molecules called cloning vectors *in vitro*, and the resulting recombinant DNA molecules are amplified by replication *in vivo* after being introduced into cells, usually bacteria, by transformation. A variety of cloning vectors have been constructed, each with advantages for certain research purposes.

DNA libraries can be constructed that contain complete sets of genomic DNA sequences or DNA copies (cDNAs) of mRNAs in an organism. Specific genes or other DNA sequences can be isolated from these libraries by genetic complementation and by hybridization to labeled nucleic acid probes containing DNA sequences of known function.

Cloned genes and other DNA sequences can be manipulated almost at will in the test tube. Phagemids provide a powerful tool for purifying the single strands of a cloned DNA sequence. Transcription vectors allow scientists to express genes *in vitro* by using coupled transcription and translation systems. DNA fragments with different termini can be joined *in vitro* with synthetic oligonucleotides called linkers and adapters. The nucleotide sequences of cloned genes can be changed as desired by *in vitro* site-specific mutagenesis.

DNA restriction fragments and other small DNA molecules can be separated by agarose and acrylamide gel electrophoresis and transferred to solid supports, usually nitrocellulose or nylon membranes, to produce DNA gel blots called Southern blots. The DNAs on these blots can be hybridized to labeled DNA probes to detect sequences of interest by autoradiography. The same procedure can be applied to RNA molecules separated by gel electrophoresis to produce RNA gel blots called northern blots. When proteins are transferred from gels to membranes and detected with antibodies, the products are called western blots.

The polymerase chain reaction can be used to amplify specific DNA sequences *in vitro* a millionfold or more. Detailed physical maps of DNA molecules can be prepared by identifying the sites that are cleaved by various restriction endonucleases. The nucleotide sequences of DNA molecules can be determined by the Maxam and Gilbert or Sanger procedures; these nucleotide sequences provide the ultimate physical maps of genes and chromosomes.

ANSWERS TO QUESTIONS AND PROBLEMS

1) restriction endonuclease 2) complementary DNA or cDNA 3) A plasmid such as pBR322 4) YACs and BACs 5) Plasmid-based vectors such as pBR322 6) forced cloning 7) Hamilton Smith, Daniel Nathans and Werner Arber 8) shuttle vector 9) Paul Berg 10) site-specific mutagenesis 11) Kerry Mullis 12) western blotting 13) phagemids 14) Southern blotting 15-18) see *Approaches to Problem Solving* section

Ideas on thought challenging exercise:

Most of the scientific discoveries presented in this chapter have made possible the array of molecular techniques now available for cloning and manipulating DNA. Yet, most were the rewards of excellent basic research. A case in point is the discovery of restriction enzymes. Their discovery came not because the scientists set out to find such enzymes, but rather were discovered during basic research on bacteria and bacteriophages.

APPROACHES TO PROBLEM SOLVING

15. To approach this problem, first indicate the sense DNA strand (complementary to the RNA) and the nonsense strand of the double helix for the indicated RNAs. When this is done, it is apparent that (b) contains a palindrome.

RNA	C C U A U A G G
DNA	**G G A T** A T C C
DNA	C C T A **T A G G**

16. To solve this problem, use the formula in your text $N = \ln(1 - P)/\ln(1 - f)$, P = probability, and f = fraction of genome present in a single clone. Thus, for this problem $P = 0.99$ and $f = 10^5/10^9 = 0.0001$.

$$N = \ln[1 - 0.99]/\ln[1 - 0.0001] = 46{,}049 \text{ clones}$$

17. The smaller fragments migrate faster in the gel. Since the polymerization stops with the addition of a dideoxyribonucleotide to the growing 3′ end, position 10 is the 3′ end and position 1 represents the 5′ end of the polynucleotide being synthesized.

lane 1 ddA	lane 2 ddT	lane 3 ddG	lane 4 ddCC	Direction of migration	Base at each position in the polypeptide
___				10 ↓	A
		___		9	G
		___		8	G
			___	7	C
	___			6	T
___				5	A
___				4	A
		___		3	G
			___	2	C
			___	1	C

The 5′ → 3′ sequence of nucleotides in the strand sequenced is complementary to the strand synthesized and is, therefore, TCCGATTCGG

18. To approach this problem, first diagram the possible restriction map for *Eco*RI fragments. The two possibilities in kb pairs are:

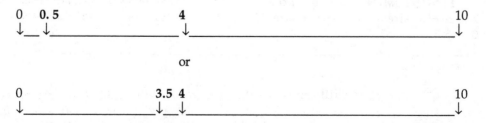

Next the restriction fragment of *Bam*HI and of the codigests of *Bam*HI and *Eco*RI are analyzed. *Bam*HI has one cutting site that results in a 2kb fragment and a 8kb fragment. Superimposing these on the *Eco*RI map diagrammed above, and noting that restriction maps are additive, the restriction map is:

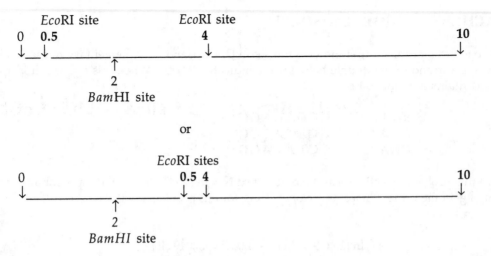

or

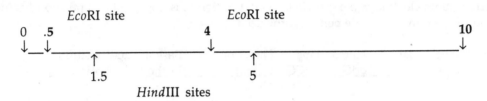

The next step is to analyze *Eco*RI and *Hind*III restriction sites. Again, this is done relative to *Eco*RI sites.

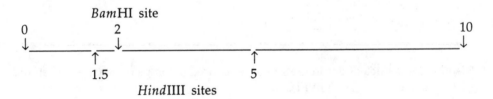

These data limit the position of the restriction site generating the 500 bp *Eco*RI fragment as indicated above. This is the only order that allows the 2500 bp *Eco*RI + *Hind*III fragment to be formed.

*Hind*II and *Bam*HI restriction sites considered together map as follows:

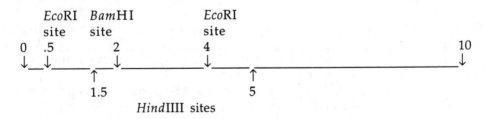

The final map pieces together as follows:

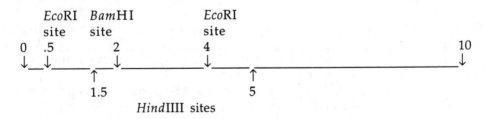

The reason that the restriction fragments don't add up to 10000 in some of the digestions is that some of the restriction cleavages produce fragments of the same size.

20

Molecular Analysis of Genes and Gene Products

MAP POSITION-BASED CLONING OF GENES
    Restriction Fragment-Length Polymorphism Maps
    Chromosome Walks and Jumps
    Physical Maps and Clone Banks

USE OF RECOMBINANT DNA TECHNOLOGY TO IDENTIFY HUMAN GENES
    Huntington's Disease
    Cystic Fibrosis
    Duchenne's Muscular Dystrophy

MOLECULAR DIAGNOSIS OF HUMAN DISEASES

THE HUMAN GENOME PROJECT

HUMAN GENE THERAPY

DNA FINGERPRINTS
    Paternity Tests
    Forensic Applications

PRODUCTION OF EUKARYOTIC PROTEINS IN BACTERIA
    Human Growth Hormone
    Proteins With Industrial Applications

TRANSGENIC PLANTS AND ANIMALS
    Transgenic Animals: Microinjection of DNA into Fertilized Eggs
    Transgenic Plants: The Ti Plasmid of *Agrobacterium tumefaciens*
    Herbicide-Resistant Plants
    Use of Antisense RNAs to Block Gene Expression

IMPORTANT CONCEPTS

A. The development of recombinant DNA and gene cloning technologies has allowed for many new types of genetic analysis and manipulation.
    1. Essentially any gene of any organism can be isolated, dissected, modified, and manipulated.
    2. Synthetic, modified, or other foreign genes can be introduced into microorganisms, plants, and animals.
    3. Recombinant DNA technologies have resulted in the explosive accumulation of new information in virtually every area of biology.
        a. To date, the most spectacular achievements of these technologies have been in furthering our knowledge of basic biology.

B. Restriction fragment length polymorphisms (RFLPs) are extremely valuable tools.
    1. RFLPs are commonly used for increasing the density of markers on genetic maps.
    2. RFLPs have been used to identify the genes responsible for several major human diseases by positional cloning, i.e., by cloning genes based on their chromosomal locations.
        a. Positional cloning of genes is greatly assisted by the preparation of physical maps of chromosomes, e.g., molecular maps prepared from the restriction maps of overlapping clones (contigs).

b. The human genes responsible for Huntington's disease, cystic fibrosis, and Duchenne muscular dystrophy were all identified by positional cloning.

(1) The identification and characterization of these genes have provided new information about the molecular and cellular basis of these diseases, made DNA diagnosis of the defective genes possible, and raised hope for successful treatment of the diseases by somatic-cell gene therapy, i.e., providing mutant cells with functional copies of the defective gene.

C. DNA fingerprinting provides valuable evidence in establishing individual identity, especially in paternity and forensic cases.

D. Transgenic microbes, plants, and animals are now being used to produce valuable proteins.

1. Transgenic microbes are used in commercial production of specific gene products such as human insulin, human growth hormone (HGH), and proteins with industrial applications such as protease, amylase and rennin.

2. Transgenic animals are routinely produced by microinjecting DNA into fertilized eggs.

3. Transgenic plants are produced by using the Ti plasmid of *Agrobacterium tumefaciens* as a gene-transfer vector or by a procedure called microprojectile bombardment, which simply involves shooting DNA coated-tungsten particles into nuclei.

4. The expression of an endogenous gene can sometimes be shut off or reduced by the presence of antisense RNA.

a. The antisense RNA is complementary to mRNA and blocks the translation of the mRNA (sense RNA).

5. Patterns of gene expression and regulatory sequences of the genes of interest can be investigated in transgenic organisms that harbor reporter genes composed of the regulatory sequences of the genes of interest and the coding sequence of a gene that produces an easily assayed product.

a. The *E. coli* DNA sequences that encode β-galactosidase and β-glucuronidase are commonly used in the construction of the reporter genes used in animals and plants, respectively.

E. The Human Genome Project was established with the goals of mapping all human genes and sequencing all 3×10^9 nucleotide pairs in the human genome

IMPORTANT TERMS

In the space allotted, concisely define each term.

Tay-Sach's disease:

position cloning:

variable number tandem repeats (VNTRs):

chromosome walking:

chromosome jumping:

contigs:

anchor genes:

sequence-tagged sites (STSs):

expressed sequence tags (ESTs):

Huntington's disease:

exon amplification:

cystic fibrosis:

CpG islands:

cystic fibrosis transmembrane conductance regulator:

Duchenne's muscular dystrophy:

dystrophin:

apodystrophins:

human genome project:

human genome organization (HUGO):

gene therapy:

transgene:

adenosine deaminase-deficient severe combined immunodeficiency disease (ADA-SCID):

gene addition:

gene replacement:

DNA fingerprints:

human growth hormone:

microprojectile bombardment:

electroporation:

Agrobacterium tumefaciens-mediated transformation:

totipotency:

Ti plasmid:

T-DNA:

vir region:

chimeric selectable marker genes:

antisense RNA:

antisense gene:

IMPORTANT NAMES

In the space allotted, concisely state the major contribution made by the individual or group.

James Gusella and Nancy Wexler:

Francis and Lap-Chee Tsui:

James Watson:

TESTING YOUR KNOWLEDGE

*In this section, answer the questions, fill in the blanks, and solve the problems in the space allotted. Problems noted with an * are solved in the Approaches to Problem Solving section at the end of the chapter.*

1. A cloning procedure that depends upon the availability of a detailed map of the region of the chromosome where the gene of interest resides is called _____.

2. A physical map constructed by ordering overlapping genomic DNA fragments is called a _____.

3. Three human disease-causing mutant genes, discussed in the text, which were cloned by positional cloning are _____ , _____ , and _____.

4. In humans, a very useful class of RFLP markers involving short tandemly repeated sequences that vary in number is called _____.

5. Restriction maps of genomic clones may be computer analyzed and organized into overlapping sets of clones called _____.

6. Clusters of cytosines and guanines that often precede human genes are called _____.

7. The largest known human gene, DMD, contains 79 exons, spans 2500 kb, and produces a protein of 3685 amino acids called _____.

8. The first four human diseases shown to be associated with an unstable trinucleotide repeats were _____, _____, _____, and _____.

9. What is the most common inherited disease in humans of Northern European heritage?

10. Genes that are mapped both genetically and physically are called _____.

11. An organism carrying a gene introduced by genetic engineering is called _____.

12. PCR-amplified short unique genomic sequences that have been assigned to contigs by Southern blotting and to chromosomal positions by *in situ* hybridization are called _____.

13. Which human genetic disease was the first to be treated by gene therapy?

14. Specific banding patterns on Southern blots of genomic DNA that has been cleaved with specific restriction enzymes and hybridized with appropriate DNA probes are called _____.

15. The first commercial success of recombinant DNA technology was the mass-production of the _____ protein by genetically engineered bacteria.

16. The most common form of transformation in dicots is mediated by _____.

*17. In a search for the gene for a genetic disease using restriction fragment length polymorphisms, DNA was isolated from five people with the disease, and five people without the disease. The DNAs were treated with a restriction endonuclease and the digested fragments separated by agarose electrophoresis. Which band probably represents the fragment of DNA most closely linked to the gene for the disease? Why? (This problem is modified from problem 10.7 in N. L. Pruitt, Problem Book and Study Guide for Cell and Molecular Biology by G. Karp, 1996).

<u>People without disease</u>   <u>People with disease</u>

— — — — —   1   — — — — —

— — — — —   2   — — — — —

— — — — —   3   — — — — —

— — — — —   4.   — — — — —

— — — — —   5.   — — — — —

*18. A man is about to go to trial for allegedly cutting the lock on a driveway gate and burgling a house. Although much circumstantial evidence suggests that the man committed the crime, the prosecuting attorney does not feel he can obtain a conviction unless it can be proven that the man was at the scene of the crime. Upon analyzing evidence provided by the police investigator he noticed a leaf was found that was lodged in the cuff of the pants worn by the defendent on the day of the crime. The prosecuting attorney remembered from his college botany class that the leaf was from a tree that was rarely found in his region of the country. Only four trees were known to exist in the city of the crime; one along the driveway of the burgled house and the others in city parks. After reviewing evidence, the defendant's lawyer insisted that the leaf must be from some source other than the burgled home. The prosecuting attorney, however, had a molecular genetics laboratory isolate and fingerprint the DNA of leaves of the four trees, and of the leaf found in the defendant's car. The resulting banding patterns (fingerprints) of the DNA from the leaves were as follows.

	Source of leaves			
defendant's pants	tree 1	tree 2	tree 3	burgled house

Based upon the DNA fingerprints of the leaves, should the prosecution of the case continue? Why or why not?

THOUGHT CHALLENGING EXERCISE

Recently, some consumer groups have been demanding that all food products from genetically engineered organisms, e.g., *flavr savr* tomato, be labeled as such. Is there a scientific basis for this concern? Do you feel that there are cases where genetically engineered foods should be labeled?

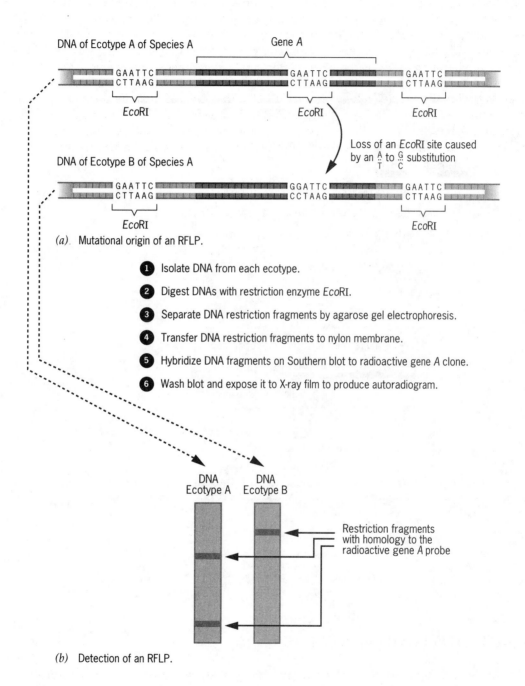

(a). Mutational origin of an RFLP.

1 Isolate DNA from each ecotype.

2 Digest DNAs with restriction enzyme *Eco*RI.

3 Separate DNA restriction fragments by agarose gel electrophoresis.

4 Transfer DNA restriction fragments to nylon membrane.

5 Hybridize DNA fragments on Southern blot to radioactive gene *A* clone.

6 Wash blot and expose it to X-ray film to produce autoradiogram.

(b) Detection of an RFLP.

Figure 20.1 The mutational origin (a) and detection (b) of RFLPs in different ecotypes of a species. In the example shown, an AT → GC base-pair substitution results in the loss of the central *Eco*RI recognition sequence present in gene A of the DNA of ecotype A. This mutation could have occurred in a common progenitor of the two ecotypes or in ecotype B ancestor during the early stages of its divergence from ecotype A. In ecotype A, gene A sequences are present on two *Eco*RI restriction fragments, whereas in ecotype B, all A gene sequences are present on one large *Eco*RI restriction fragment.

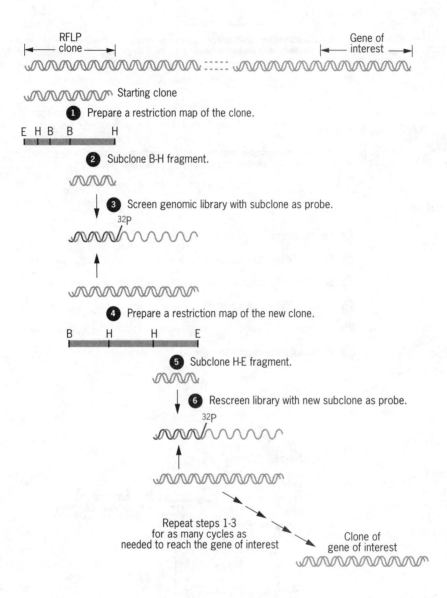

Figure 20.5 Positional cloning of a gene by chromosome walking. A walk starts with the identification of the molecular marker — such as the RFLP shown — close to the gene of interest. A restriction map is prepared for the initial RFLP clone, and the restriction fragment closest to the gene of interest is used as a probe to screen a genomic library for overlapping clones. Restriction maps are then prepared for the new genomic clones (for simplicity, only one is shown), and, again, the restriction fragment proximal to the gene of interest is subcloned and used as a probe to isolate a second set of overlapping genomic clones. Restriction maps of the new clones are prepared, and the process is repeated until the walk reaches the gene of interest.

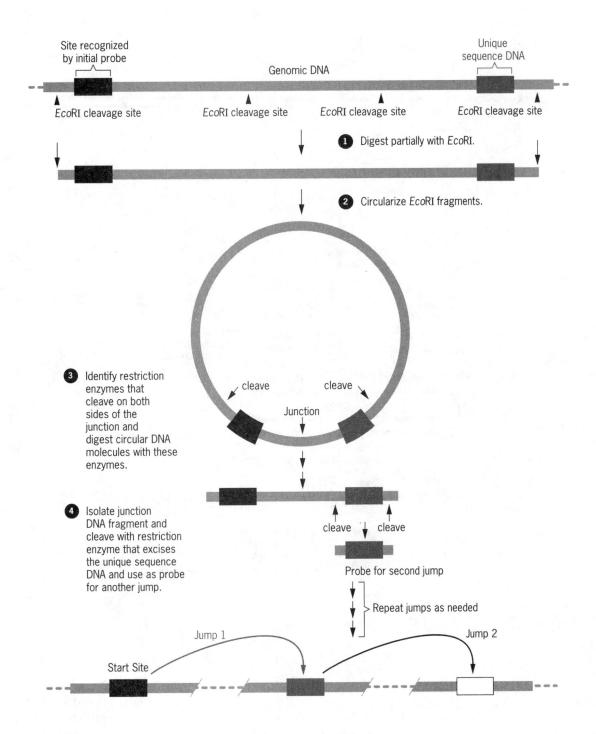

Figure 20.6 Chromosome jumping as a shortcut method for long chromosome walks. This procedure can also be used to jump over repetitive DNA sequences that block chromosome walks.

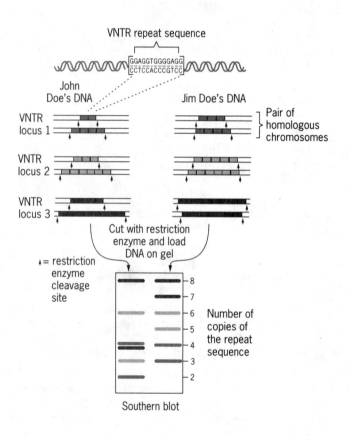

Figure 20.19 Simplified diagram of the use of variable number tandem repeats in preparing DNA fingerprints.

SUMMARY OF KEY POINTS

Detailed genetic and physical maps of genomes allow genes to be identified, cloned, and characterized based on their chromosomal locations. A closely linked molecular marker is used to initiate walks or jumps along the chromosome to a gene of interest. Positional cloning of genes can be done without laborious walks or jumps once complete physical maps and clone banks of chromosomes are available.

The human genes that cause Huntington's disease, cystic fibrosis, Duchenne's muscular dystrophy, and several other disorders have been identified by positional cloning. The nucleotide sequences of these genes were used to predict the amino acid sequences of their polypeptide products and to obtain valuable information about the functions of the gene products. The identification of these genes led directly to new methods of treating the diseases, possible approaches to gene therapy, and DNA tests for the mutations that cause the diseases.

The mutant genes responsible for several inherited human diseases can be accurately diagnosed by screening genomic DNAs for the genetic defect. The results of these tests provide information that allows genetic counselors to inform families of the risk of an affected child.

Important goals of the Human Genome Project are to map all 50,000 to 100,000 human genes and to determine the complete nucleotide sequence of all 24 human chromosomes.

Gene therapy involves the addition of a normal (wild-type) copy of a gene to the genome of an individual who carries defective copies of the gene. Physicians are currently testing the effectiveness of somatic-cell gene therapy in the treatment of patients with adenosine deaminase-deficient severe combined immunodeficiency disease. Although many technical details must still be worked out, somatic cell gene therapy holds great promise for the treatment of many inherited human diseases. In the future, somatic-cell gene therapy will probably involve gene replacements, rather than gene additions.

DNA fingerprints detect and record polymorphisms in the genomes of individuals. DNA prints provide strong evidence of individual identity, evidence that may be extremely valuable in paternity and forensic cases.

Valuable proteins that could be isolated from eukaryotes only in small amounts and at great expense can now be produced in large quantities in genetically engineered bacteria. Proteins such as human insulin and human growth hormone are valuable pharmaceuticals used to treat diabetes and pituitary dwarfism, respectively. Other proteins have important industrial applications.

Synthetic, modified, or foreign genes can now be introduced into most plant and animal species. The resulting transgenic organisms provide valuable systems in which to study the functions of the introduced genes. The Ti plasmid of *Agrobacterium tumefaciens* is an important tool for transferring genes into plants. Antisense RNA, which is complementary to sense RNA (mRNA), can be used to shut off or reduce the expression of individual genes.

ANSWERS TO QUESTIONS AND PROBLEMS

1) positional cloning 2) contiguous clone map 3) Huntington's disease, cystic fibrosis, and Duchenne muscular 4) VNTRs (variable number of tandem repeats) 5) contigs 6) CpG islands 7) dystrophin 8) fragile-X syndrome, spino-bulbular muscular atrophy, myotonic dystrophy, and Huntington's disease 9) cystic fibrosis 10) anchor genes 11) transgenic 12) sequence tagged sites (STSs) 13) severe combined immunodeficiency syndrome (SCID) 14) DNA fingerprints 15) insulin 16) *Agrobacterium tumefaciens* 17 and 18) see *Approaches to Problem Solving* section.

Some ideas concerning the thought challenging exercise:

For the most part, there should be no extraordinary concern about food derived from a genetically engineered organism, particularly if no new compound is appearing in the food. Some misinformed individuals believe that a conventionally-bred organism is safer as a food source than an organism that is genetically engineered. There is no *a priori* reason to believe that natural or conventionally bred organisms produce safer food. Some plants, conventionally bred for food products, retain undesirable or suspected carcinogenic compounds. Certainly, we all agree that there should be regulations concerning the marketing of food products derived from genetically engineered organisms, but labeling all products derived from a genetically engineered organism would not be based on scientific knowledge.

APPROACHES TO PROBLEM SOLVING

17. To solve this problem, look for a pair of bands that is associated with individuals with the disease, but is polymorphic among normal individuals. One form of band 4 is associated with the disease and may be closely linked to the gene for the disease. Since there is only a sample of five individuals, more sampling would be necessary to support this hypothesis.

18. Since the banding pattern of the DNA of the leaf found on the defendant's pants matches exactly that of the tree found in the driveway of the burgled house, but not that of the other four trees of the same kind growing in the town, this would be evidence that the defendant was at the scene of the crime. However, because a leaf could be carried by wind or other vectors to other locations, this would be circumstantial evidence that could not establish absolute proof that the defendant was at the scene of the crime. Whether the prosecution continued would depend upon how strong a case could be made on the cumulative evidence collected.

21

Regulation of Gene Expression in Prokaryotes

CONSTITUTIVE, INDUCIBLE, AND REPRESSIBLE GENE EXPRESSION

OPERONS: COORDINATELY REGULATED UNITS OF GENE EXPRESSION

THE LACTOSE OPERON IN *E. COLI*: INDUCTION AND CATABOLIC REPRESSION
    Induction
    Catabolite Repression

THE TRYPTOPHAN OPERON IN *E. COLI*: REPRESSION AND ATTENUATION
    Repression
    Attenuation

THE ARABINOSE OPERON IN *E. COLI*: POSITIVE AND NEGATIVE CONTROLS

BACTERIOPHAGE LAMBDA: LYSOGENY OR LYSIS
    Repression of Lambda Lytic Pathway Genes During Lysogeny
    The Lambda Lytic Regulatory Cascade
    The Lambda Switch: Lytic Development or Lysogeny

TEMPORAL SEQUENCES OF GENE EXPRESSION DURING PHAGE INFECTION

TRANSLATIONAL CONTROL OF GENE EXPRESSION

POST-TRANSLATIONAL REGULATORY MECHANISMS

IMPORTANT CONCEPTS

  A. Gene expression is frequently under the control of regulator genes.
    1. Regulator genes act primarily at the level of transcription, often affecting the ability of RNA polymerase to bind to promoter sequences.
    2. The effects of regulator gene products, in turn, are controlled by the presence or absence of specific effector molecules in the environment.
    3. In prokaryotes, genes with related functions are frequently present in coordinately regulated units called operons.
      a. Each operon is one unit of transcription, i.e., a single mRNA carries the coding sequences of all the genes in the operon.
    4. Certain regulatory genes encode proteins called repressors, which function by means of their sequence-specific binding to DNA.
      a. Some repressors (those for repressible operons) bind to DNA only in the presence of effector molecules called co-repressors; others (those for inducible operons) bind to DNA only in the absence of effector molecules called inducers.
      b. Repressors are allosteric proteins, i.e., proteins that undergo conformational shifts and correlated changes in activity in response to the binding of specific effector molecules.
        (1) Repressors bind to their DNA binding sites in one conformation, but not in other conformations.
      c. Repressors act by binding to DNA sequences called operators, which are located adjacent to the structural genes whose transcription they control.

    (1) When a repressor, or the complex of repressor and co-repressor, is bound to the operator sequence of an operon, it prevents RNA polymerase from binding to the contiguous promoter sequence and initiating the transcription of the operon.

  5. Operons containing genes that encode enzymes involved in catabolic pathways often are under positive control by the catabolite activator protein (CAP) and cyclic AMP (cAMP).

   a. The binding of the CAP-cAMP complex to a site within the promoter region of an operon is required for the efficient binding of RNA polymerase to its binding site in the promoter region.

    (1) High concentrations of glucose result in low intracellular concentrations of cAMP.

    (2) Such operons cannot be induced in the presence of high concentrations of glucose, a phenomenon known as catabolite repression.

  6. Operons controlling enzymes involved in amino acid biosynthetic pathways are frequently controlled by attenuation.

   a. Attenuation occurs by premature termination of transcription at an attenuator site located within the mRNA leader sequence.

    (1) Attenuator sites contain transcription-termination sequences; however, the ability of the attenuator RNA sequences to form the secondary structures that lead to termination of transcription depends on the presence of the amino acid that is the end-product of the pathway controlled by the operon in question.

B. A temperate bacteriophage such as lambda can follow either the lytic cycle, during which it reproduces and lyses the host cell, or the lysogenic cycle, during which its chromosome exists as a dormant prophage inserted into the chromosome of the host bacterium.

  1. During lysogeny, the lytic genes of the prophage are kept turned off by a repressor-operator-promoter-circuit similar to those of bacterial operons.

  2. The decision between lysogeny and lytic development involves an elegant genetic switch that is controlled by two regulatory proteins, λ repressor and Cro protein, which bind to operator sites that control transcription of the λ genome.

   a. If repressor is bound at these sites, lysogeny ensues.

    (1) The lysogenic state is maintained by the autogenous control of the λ repressor.

   b. If Cro protein occupies these sites, lytic development occurs.

    (2) Lytic development is controlled by a regulatory cascade in which transcriptional antiterminators play a central role.

C. Preprogrammed sequences of viral gene expression occur in bacteriophage-infected cells.

  1. The products of genes expressed early after infection interact with RNA polymerase to change its promoter specificity and switch transcription to a second set of genes.

   a. In the case of some of the more complex bacteriophages, this switching process may be repeated several times during the life cycle of the virus.

IMPORTANT TERMS

In the space allotted, concisely define each term.

constitutive genes:

inducible genes:

catabolic pathways:

repression:

derepression:

anabolic pathways:

regulator gene:

repressor:

operator:

operon:

effector molecules:

inducers:

co-repressors:

allosteric transitions:

repressor-inducer complex:

negative control system:

positive control system:

catabolite repression:

CAP:

cyclic AMP (cAMP):

attenuation:

attenuator:

transcription-termination signals:

*ara*C protein:

DNA binding domain:

connector domain:

N protein:

Q protein:

Cro protein:

transcriptional antiterminator:

negative self-regulation:

negative autogenous regulation:

feedback inhibition:

end product or regulatory binding site:

IMPORTANT NAMES

In the space allotted, concisely state the major contribution made by the individual or group.

Felix d'Herelle:

Frederick W. Twort:

Francois Jacob and Jacques Monod:

Charles Yanofsky:

TESTING YOUR KNOWLEDGE

In this section, answer the questions, fill in the blanks, and solve the problems in the space allotted. Problems noted with an * are solved in the Approaches to Problem Solving section at the end of the chapter.

1. β-galactosidase cleaves lactose to form _____ and _____.

2. In inducible bacterial operons the inactivation of induction by elevated levels of glucose, even though the inducer is present, is called _____.

3. The two individuals who received a Nobel Prize for their research leading to the operon model of gene regulation in bacteria were _____ and _____.

4. Name the two protein complexes that bind to the promoter of the *lac* operon.

5. What would be the effect of a deletion of the promoter of the *lac* operon?

6. What would be the effect of a deletion of the promoter of the *i* gene of the *lac* operon?

7. Whether or not a repressor will bind to the operator and turn off the transcription of the structural genes in the operon is determined by the presence or absence of molecules called _____.

8. The enzymes produced by the coordinate regulation of the three genes in the *lac* operon are _____, _____, and _____.

9. The effector molecule of the *lac* operon is _____.

10. To bind to the promoter of the *lac* operon, the catabolite activator protein must be complexed with _____.

11. The best understood repressible operon of *E. coli* is the _____ operon.

12. A nucleotide sequence in the 5′ region of a prokaryotic gene (or in its RNA) that causes premature termination of transcription is called a (an) _____.

13. In the *trp* operon, the co-repressor is _____.

14. The induction of the arabinose operon of bacteria depends on the positive regulatory effects of two proteins, _____, and _____.

15. Which pathway, lysogeny or lytic development will occur if the C_1 repressor is bound to both the O_L and O_R operator regions of the λ prophage?

16. For the λ bacteriophage, the determination of lysogeny or lytic growth depends upon a balance between two proteins called _____ and _____.

17. The inhibition of translation of an mRNA molecule by one of the products that it encodes is called _____.

18. The inhibition of the activity of an enzyme by an end product of the biosynthetic pathway is called _____.

19. The enhancement of enzyme activity by a substrate or other effector molecule is called _____.

*20. For each of the following merozygotes indicate the conditions under which β-galactosidase is produced.

(a) $i^-p^+o^+z^+ / i^+p^+o^+z^-$

(b) $i^+p^+o^+z^+ / i^+p^+o^cz^+$

(c) $i^+p^+o^cz^- / i^+p^+o^+z^+$

(d) $i^s p^+ o^+ z^+$ / $i^+ p^+ o^+ z^+$

(e) $i^s p^+ o^+ z^+$ / $i^+ p^+ o^c z^+$

THOUGHT CHALLENGING EXERCISE

You are given a *lac⁻* mutant of *E. coli*. How could you determine if the uninducible phenotype is due to a defective repressor (i^s), a defective promoter, or to a defective structural gene?

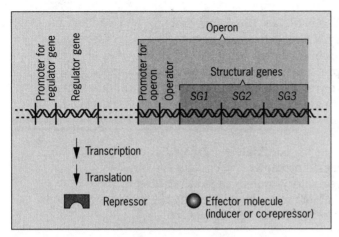

(a) The operon: components.

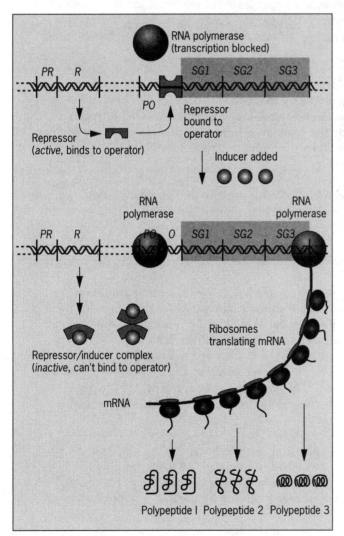

(b) The operon: induction.

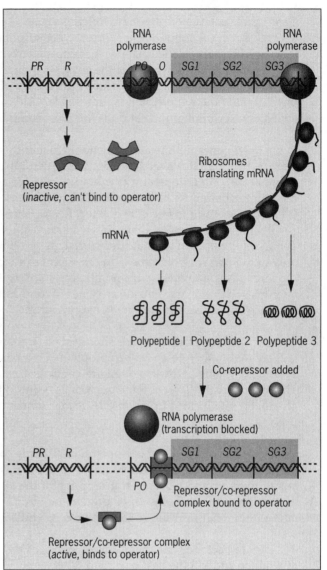

(c) The operon: repression.

Figure 21.3 Regulation of gene expression by the operon mechanism. (a) Essential components of an operon: one or more structural genes (three, SG1, SG2, and SG3, are shown) and the adjoining operator and promoter sequences. The transcription of gene (R) is initiated by RNA polymerase, which binds to its promoter (PR). When the repressor is bound to the operator, it sterically prevents RNA polymerase from initiating transcription of the structural genes. The difference between an inducible operon (b) and a repressible operon (c) is that free repressor binds to the operator of an inducible operon, whereas the repressor-effector molecule complex binds to the operon of a repressible operon. Thus an inducible operon is turned off in the absence of the effector (inducer) molecule, and a repressible operon is turned on in the absence of the effector (co-repressor) molecule.

SUMMARY OF KEY POINTS

Genes that specify housekeeping functions such as rRNAs, tRNAs, ribosomal proteins, and DNA and RNA polymerase subunits are expressed constitutively. Other genes usually are expressed only when their products are needed. Genes that encode enzymes in catabolic pathways are often expressed only in the presence of the substrates of the enzyme; their expression is inducible. Genes that encode enzymes involved in anabolic pathways usually are turned off or repressed in the presence of the end product of the pathway; their expression is repressible. Although gene expression can be regulated at many levels, transcriptional regulation is the most common.

In bacteria, genes with related functions frequently occur in coordinately regulated units called operons. Each operon contains a set of contiguous structural genes, a promoter (the binding site for RNA polymerase) and an operator (the binding site for a regulatory protein called a repressor). When the repressor is bound to the operator, RNA polymerase cannot transcribe the structural genes in the operon. When the operator is free of repressor, RNA polymerase can transcribe the operon.

The *E. coli lac* operon is an inducible system; the three structural genes of the *lac* operon are transcribed only in the presence of lactose. In the absence of lactose, the *lac* repressor binds to the *lac* operator and prevents RNA polymerase from initiating transcription of the operon. Catabolic repression keeps operons such as *lac* that encode enzymes involved in carbohydrate catabolism from being induced in the presence of glucose, the preferred energy source.

The *E. coli trp* operon is a repressible system; transcription of the five structural genes in the *trp* operon is repressed in the presence of tryptophan. Operons such as *trp* that encode enzymes involved in amino acid biosynthetic pathways often are controlled by a second regulatory mechanism called attenuation. The level of expression of these operons is reduced by the premature termination of transcription at the attenuator site located in the mRNA leader sequence when the amino acid produced by the pathway is present.

Transcription of the arabinose operon of *E. coli* is controlled by both positive and negative regulatory mechanisms. In the absence of arabinose, the *ara*C protein acts as a negative regulator by binding to the *ara*O$_2$ operator and blocking transcription of the *ara*B, *ara*A, and *ara*D structural genes. In the presence of arabinose, the arabinose-*ara*C protein and cAMP-CAP complexes function as positive regulators by stimulating transcription of the *ara*B, *ara*A, and *ara*D structural genes.

Temperate bacteriophages can follow either of two pathways: (1) lytic growth, during which they reproduce and kill the host cells, and (2) lysogeny, during which their chromosomes exist as a dormant prophage covalently inserted in the chromosomes of the host bacteria. During lysogeny, the lytic genes of the prophage are kept turned off by a repressor-operator-promoter circuit similar to those of bacterial operons. Whether a lambda prophage will enter the lysogenic state or undergo lytic development is determined by which of two regulatory proteins, λ repressor or Cro protein, occupies key operator sites that control transcription of the λ genome. The lysogenic state is maintained by the autogenous control of λ repressor, whereby lytic development is controlled by a regulatory cascade in which transcriptional antiterminators play a major role.

Preprogrammed temporal sequences of the viral gene expression occur in bacteriophage-infected cells. The first viral genes expressed in an infected cell are transcribed by unmodified bacterial RNA polymerase. Subsequent sets of expressed viral genes are transcribed either by an RNA polymerase encoded by the phage genome or by bacterial RNA polymerase modified by the addition of viral protein(s).

Regulatory fine tuning often occurs at the level of translation by modulation of the rate of either polypeptide chain initiation or chain elongation. Sometimes regulation occurs by the differential degradation of specific regions of polygenic mRNAs. The inhibition of translation of a specific mRNA by a protein that it encodes is called negative autogenous regulation.

The end product of a biosynthetic pathway often inhibits the activity of the first enzyme in the pathway, rapidly shutting off the synthesis of the product. This regulatory mechanism is called feedback inhibition. Enzyme activation occurs when a substrate or other effector molecule enhances the activity of an enzyme.

ANSWERS TO QUESTIONS AND PROBLEMS

1) galactose and glucose **2)** catabolite repression **3)** Jacob and Monod **4)** RNA polymerase and cAMP-CAP **5)** uninducibility of the operon **6)** constitutive synthesis of *lac* operon proteins **7)** effectors **8)** β-galactosidase, β-galactoside permease, β-galactoside transacetylase **9)** allolactose **10)** cyclic AMP **11)** tryptophan **12)** attenuator **13)** tryptophan **14)** *ara*C and cAMP-CAP **15)** lysogeny **16)** λ repressor and Cro protein **17)** negative self-regulation or negative autogenous regulation **18)** feedback (end-product) inhibition **19)** substrate activation **20)** see *Approaches to Problem Solving* section below.

Some ideas concerning the thought challenging exercise:

First of all, it would be best to assay for the inducibility of the other two genes of the operon, i.e., permease and transacetylase. If the other two genes are inducible, but the i gene is not, then the z gene is defective. Then, if all three structural genes of the operon were uninducible, it could be due to either a mutation of the repressor gene or to a mutant promoter. It could be determined if the lac⁻ mutant was defective in the i gene by forming a merozygote between the mutant and an $i^- p^+ o^+ z^+$ cell. If the production of β-galactosidase is lacking in the merozygote and cannot be induced, then the mutant is most likely a mutant of the i gene, i.e., i^s. If all three genes of the operon are uninducible, and it is not due to a defective regulator gene, then the mutant is either a defective promoter or a mutation or deletion of the structural gene(s). The simplest approach to distinguish between these two possibilities would be to genetically map the position of the mutant gene.

APPROACHES TO PROBLEM SOLVING

20. The key to solving problems involving operons is first to fully learn how the different components interact. The operon model is presented in Key Figure 21-3.

(a) $i^- p^+ o^+ z^+ / i^+ p^+ o^+ z^-$

In this merozygote, the DNA strand indicated on the left has a mutant regulator gene i^-, and the other genes are wild-type. This would show constitutive synthesis of β-galactosidase, but i^+ is dominant to i^-; it produces a normal repressor that will fuse to o^+ unless lactose is added. The DNA strand on the right does not produce β-galactosidase because of the mutant gene z^-. This merozygote produces β-galactosidase only inducibly.

(b) $i^+ p^+ o^+ z^+ / i^+ p^+ o^c z^+$

In this merozygote, all genes are wild-type except for the operator gene o^c of the DNA strand indicated on the right. Due to the o^c gene, this merozygote will produce β-galactosidase constitutively. Since the DNA on the left is normal, it can be induced by lactose. Therefore, although the cell has constitutive production of β-galactosidase, the amount of β-galactosidase is even higher in induced cells.

(c) $i^+p^+o^cz^-$ / $i^+p^+o^+z^+$

The DNA strand on the left has an o^c gene and a z^- gene. No β-galactosidase will ever be produced following transcription of this strand. Remember, operator genes are binding sites, and therefore can only affect structural genes of the operon in the cis position. The DNA on the right is wild-type. It will produce β-galactosidase when induced.

(d) $i^sp^+o^+z^+$ / $i^+p^+o^+z^+$

This merozygote is normal except for the presence of a mutant gene i^s that produces a superrepressor. The superrepressor cannot be removed from the operator gene by the inducer and therefore this merozygote is permanently suppressed.

(e) $i^sp^+o^+z^+$ / $i^+p^+o^cz^+$

This merozygote has a superrepressor mutant of the i gene in the DNA on the left. The operon contained in this DNA molecule is permanently suppressed. However, the superrepressor cannot bind to the mutant operator o^c of the DNA strand on the right. The constitutive operator gene is epistatic to the i^s mutant. Therefore, this strand is transcribed all of the time and the production of β-galactosidase is constitutive.

22

Gene Regulation in Eukaryotes and the Genetic Basis of Cancer

---

IMPORTANT CONCEPTS

A. Gene expression is spatially and temporally regulated in multicellular eukaryotes; it can occur in the nucleus or in the cytoplasm.
   1. In the nucleus, gene expression can be regulated at transcription or by RNA processing.
   2. In the cytoplasm, it can be regulated by mRNA stability.
   3. The transcription of some eukaryotic genes is induced by environmental stimuli such as heat shock or light.
   4. Other eukaryotic genes are transcribed in response to signaling molecules such as hormones and growth factors.

a. These factors are mediated by protein transcription factors that bind to enhancer or silencer sequences located in the vicinity of a gene.

    (1) Transcription factors have characteristic structural motifs such as the zinc finger, leucine zipper, helix-turn-helix, or helix-loop-helix.

5. Alternate splicing of eukaryotic gene transcripts provides a way for a single gene to encode several different polypeptides.

6. The stability of a mRNA depends on sequences in the 5' untranslated region and on polyadenylation at the 3' end.

B. Eukaryotic gene expression is influenced by chromosome organization.

1. Transcription typically occurs in open chromatin configurations such as amphibian lampbrush chromosome loops or dipteran polytene chromosome puffs.

2. Transcriptionally active DNA is more sensitive to digestion with the enzyme DNase I.

3. Chromosomes may be organized into domains of gene expression by specialized chromatin structure.

    a. The transcription of eukaryotic genes that have been transposed into heterochromatin is repressed, causing the phenomenon called position-effect variegation.

    b. Some genes such as those encoding the ribosomal RNAs in amphibian oocytes are amplified to increase their expression.

    c. Large arrays of genes on an entire X chromosome are regulated to achieve equal X-linked gene expression in males and females.

        (1) These genes may be coordinately inactivated, hyperactivated or hypoactivated by RNAs or proteins that associate with the X chromosome.

C. Abnormal gene expression is the basis for many forms of cancer.

1. Cancer-causing genes, or oncogenes, were originally discovered in the genomes of retroviruses.

    a. These *v-onc* genes have normal cellular counterparts, the *c-onc* genes which play important roles in regulating cell growth and differentiation.

2. Cancer can be caused by mutant cellular oncogenes, or by the inappropriate expression of normal cellular oncogenes that have been relocated by chromosomal rearrangements.

3. Cancer can result by inactivation of tumor-suppressor genes.

4. Some forms of cancer are inherited; other forms appear to be caused by the occurrence of a series of mutations in the somatic tissues.

IMPORTANT TERMS

In the space allotted, concisely define each term.

pseudogene:

alternate splicing:

inducer:

heat-shock proteins:

heat-shock transcription factor (HSTF):

heat-shock response elements (HSREs):

peptide hormones:

membrane-bound receptor proteins:

signal transduction:

hormone response elements (HREs):

general transcription factors:

special transcription factors:

enhancers:

silencers:

zinc finger:

homeodomain:

leucine zipper:

helix-loop-helix:

chromomeres:

puffs:

ecdysone:

DNase I hypersensitivity:

specialized chromatin structures (scs and scs'):

heterochromatin:

position-effect variegation:

gene amplification:

nucleolar organizer:

homogeneously staining region (HSR):

double minutes (Dms):

dosage compensation:

X inactivation center (XIC):

facultative heterochromatin:

malignant:

benign:

metastasis:

Rous sarcoma virus:

oncogenes:

proto-oncogenes (normal cellular oncogenes):

Philadelphia chromosome:

Burkitt's lymphoma:

retinoblastoma:

tumor-suppressor gene:

IMPORTANT NAMES

In the space allotted, concisely state the major contribution made by the individual or group.

Pamela Geyer and Victor Corces:

M. Groudine and H. Weintraub:

Paul Schedl and Rebecca Kellum:

Peyton Rous:

Robert Weinberg:

Alfred Knudson:

TESTING YOUR KNOWLEDGE

In this section, answer the questions, fill in the blanks, and solve the problems in the space allotted. Problems noted with an * are solved in the Approaches to Problem Solving section at the end of the chapter.

1. The tissue-specific expression of genes results from _____ regulation.

2. The _____ regulation of genes results in different genes being expressed at different times during development.

3. A human cancer that is associated with a reciprocal translocation between chromosomes 8 and 14 and overexpression of the *c-myc* gene is called _____.

4. The reciprocally translocated chromosome (22/9) that is associated with myelogenous leukemia in humans is called the _____ chromosome.

5. A mutant, nonfunctional duplicate of a normal gene is called a (an) _____.

6. The different rat troponin T proteins result from a phenomenon called _____.

7. A mutant gene that results in the transformation of a normal animal cell to a cancerous cell is called a (an) _____.

8. Name the three levels at which eukaryotic gene expression can be regulated.

9. In animals the small lipid-soluble molecules derived from cholesterol, e.g., estrogen, progesterone, and testosterone, form a class of hormones called _____.

10. For estrogen to enter the nucleus and act as a transcription factor it first must bind to a cytoplasmic protein called a (an) _____.

11. A class of hormones consisting of polypeptides, e.g., insulin, is called _____.

12. The process of transmitting a hormonal signal through the cell and into the nucleus is called _____.

13. Proteins that bind to a sequence within a promoter and facilitate the proper alignment of the RNA polymerase are called _____.

14. What are specific DNA sequences that mediate hormone-induced gene expression called?

15. Special transcription factors such as those involved in the regulation of the heat, light, and hormone inducible genes bind to response elements in the DNA called _____ and _____.

16 Peptide hormones are transmitted through the cell membrane by _____.

17. DNA sequences that can act over relatively large distances and influence gene regulation independent of orientation and position are called _____.

18. A structural motif found in some eukaryotic transcription factors that plays an important role in DNA binding, consisting of a protein loop formed when two cysteines in one part of the polypeptide and two histidines in another part nearby jointly bind to a zinc ion, is called a (an) _____ _____.

19. The highly conserved region of approximately 60 amino acids in many transcription factors that coincides with a helix-turn-helix motif is called a _____.

20. A steroid hormone that has been demonstrated to induce puffing of polytene chromosome bands in *Drosophila* is called _____.

21. A mutant that results in a body part developing in an inappropriate position is called a (an) _____ mutant.

22. Chromatin staining darkly even during interphase, often containing repetitive DNA, and with few genes is called _____.

23. A structural motif found in some transcription factors that consists of a stretch of amino acids with leucine at every seventh position is called a _____.

24. The region of eukaryotic chromosomes that contains numerous 18S-28S ribosomal RNA genes is called the _____.

25. A unique feature of tumorogenic Rous sarcoma virus is the presence of the oncogene *v-src* that encodes a polypeptide with _____ activity.

*26. Why do familial forms of retinoblastoma generally involve both eyes and occur early in life, whereas, non-familial forms generally involve only one eye and occur later in life?

THOUGHT CHALLENGING EXERCISE

Discuss the concept that cancer has been called a genetic disease, even though a high percentage of cancer is believed to be environmentally induced.

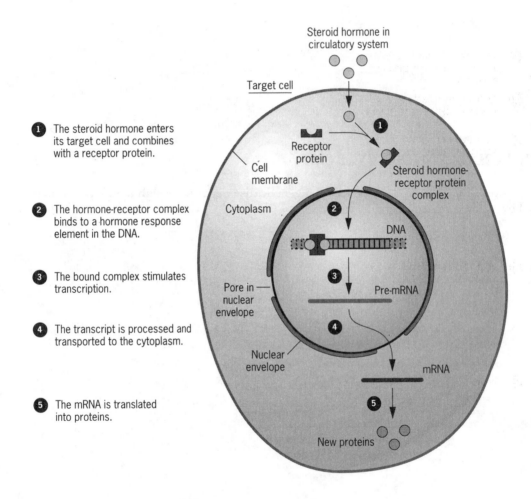

1. The steroid hormone enters its target cell and combines with a receptor protein.

2. The hormone-receptor complex binds to a hormone response element in the DNA.

3. The bound complex stimulates transcription.

4. The transcript is processed and transported to the cytoplasm.

5. The mRNA is translated into proteins.

Figure 22.8 Regulation of gene expression by steroid hormones. The hormone interacts with a receptor inside its target cell, and the resulting complex moves into the nucleus where it activates the transcription of particular genes.

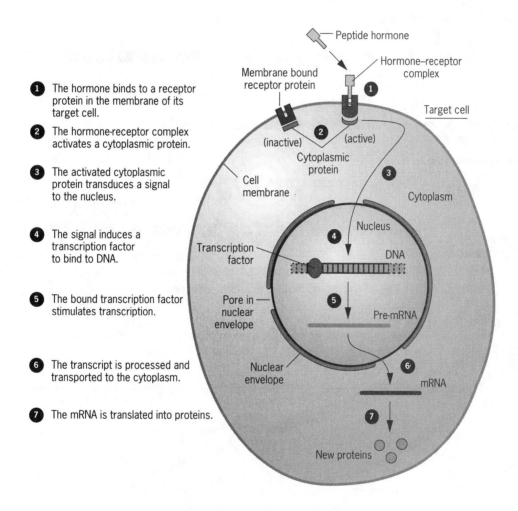

1. The hormone binds to a receptor protein in the membrane of its target cell.

2. The hormone-receptor complex activates a cytoplasmic protein.

3. The activated cytoplasmic protein transduces a signal to the nucleus.

4. The signal induces a transcription factor to bind to DNA.

5. The bound transcription factor stimulates transcription.

6. The transcript is processed and transported to the cytoplasm.

7. The mRNA is translated into proteins.

Figure 22.9   Regulation of gene expression by peptide hormones. The hormone (an extracellular signal) interacts with a receptor in the membrane of its target cell. The resulting hormone-receptor complex activates a cytoplasmic protein that triggers a cascade of intracellular changes. These changes transmit the signal into the nucleus, where a transcription factor stimulates the expression of particular genes.

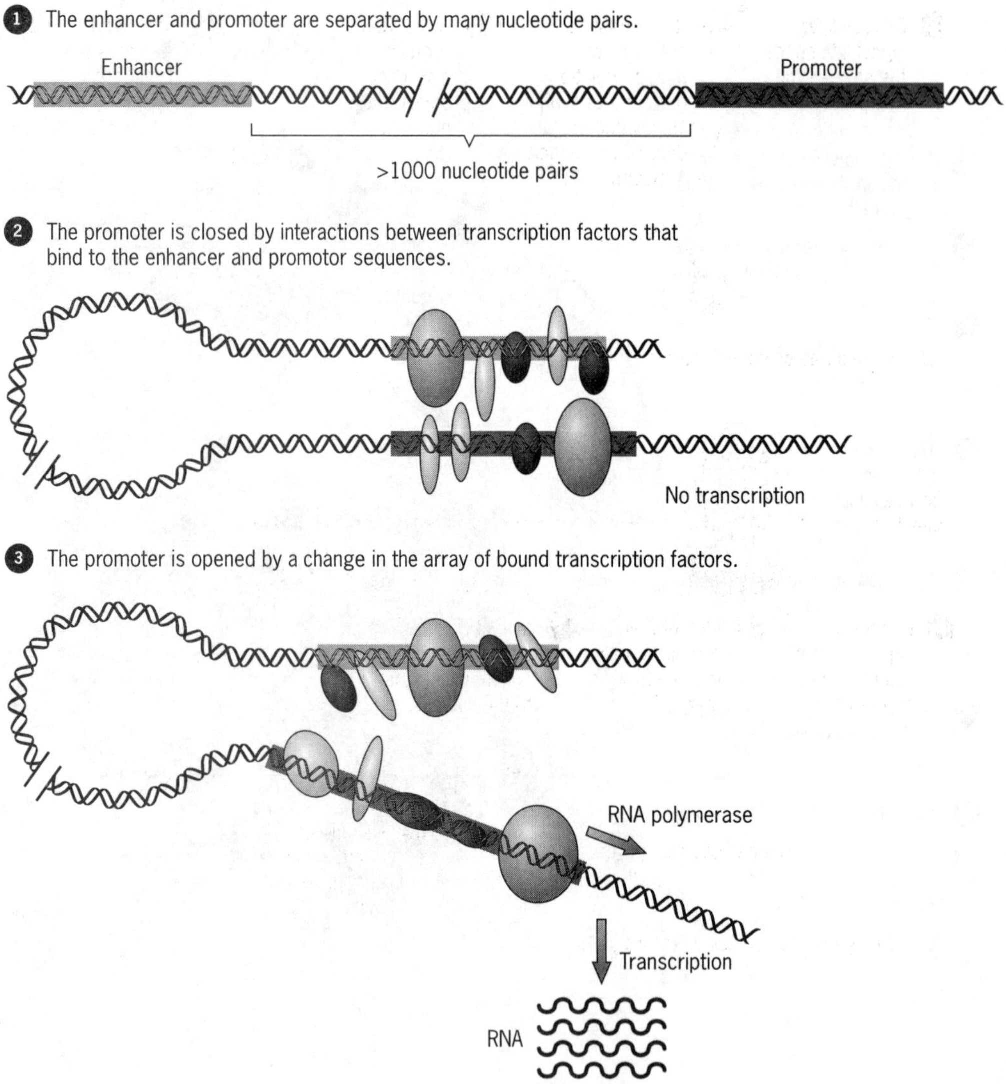

① The enhancer and promoter are separated by many nucleotide pairs.

Enhancer   Promoter

>1000 nucleotide pairs

② The promoter is closed by interactions between transcription factors that bind to the enhancer and promotor sequences.

No transcription

③ The promoter is opened by a change in the array of bound transcription factors.

RNA polymerase

Transcription

RNA

Figure 22.12 Model for the control of transcription by proteins bound to the enhancer and a promoter.

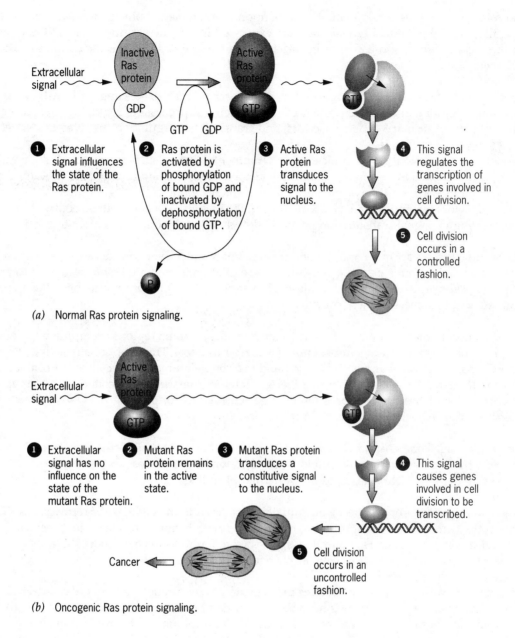

(a) Normal Ras protein signaling.

1 Extracellular signal influences the state of the Ras protein.

2 Ras protein is activated by phosphorylation of bound GDP and inactivated by dephosphorylation of bound GTP.

3 Active Ras protein transduces signal to the nucleus.

4 This signal regulates the transcription of genes involved in cell division.

5 Cell division occurs in a controlled fashion.

(b) Oncogenic Ras protein signaling.

1 Extracellular signal has no influence on the state of the mutant Ras protein.

2 Mutant Ras protein remains in the active state.

3 Mutant Ras protein transduces a constitutive signal to the nucleus.

4 This signal causes genes involved in cell division to be transcribed.

5 Cell division occurs in an uncontrolled fashion.

Cancer

Figure 22.26 Ras protein signaling and cancer. (a) The normal protein product of the *ras* gene alternates between inactive and active states, depending on whether it is bound to GDP or GTP. Extracellular signals, such as growth factors, stimulate the conversion of inactive Ras to active Ras. Through active Ras, these signals are transmitted to other proteins and eventually to the nucleus, where they induce the expression of genes involved in cell division. Because this signaling is intermittent and regulated, cell division occurs in a controlled manner. (b) Mutant Ras protein exists mainly in the active state. These proteins transmit their signals more or less constantly, leading to uncontrolled cell division, the hallmark of cancer.

SUMMARY OF KEY POINTS

In eukaryotes, gene expression is spatially and temporally regulated. Some genes, such as those encoding the α and β tubulins in *Arabidopsis*, are expressed in a tissue-specific manner. Other genes, such as those coding for α and β globins in vertebrates, are expressed in a specific temporal pattern during development.

Eukaryotic gene expression can be regulated at the transcriptional, processing, or translational levels. Transcriptional regulation involves protein-DNA interactions, processing regulation involves the alternate splicing of primary gene transcripts, and translational regulation involves mRNA stability.

Eukaryotic gene expression can be induced by environmental factors such as heat and light, and by signaling molecules such as hormones and growth factors. Hormone-induced gene expression is mediated by proteins that interact with the hormones. Some of the hormone receptors act directly as transcription factors by binding to DNA sequences in the vicinity of a gene; others control transcription indirectly through a signal transduction pathway that targets transcription factors to a gene.

Enhancers and silencers act in an orientation-independent manner over considerable distances to regulate transcription from a gene's promoter. This regulation is mediated by different types of transcription factor proteins that bind to these DNA sequences. Different motifs have been identified in the amino acid sequences of these proteins.

Transcription occurs preferentially in loosely organized chromosome regions exemplified by loops of lampbrush chromosomes and the puffs of polytene chromosomes. Transcriptionally active DNA tends to be more sensitive to digestion with DNase I and may be delimited by specialized chromatin structures that insulate it from neighboring DNA. Heterochromatin is associated with the repression of transcription. Increased gene expression may be achieved by amplifying the DNA, either within chromosomes or on extrachromosomal DNA molecules.

In mammals, dosage compensation is achieved by inactivating one of the X chromosomes in XX females, in *Drosophila* by hyperactivating the single X chromosome in XY males, and in *Caenorhabditis*, by hypoactivating the two X chromosomes in XX hermanphrodites.

Cancer-causing agents, or oncogenes, were initially discovered in the genomes of retroviruses. These genes encode a variety of different proteins, including growth factors, growth-factor receptors, tyrosine kinases, and transcription factors. Later, these genes were found to have normal cellular homologs, but unlike their viral counterparts, these cellular oncogenes possess introns.

Some cancers are caused by mutant proteins that function as dominant activators of cell division. Mutations that produce these proteins have been identified by transfecting cultured cells with DNA isolated from tumor cells. Other cancers, such as Burkitt's lymphoma and chronic myelogenous leukemia, are associated with chromosomal translocations that alter the expression of a particular cellular oncogene. Still other cancers, such as retinoblastoma, are associated with deletions of genes that suppress tumor formation. Such a deletion may predispose an individual to develop a cancer if a spontaneous mutation eliminates the other copy of the gene.

ANSWERS TO QUESTIONS AND PROBLEMS

1) spatial 2) temporal 3) Burkitt's lymphoma 4) Philadelphia 5) pseudogene 6) alternate splicing 7) oncogene 8) trancriptional, processing, translational 9) steroid hormones 10) hormone receptor 11) peptide hormones 12) signal transduction 13) general transcription factors 14) hormone response elements 15) enhancers and silencers 16) membrane-bound receptor proteins 17) enhancers 18) zinc finger 19) homeodomain 20) ecdysone 21) homeotic 22) heterochromatin

23) leucine zipper **24)** nucleolus organizer **25)** protein kinase **26)** see *Approaches to Problem Solving* section

Ideas concerning thought challenging exercise:

Although environmental carcinogens (mutagens) are believed to be the major cause of cancer, their mode of action is likely through the induction of mutations. The current concept of cancer development is that it results from an accumulation of mutations and/or chromosomal aberrations. Cancer is considered to develop after a critical number of specific mutations has occurred. In inherited forms of cancer, the individual receives mutant allele(s) from a parent(s).

APPROACHES TO PROBLEM SOLVING

26. Retinoblastoma is a cancer that is determined by homozygosity for a recessive allele. Individuals who inherit one recessive allele from their parents require only one more mutation for retinoblastoma to develop. Since there are millions of cells in a retina, there is a high probability that the additional mutation will occur in at least one cell in each retina. Therefore, retinoblastoma of the familial type generally affects both eyes and at a young age. Individuals who do not inherit a mutant allele will still have a high probability of mutations occurring in retinal cells. In this case, for retinoblastoma to develop two mutations of the allele must occur in the same cell, or in descendants of a cell carrying the first mutant. Consequently, retinoblastoma of the nonfamilial type is rare, generally affects only one eye, and occurs later in life.

23

The Genetic Control of Animal Development

IMPORTANT CONCEPTS

A. Development of an animal results from cell divisions and differentiation that are directed by differential gene activity.
1. The blastula of an animal embryo is formed by cleavage divisions in a zygote.
   a. This sphere of cells is reorganized by cell movements (gastrulation) to produce a body with primitive tissues (ectoderm, endoderm, and mesoderm).
   b. During morphogenesis, ectoderm, endoderm, and mesoderm differentiate into mature tissues and organs.
B. The genetic analysis of animal development has concentrated on two model organisms, the fruit fly *Drosophila melanogaster* and the nematode *Caenorhabditis elegans*.
1. In these organisms, geneticists are able to dissect developmental pathways by identifying genes whose products are involved in differentiation.
   a. Some of the genes involved in sexual differentiation have been identified in both species.
2. The analysis of genetic mosaics may allow a researcher to trace cell lineages during development, and to ascertain if a gene's function is cell-autonomous.

3. Cloning a gene can provide important clues about its role in development.
4. The course of development can be influenced by maternal-effect products that are transported into the egg during oogenesis.
   a. In *Drosophila*, maternal-effect products are responsible for establishing the dorsal-ventral and anterior-posterior axes of the embryo.
      (1) Subsequent embryonic development, including the formation of body segments and the differentiation of specific cell types, requires the expression of zygotic genes.
C. Genes that are involved in vertebrate development have been identified by virtue of their homology to invertebrate genes.
   1. Mutations that affect vertebrate development have been studied in the mouse.
   2. Transgenic mice, which are useful in the study of development, can be created by injecting DNA into developing embryos.

IMPORTANT TERMS

In the space allotted, concisely define each term.

blastula:

gastrula:

gastrulation:

ectoderm:

mesoderm:

endoderm:

differentiation:

cleavage divisions:

determination:

primary germ layers:

morphogenesis:

chorion:

micropyle:

syncytium:

cellular blastoderm:

pole cells:

imaginal disks:

numerator elements:

denominator elements:

genetic mosaic:

twin spot:

transgene:

maternal-effect genes:

segments:

homeotic mutations:

homeosis:

Bithorax complex:

selector genes:

segmentation genes:

gap genes:

pair-rule genes:

segment-polarity genes:

embryonic stem cells:

targeting:

"knockout mutation":

IMPORTANT NAMES

In the space allotted, concisely state the major contribution made by the individual or group.

T. H. Morgan:

Christiane Nusslein-Volhard, Eric Weischaus, Trudi Schupbach, and Gerd Jurgens:

Calvin Bridges:

Edward Lewis:

Thomas Kaufman and Matthew Scott:

Gerald Rubin:

Walter Gehring:

TESTING YOUR KNOWLEDGE

In this section, answer the questions, fill in the blanks, and solve the problems in the space allotted. Problems noted with an * are solved in the Approaches to Problem Solving section at the end of the chapter.

1. The sphere of relatively nondescript cells that result from a rapid series of mitotic divisions of the zygote is called a _____.

2. The process in which undifferentiated cells are assigned developmental fates is called _____.

3. The combination of cell movements and differentiation through which the body of an animal takes shape is called _____.

4. The phenomenon of cells migrating to new positions in the embryo, therefore converting the nondescript blastula into an embryo with three distinct cell layers is called _____.

5. The single layer of cells formed at the surface of a *Drosophila* embryo after 13 cycles of mitosis that is equivalent to the blastula of vertebrates is called the _____.

6. X-linked genes which code for proteins in the *Drosophila* embryo that "count" the number of X chromosomes present are called _____.

7. Genes located on autosomes in *Drosophila* that affect the nominator of the X:A ratio are called _____.

8. The X-linked _____ gene is the master regulator of the sex-determination pathway in *Drosophila*.

9. The SLX protein allows the synthesis of a functional _____ protein in XX embryos but not in XY embryos.

10. The autosomal gene in *Drosophila* that is regulated by a protein encoded by the *transformer* 2 (*tra*2) gene and produces two proteins (through alternate splicing of its RNA) which determine whether male or female development will proceed is called the _____ gene.

11. What is the phenotype of embryos that are homozygous for loss-of-function mutants in *Sxl*?

12. What is the sex of an XY fly that is hemizygous for a loss-of-function mutation of the *Sxl* gene?

13. What is the sex of flies homozygous for loss-of-function mutants of *tra* or *tra* 2?

14 Genes that are expressed independently of the genotype of adjacent cells, e.g., *yellow* and *singed*, are called _____ genes.

15. Mutations in genes that contribute to the formation of healthy eggs, but have no effect on the phenotype of the female making the eggs are called _____ mutations.

16. In *Drosophila*, the formation of the dorsal-ventral axis is determined by the selective induction and repression of genes by action of a maternally synthesized transcription factor encoded by the _____ gene.

17. In *Drosophila*, the asymmetric synthesis of a protein coded by the *hunchback* gene determines the _____ axis of the embryo.

18. The *hunchback* gene is activated by a transcription factor encoded by the _____ gene.

19. Genes that encode transcription factors and define segmental regions in the early embryo, e.g., *Kruppel*, *giant*, *hunchback* and *knirps*, are called _____ genes.

20. Genes that are regulated by gap genes, e.g., *tarazu* and *even-skipped*, and define a pattern of segments within the embryo are called _____ genes.

21. Genes that define the anterior and posterior compartments of individual segments along the anterior-posterior axis, e.g., *gooseberry*, *wingless*, and *engrailed*, are called _____ genes.

22. Masses of cells in the larvae of *Drosophila* that give rise to specific adult organs such as antennae, eyes, and wings are called _____.

*23. A geneticist performed the *Drosophila* mating, $y^+sn^+/y^+sn^+ \times y\ sn\ \bigcirc$, and irradiated larvae to induce somatic recombination. Upon examination of the $y^+sn^+/y\ sn$ female progeny, single yellow spots, single singed spots and single yellow, singed spots were observed. A few twin spots, yellow and singed, were also observed. Illustrate the events involved in the formation of these twin spots.

THOUGHT CHALLENGING EXERCISE

What are some major features that differ between animal and plant development?

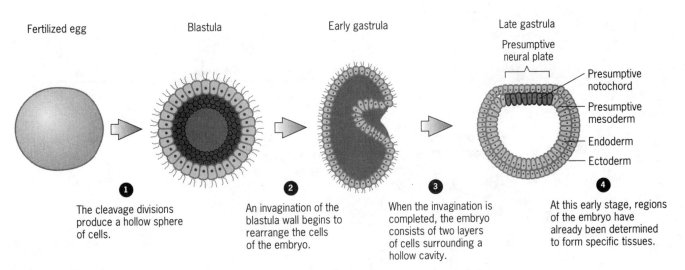

Fertilized egg

Blastula

Early gastrula

Late gastrula

Presumptive
neural plate

Presumptive
notochord

Presumptive
mesoderm

Endoderm

Ectoderm

1 The cleavage divisions produce a hollow sphere of cells.

2 An invagination of the blastula wall begins to rearrange the cells of the embryo.

3 When the invagination is completed, the embryo consists of two layers of cells surrounding a hollow cavity.

4 At this early stage, regions of the embryo have already been determined to form specific tissues.

(a) Early stages of development in *Amphioxus*.

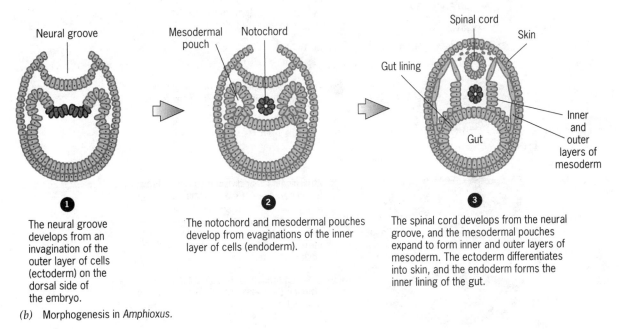

Neural groove

Mesodermal pouch

Notochord

Spinal cord

Skin

Gut lining

Inner and outer layers of mesoderm

Gut

1 The neural groove develops from an invagination of the outer layer of cells (ectoderm) on the dorsal side of the embryo.

2 The notochord and mesodermal pouches develop from evaginations of the inner layer of cells (endoderm).

3 The spinal cord develops from the neural groove, and the mesodermal pouches expand to form inner and outer layers of mesoderm. The ectoderm differentiates into skin, and the endoderm forms the inner lining of the gut.

(b) Morphogenesis in *Amphioxus*.

Figure 23.2   Embryonic development of the marine animal *Amphioxus*, a protochordate. (a) Early stages of embryonic development. (b) Morphogenesis.

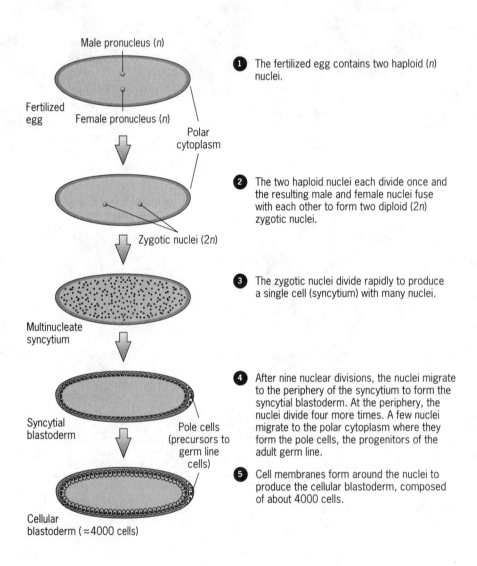

Male pronucleus (*n*)

Fertilized egg

Female pronucleus (*n*)

Polar cytoplasm

Zygotic nuclei (2*n*)

Multinucleate syncytium

Syncytial blastoderm

Pole cells (precursors to germ line cells)

Cellular blastoderm (≈4000 cells)

1 The fertilized egg contains two haploid (*n*) nuclei.

2 The two haploid nuclei each divide once and the resulting male and female nuclei fuse with each other to form two diploid (2*n*) zygotic nuclei.

3 The zygotic nuclei divide rapidly to produce a single cell (syncytium) with many nuclei.

4 After nine nuclear divisions, the nuclei migrate to the periphery of the syncytium to form the syncytial blastoderm. At the periphery, the nuclei divide four more times. A few nuclei migrate to the polar cytoplasm where they form the pole cells, the progenitors of the adult germ line.

5 Cell membranes form around the nuclei to produce the cellular blastoderm, composed of about 4000 cells.

Figure 23.4 **Early embryonic development in *Drosophila*.**

SUMMARY OF KEY POINTS

The development of a fertilized egg into a multicellular animal involves cell division, the assignment of fates to individual cells (determination), and the subsequent realization of those fates (differentiation). Key materials for the early phases of this process are transferred into the egg during oogenesis. After fertilization, the egg proceeds through several cleavage divisions to form a sphere of cells (the blastula), which is subsequently reorganized by cell movements (gastrulation) to form the primary germ layers (ectoderm, endoderm, and mesoderm). These primary tissues eventually produce mature tissues and organs.

The fruit fly *Drosophila melanogaster* and the hermaphroditic nematode *Caenorhabditis elegans* are model organisms with features that make them ideal for studies of the genetic control of development.

The genetic dissection of a developmental pathway consists of analyzing genes whose products are involved in the differentiation of specific tissues or organs. In *Drosophila*, for example, the pathway that controls sexual differentiation into male or female involves some genes that ascertain the X:A ratio, some that convert this ratio into a developmental signal, and others that respond to the signal by producing either male or female structures. In *Caenorhabditis*, the sexual-differentiation pathway involves genes that encode signaling proteins, their receptors, and transcription factors.

Genetic mosaics caused by mitotic chromosome loss or somatic recombination can be used to trace cell lineages during development and to ascertain whether a gene functions autonomously within cells. In *Drosophila*, such mosaics can be created by X chromosome loss or by radiation-induced somatic recombination.

Cloning a gene is fundamental to determining its role in development. Genetically transformed, or transgenic, animals can provide information on how a cloned gene functions *in vivo*.

Materials transported into the egg during oogenesis play a major role in its development. These are the products of maternal-effect genes such as *dorsal*, *bicoid*, and *nanos* in *Drosophila*, all of which function in the determination of the embryonic axes. Recessive mutations in maternal effect genes are expressed only in embryos produced by homozygous females.

The zygotic genes are activated after fertilization in response to maternal products. In *Drosophila*, a cascade of zygotic gene activity brings about the differentiation of the dorsal-ventral and anterior-posterior dimensions of the embryo. Ultimately, this activity subdivides the embryo into a series of segments. The identity of each segment is determined by homeotic genes of the Bithorax and Antennapedia complexes, acting in response to a regulatory hierarchy of segmentation genes. Once segmental identities have been established, specific cell types differentiate. The formation of an entire organ may depend on the product of a master regulatory gene, such as *eyeless* in *Drosophila*.

Many vertebrate genes have been identified by homology to genes isolated from model organisms such as *Drosophila* and *Caenorhabditis*. For example, the mammalian *Hox* genes are structurally and functionally similar to the homeotic genes of *Drosophila*. Among vertebrates, the mouse provides opportunities to study mutations that affect development. Transgenic mice can be created by injecting DNA into eggs or embryos, or by inserting transfected embryonic stem cells into developing embryos. These mice can be a source of insertional mutations, including knockout mutations, in genes that have a developmental significance.

ANSWERS TO QUESTIONS AND PROBLEMS

1) blastula 2) gastrulation 3) morphogenesis 4) determination 5) cellular blastoderm 6) numerator elements 7) denominator elements 8) *Sex-lethal (SXL)* 9) transformer 10) *double sex (dsx)* 11)

males that die as embryos **12)** male, because SXL protein is normally not made in males **13)** both XX and XY embryos develop into males **14)** cell-autonomous **15)** maternal-effect **16)** *dorsal* **17)** anterior-posterior **18)** *bicoid* **19)** gap **20)** pair-rule **21)** segment-polarity **22)** imaginal disks **23)** see *Approaches to Problem Solving* section

Ideas concerning thought challenging exercise:

There are several major features that distinguish plant and animal development.   One major feature is that animal embryogenesis is characterized by cell migration; plant embryogenesis lacks cell migration. Another major difference is that differentiation in plants occurs from meristems that may be active for the life of the plant. A plant meristem can differentiate shoots and leaves throughout much of the life of a plant and then switch to a floral meristem and differentiate flowers. Animal germ lines are set aside early in development. Although both plant and animal development is influenced by hormones, animal hormones are often synthesized in specialized cells or tissues. Plant hormones are not synthesized in specialized cells that function solely for the purpose of hormone production. Of course, the evolution of much more complex organs and tissues has occurred during animal, relative to plant evolution.

APPROACHES TO PROBLEM SOLVING

23. To obtain a singed and yellow twin spot on a fly of the genotype $y^+sn^+ / y\,sn$ would **require a four-strand double crossover between the *singed* and *yellow* loci, and a third crossover between *singed* and the centromere as illustrated below:

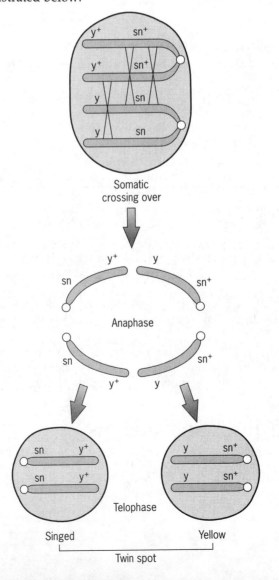

24

The Genetic Control of the Vertebrate Immune System

IMPORTANT CONCEPTS

A. The immune system of vertebrate animals is complex and protects them from invasion by pathogenic microorganisms and other foreign substances.

    1. Three different types of white blood cells are involved in the immune system.

        a. B lymphocytes differentiate into antibody-producing plasma cells.

        b. T lymphocytes develop into killer T cells that produce T-cell receptors and use them to seek out and destroy cells carrying foreign antigens.

        c. Macrophages carry out phagocytosis of antibody-antigen complexes, viruses, bacteria, fungi, and protozoa.

    2. T-cell recognition of cells that contain foreign antigens requires the presence of specific histocompatibility antigens encoded by the major histocompatibility complex (the HLA locus in humans).

B. A seemingly limitless variety of antibodies and T cell receptor proteins can be synthesized in response to antigens that the animal has not previously encountered.

    1. It is now understood how this phenomenal diversity of antibody specificity and T-cell receptor diversity are produced.

a. The genetic information encoding antibody and T-cell receptor chains is stored in several sets of gene segments, and the segments are put together in the appropriate sequences by genome rearrangements that occur during the development of the antibody-producing plasma cells.

C. Antibodies are composed of two light chains (either kappa or lambda) and two heavy chains.
 1. Each antibody chain contains a variable region that forms the antigen-binding site and a constant region that anchors the antibody to the cell surface in the case of membrane-bound antibodies.
  a. The variable regions of antibodies are encoded by gene segments that are present in the germ-line DNA in multiple copies.
  b. The constant region of antibodies is encoded by gene segments that are present in the genome in only one or a few copies per cell.
 2. Kappa light chain genes are assembled by genome rearrangements that occur during B cell differentiation from about 300 *LKVK* gene segments, 5 *JK* gene segments, and one *CK* gene segment.
 3. Functional heavy chain genes are similarly assembled from about 300 *LHVH* gene segments, 10 *D* segments, 4 *JH* gene segments, and 8 *CH* gene segments.
 4. The use of alternate sites of gene segment joining during the genomic rearrangement events and somatic hypermutation within the variable region gene segments contributes to the production of additional antibody diversity.

D. Antibodies are of five different immunoglobulin classes: IgM, IgD, IgG, IgE, and IgA.
 1. The class of antibody is determined by its heavy chain constant region, which in turn is determined by the C segment that was expressed during its synthesis.
 2. Class switching occurs when a B lymphocyte stops synthesizing one class of antibody and begins synthesizing another class of antibody with the same antigen specificity.
  a. Class switching involves the expression of the same variable region gene segments but a different heavy chain constant region gene segment.
  b. Class switching most often occurs by further genomic rearrangements similar to those that resulted in the synthesis of the original antibody chains.
  c. Class switching can also occur by alternate patterns of transcript splicing.

E. The productive rearrangement of antibody gene segments during B cell differentiation activates transcription of the assembled gene by bringing the promoter located upstream from the *LV* gene segment into the range of influence of a tissue-specific enhancer located in the intron between the J segment cluster and the C gene segment.

F. Clonal selection explains the production of large numbers of plasma cells all synthesizing the same antibody specific for a particular antigen present in the circulatory system.
 1. Only one productive light chain genomic rearrangement and one productive heavy chain genomic rearrangement can occur in a given B lymphocyte. This is known as allelic exclusion.
  a. The molecular basis of allelic exclusion remains unknown, but must be controlled by some type of feedback mechanism.

G. The class I antigens produced by the major histocompatibility genes (HLA genes in humans) usually are responsible for tissue rejections during tissues - and organ - transplant operations.
 1. The genes encoding class I histocompatibility antigens are highly polymorphic, thus different individuals (other than identical twins) are highly unlikely to carry histocompatibility antigens that are identical.

H. There are both inherited and acquired forms of immunodeficiences.
 1. Inherited immunodeficiences such as X-linked agammaglobulinemia and severe combined immunodeficiency syndrome (SCID) are usually fatal early in life if untreated.
  a. The adenosine deaminase deficient type of SCID was the first human disease to be treated by somatic cell therapy.
 2. The most severe of the acquired immunodeficiences is acquired immunodeficiency syndrome (AIDS), which is caused by a retrovirus, the human immunodeficiency virus (HIV).

a. HIV destroys the helper T cells of the infected individuals so that they are unable to combat infections by microorganisms.

IMPORTANT TERMS

In the space allotted, concisely define each term.

antigen:

immunogen:

stem cell:

B lymphocytes (B cells):

T lymphocytes (T cells):

phagocytes:

memory T cells:

cytotoxic (killer) T cells:

helper T cells:

suppressor T cells:

antibodies:

immunoglobulins:

light chains:

heavy chains:

variable region:

constant region:

domains:

antigen-binding sites:

effector function domain:

kappa chains:

lambda chains:

class switching:

T-cell receptors:

histocompatibility antigens:

major histocompatibility complex (MHC):

HLA antigens:

HLC locus:

HLA class I genes:

transplantation antigens:

HLC class II genes:

HLC class III genes:

complement:

antibody-mediated (humoral) immune response:

epitope:

T-cell-mediated (cellular) immune response:

primary immune response:

secondary immune response:

activated B or T cells:

joining sequence:

somatic hypermutation:

allelic exclusion:

monoclonal antibody:

autoimmune disease:

juvenile diabetes:

systemic lupus erythamatosus:

SCID:

X-linked agammaglobulinemia:

DiGeorge syndrome:

AIDS:

HIV:

IMPORTANT NAMES

In the space allotted, concisely state the major contribution made by the group.

Nobumichi Hozumi and Susumu Tonegawa:

TESTING YOUR KNOWLEDGE

*In this section, answer the questions, fill in the blanks, and solve the problems in the space allotted. Problems noted with an * are solved in the Approaches to Problem Solving section at the end of the chapter.*

1. The cells of the immune system that respond to the display of an antigen by a macrophage by stimulating B lymphocytes to produce antibodies and T lymphocytes to produce T-cell receptors are called _____.

2. The undifferentiated bone marrow cells that give rise to the various cells of the immune system are called _____.

3. Lymphocytes that carry T-cell receptors and kill cells displaying the recognized antigens are called _____.

4. Cells of the immune system that differentiate in the thymus are _____.

5. Antibody-producing white blood cells derived from B lymphocytes are called _____.

6. T cells that assist in down regulating the production of antibodies and T-cell receptors by B cells and T cells, respectively, are called _____.

7. Large cells that capture, ingest, and destroy invading foreign agents such as viruses, bacteria, and fungi are called _____.

8. The substances that trigger an immune response are called _____.

9. Proteins produced by killer T cells that bind antigens with the help of major histocompatibility antigens are called _____.

10. Cells that differentiate in the bone marrow to form antibody-producing plasma cells and memory B cells are called _____.

11. T cells that facilitate the rapid production of a given T-cell receptor in repeated encounters with an antigen are called _____.

12. Cell surface proteins that allow immune system cells to distinguish foreign substances from self and facilitate communication between cells of the immune system are called _____.

13. Phagocytic cells that ingest antigens and display them on the cell surface for interactions with other cells of the immune system are called _____.

14. The most predominant class of antibodies present in humans is _____.

15. The second most abundant class of human immunoglobulins that are secreted in milk, saliva and tears is _____.

16. An abnormal amount or functioning of which immunoglobin class is related to allergies in humans?

17. What is a group of approximately twenty soluble proteins called that circulate in the bloodstream and form large protein complexes, which in turn kill the cells involved in antibody-antigen binding?

18. The first class of antibody produced in a developing B cell is always _____.

19. The two types of light antibody chains are _____ and _____.

*20. A given mouse has 300 *LH-VH* gene segments, 10 *D* gene segments, 4 *JH* gene segments, and 8 *CH* gene segments on chromosome 12. How many different heavy antibody genes could be formed by this mouse?

*21. Assume that the above mouse also has 290 *LK-VK* gene segments, 5 *JK* gene segments, and 1 *CK* region. How many different kappa light antibody genes could be assembled by this mouse?

*22 Using the numbers calculated for problems 21 and 22, how many different antibodies could be formed during the differentiation of B lymphocytes?

THOUGHT CHALLENGING EXERCISE

Discuss the nature of the genetic information encoding antibody and T-cell receptor chains in light of previously presented concepts on the nature of a gene.

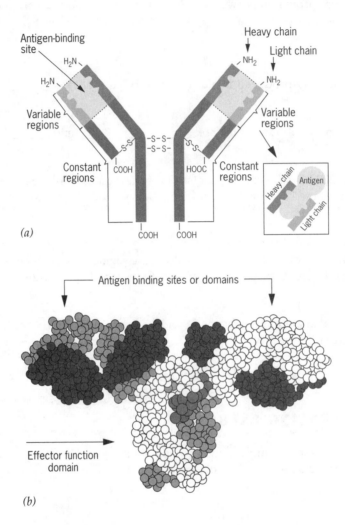

Figure 24.2   (a) Diagram and (b) space-filling model of antibody structure. Each antibody is a tetramer composed of four polypeptide chains: two identical light-chains and two identical heavy chains. (a) Each chain consists of a variable region and a constant region, and each antibody has two antigen-binding sites, formed by heavy and light chain regions. (b) The structure shown is shown is for a human IgG molecule.

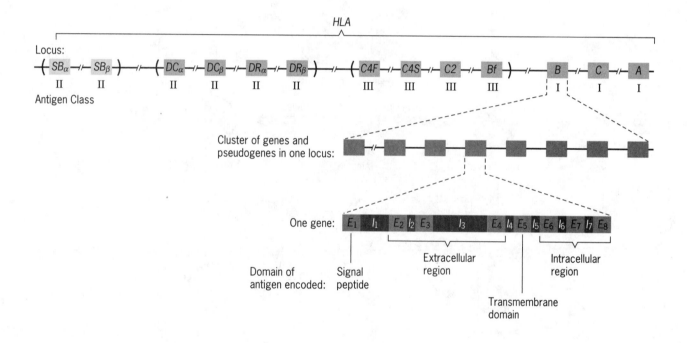

Figure 24.5 Organization of the major histocompatibility complex (HLA) on human chromosome 6. The relative positions of mapped loci within this huge gene complex are shown at the top. The order of loci enclosed in parenthesis is uncertain. The entire HLA complex is over 2×10^6 nucleotide pairs in length. The class of histocompatibility antigen encoded by genes in each locus is indicated below the map. Note that each of the loci within the HLA complex is itself a complex locus containing several genes and pseudogenes (center). The structure of a typical class I gene is shown at the bottom. Note that different exons encode different functional domains of the polypeptide gene product.

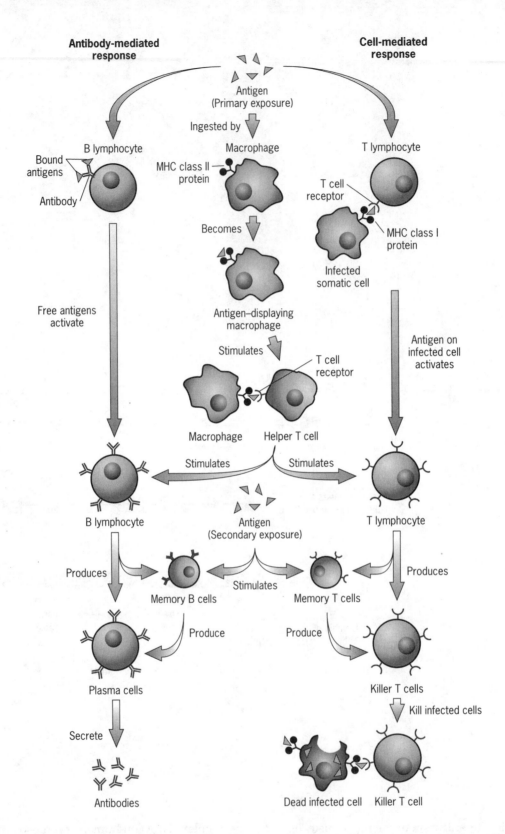

Antibody-mediated response

Cell-mediated response

Antigen (Primary exposure)

Ingested by

B lymphocyte

Bound antigens

Antibody

MHC class II protein

Macrophage

T lymphocyte

T cell receptor

MHC class I protein

Infected somatic cell

Becomes

Free antigens activate

Antigen–displaying macrophage

Stimulates

T cell receptor

Antigen on infected cell activates

Macrophage   Helper T cell

Stimulates   Stimulates

B lymphocyte

Antigen (Secondary exposure)

T lymphocyte

Produces

Stimulates

Produces

Memory B cells   Memory T cells

Produce

Produce

Plasma cells

Killer T cells

Kill infected cells

Secrete

Antibodies

Dead infected cell   Killer T cell

Figure 24.8 Summary of the immune response in mammals. The antibody-mediated (humoral) immune response, which is activated by free antigens, is carried out by B lymphocytes, and the cell-mediated (cellular) immune response, which is activated by antigens displayed on infected cells, is performed by T lymphocytes. Both responses are stimulated by activated helper T cells. The rapid immune responses that occur after a second or subsequent exposure to an antigen are mediated by memory B and T lymphocytes that rapidly produce antibody-producing plasma cells and killer T cells, respectively.

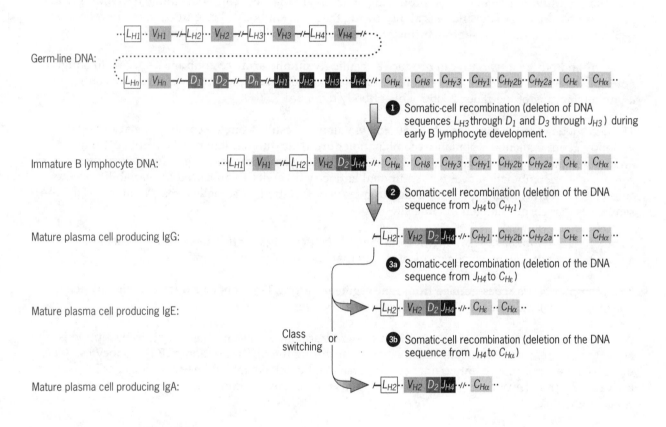

Figure 24.14. The genetic control of mouse antibody heavy chains. Each heavy chain gene is assembled from four different gene segments located on chromosome 12 by somatic recombination events that occur during B lymphocyte differentiation. $C_{H\mu}$, $C_{H\delta}$, $C_{H\epsilon}$, and $C_{H\alpha}$ encode the heavy chain constant regions of IgM, IgD, IgE, and IgA, respectively. $C_{H\gamma3}$, $C_{H\gamma1}$, $C_{H\gamma2b}$, and $C_{H\gamma2a}$, encode heavy chain constant regions of four closely related forms of IgG.

SUMMARY OF KEY POINTS

The immune response of mammals involves the coordinated activities of several specialized white blood cells. After exposure to a foreign substance — an antigen — B lymphocytes differentiate into plasma cells that produce antigen-binding proteins called antibodies. Similarly, T lymphocytes develop into killer T cells that carry T-cell antigen receptors on their surfaces and kill cells displaying the recognized antigens. Highly polymorphic, major histocompatibility antigens on cells allow killer T cells to distinguish foreign cells from "self" cells, the cells of the individual mounting the immune response.

The immune response in mammals involves three steps: (1) recognition of foreign substances, (2) communication of this recognition to the responding cells, and (3) elimination of the invading agent. Nonspecific responses include the recruitment of phagocytes to ingest and destroy viruses and microorganisms. Antigen-specific responses include antibody production and the activation of killer T cells. The production of large amounts of antibodies specific for the invading antigen results from clonal selection, the stimulation of a B cell producing an antibody that recognizes the antigen to multiply and produce a population of plasma cells all producing the same antibody. Long-lived memory cells facilitate a faster immune response during a second encounter with a foreign substance.

Humans and other vertebrates can produce a seemingly infinite array of antibodies specific to antigens not previously encountered. The germ line, somatic mutation, and minigene hypotheses all contribute to an understanding of the genetic control of antibody diversity.

The enormous variety of antibodies produced by mammals results from the combinational joining of antibody gene segments by somatic recombination during the differentiation of B lymphocytes into antibody-producing plasma cells. Signal sequences flanking antibody gene segments control these somatic-cell recombination events. Additional antibody diversity is produced by variability in the sites of gene-segment joining and by a high frequency of mutation — somatic hypermutation — in DNA sequences that encode antigen-binding sites.

Developing B lymphocytes can switch from the production of IgM antibodies to other classes of antibodies by genome rearrangements or by alternate pathways of transcript splicing.

T cell receptor genes are assembled from gene segments during T-lymphocyte differentiation by genome rearrangements analogous to those involved in antibody production.

Antibody gene segments are not transcribed at significant levels in stem cells. Their transcription is activated by the genome rearrangements that occur during the differentiation of B lymphocytes. Only one productive rearrangement occurs per cell. Transcription of assembled heavy chain genes is activated by transfer of their promoters to positions where they are controlled by a strong tissue-specific enhancer.

Survival of humans and other vertebrates depends on a functional immune system. Severe immunodeficiencies, either inherited or acquired, usually are fatal.

ANSWERS TO QUESTIONS AND PROBLEMS

1) helper T cell 2) stem cells 3) cytotoxic or killer T cells 4) T lymphocytes 5) plasma cells 6) suppressor T cell 7) phagocytes 8) antigens 9) T cell receptors 10) B lymphocytes 11) memory T cells 12) MHC antigens 13) macrophages 14) IgG 15) IgA 16) IgE 17) complement 18) IgM 19) kappa and lambda 20-22) see *Approaches to Problem Solving* section.

Ideas concerning thought challenging exercise:

The amino acid sequence of a protein is determined by a colinear nucleotide sequence in the DNA or gene. Except for allelic differences, this sequence is the same in all cells of the organism and individuals of the population. Generally, there is a one gene-one polypeptide relationship. The processing and removing of introns, if present, occurs after transcription. The final gene product of a normal gene generally has one specific function or catalytic activity. A mutation in a protein-coding gene, which will probably be harmful, will be selected at the organism level.

The genetic organization of the immune system, e.g., that of B lymphocytes, is very different compared to genes coding for most proteins. There are many representatives of different sections of the final gene sequence. These sections are processed by somatic rearrangement of DNA to generate a random combination for each allele. Since there are many millions of possible rearrangements, the combination generated in each cell will in all probability be different. Further mutational events may superimpose even more variability potential. Selection among B cells occurs in the circulatory system by a process called clonal selection.

APPROACHES TO PROBLEM SOLVING

20. A heavy chain will have 1 of the 300 LH-VH regions, 1 of 10 D regions, 1 of 4 JH regions, and 1 of 8 CH regions. Assuming that the final gene results from random assemblage of these four regions, the different number of antibody heavy chains that could be produced is (300)(10)(4)(8) = 96,000.

21. A light kappa chain will have a CK region, a JK region, and a LK-VK region. Assuming that the final gene results from random assemblage of these three regions, the different number of genes for kappa light antibody chains is (290)(5)(1) = 1450.

22. An antibody contains two identical heavy chains and two identical light chains. The number of different antibodies capable of being formed by the somatic recombinational method alone is (96,000)(1450) = 13,920,000.

25

Complex Inheritance Patterns

IMPORTANT CONCEPTS

A. The rediscovery of Mendel's principles in 1900 led to a heated debate between Mendelian geneticists and biometricians interested in the inheritance of continuously varying (quantitative) traits.

1. The biometricians did not accept Mendelian genetics based primarily on the inheritance of discontinuous traits.

   a. Francis Galton, Charles Darwin's cousin, thought that traits were inherited by genetic elements called "gemules" that were not the same as Mendelian segregating genes.

2. Through several genetic experiments including those of Johannsen, Nilssen-Ehle, and East, it was demonstrated that continuous variation could be explained by Mendelian segregating alleles at several loci (polygenes) whose expression is influenced by the environment.

3. A simple polygenic model for the inheritance of quantitatively varying traits can be summarized as follows:

   a. Alleles at two or more loci determine a quantitative trait.

   b. Each loci may have two kinds of alleles, those that contribute to the phenotype (contributing alleles) and those that are neutral (noncontributing alleles) and do not contribute to the phenotype.

   c. The effects of contributing alleles are equal and additive.

   d. The alleles of different loci are nonepistatic, and segregate independently.

   e. The expression of a polygenic trait is influenced by the environment.

4. Examples in the text of polygenic traits are DDT resistance, corolla length of tobacco flowers, kernel color of wheat, and the weight of bean seeds.
5. Some traits, such as cleft lip in humans, are polygenic with a threshold effect.

B. Statistical manipulations such as analysis of variance, correlation, and regression are used to analyze data in a population or cross involving quantitative traits.
1. Heritability is the proportion of a population's phenotypic variation attributable to genetic factors.
   a. To estimate heritability, researchers must determine the total variation in a trait, then partition the variance due to genetic differences and variance due to environmental differences.
   b. Broad-sense heritability includes all types of genetic variation including dominant alleles, epistatic alleles, and alleles having additive effects.
   c. Narrow-sense heritability is based on the variance due to additive effects.

C. Inbreeding is the preferential mating between close relatives, whereas outbreeding is the preferential mating between unrelated individuals.
1. Inbreeding leads to an increase in the frequency of homozygotes and a decrease in the frequency of heterozygotes in a population.
2. Inbreeding often leads to inbreeding depression, whereas outbreeding can lead to hybrid vigor.

IMPORTANT TERMS

In the space allotted, concisely define each term.

allozymes:

polygene:

threshold trait:

statistics:

frequency distribution:

mean:

modal class:

variance:

standard deviation:

normal distribution:

correlation coefficient (r):

covariance:

regression:

multifactorial:

heritability:

broad-sense heritability:

narrow-sense heritability:

additive:

inbreeding:

inbreeding depression:

outbreeding:

gametophytic incompatibility:

Heterosis (hybrid vigor):

quantitative trait loci (QTL's):

maturity-onset diabetes of the young:

IMPORTANT NAMES

In the space allotted, concisely state the major contribution made by the individual or group.

Charles Darwin:

Francis Galton:

W. Johannsen:

Herman Nilsson-Ehle:

Edward M. East:

James Crow:

Ronald A. Fisher:

F. D. Enfield:

R. Cooper and J. P. Zupek:

H. D. Bradshaw and F. F. Stettler:

TESTING YOUR KNOWLEDGE

In this section, answer the questions, fill in the blanks, and solve the problems in the space allotted. Problems noted by an * are solved in the Approaches to Problem Solving section at the end of the chapter.

1. A statistical association among variables is called _____.

2. Quantitative variation not represented by distinct classes that is generally binomially distributed is called _____.

3. Phenotypic variation involving distinct classes such as axial or terminally located flowers is called _____.

4. For quantitative traits, the proportion of the phenotypic variance that is attributed to genotypic variance is called _____.

5. For quantitative traits, the proportion of the phenotypic variance that is due to the additive effects of alleles is called _____.

6. A measure of variation in a population expressed as the square of the standard deviation is called the _____.

*7. Assume that three polygenes (A, B, C) control the hypocotyl length in sunflower seedlings. A plant with 4 cm hypocotyls (*aa bb cc*) is crossed with a plant that has 10 cm hypocotyls (*AA BB CC*). What are the phenotypes and their frequencies among the F_1 and F_2 progeny?

*8. A quantitative trait is controlled by four polygenes. If F_1 individuals, heterozygous for the four genes, are crossed, what proportion of the F_2 progeny will have the same phenotype as the F_1?

*9. A cross between two homozygous lines of wheat having mean seed weights of 40 mg and 60 mg, respectively, produced F_1 progeny with seeds that uniformly weighed 50 mg. The F_2 consisted of 4,080 plants with seed weights binomially distributed between 40 mg and 60 mg, and a mean of 40 mg. If one of these F_2 plants produced seeds that were 40 mg and another F_2 plant produced seed weighing 60 mg, how many pairs of polygenes segregated in the F_1?

*10. The frequencies of genotypes in a plant population involving lactic dehydrogenase alleles, A and B, are .36 AA, .48 A B, and .16 B B. What would be the frequency of the A B genotype after three generations of self-pollination?

*11. A plant breeder is selecting safflower plants for oil content in seeds. In the breeding population the mean oil content is 10%. Plants with 15% oil content were crossed. If the narrow-sense heritability is 0.1, what will the average oil content be in the plants of the next generation?

*12. The following table presents measurements of the corolla length of two inbred lines (A & B) of fox gloves, their F1, and F2 progeny.

Inbred line A								Inbred line B			

Inbred line A

8   50   49   10

Inbred line B

6   47   53   7

F1 Progeny

5   45   39   7

F2 Progeny

1   7   29   55   71   57   26   8   1

21 22 23 24 25 26 27 28 29 30 31 32 33 34 35 36

COROLLA LENGTH (mm)

Assuming that each parental variety was homozygous and that the genes are additive and equal in effect, how many genes are apparently segregating in the F$_1$?

*13. The data shown below are from a series of crosses between two homozygous rodent lines to study hair length. All animals were grown under the same uniform environmental conditions.

Line A (hair length)   line 2 (hair length)

20 mm   36 mm

F$_1$ (hair length)

28 mm

F$_2$	Hair length (mm)	# of animals
	36	2
	34	14
	32	60
	30	108
	28	140
	26	114
	24	52
	22	18
	20	2

Assume that all the contributing alleles have equal and additive effects.

(a) Determine the number of polygene loci at which the parents carry different alleles.

(b) Provide a genotype for the parents and the F$_1$.

(c) What is the phenotypic value of each contributing allele?

(d) List the expected phenotypes and their frequencies among the progeny of a cross between the F$_1$ and a homozygous 24 mm hair length F$_2$.

*14. The Jumping Frenchman of Maine disorder (caused by homozygosity for a rare recessive allele) is characterized by an abnormal, exaggerated, startle reflex reaction. It occurs in about 25 out of a million people in Maine and southeastern Canada. Study the following pedigree.

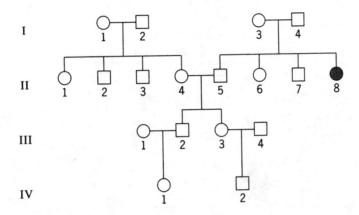

(a) If IV-1 marries at random in the population and does not marry a relative, what is the chance that her first child will be afflicted with the jumping disorder? The frequency of heterozygous individuals in the population is about 0.01.

(b) If IV-1 marries her cousin IV-2, what is the probability that her first child will be afflicted with the jumping disorder?

THOUGHT CHALLENGING EXERCISE

Intelligence is a complex polygenic trait in humans. Some psychologists and sociologists have been interested in differences in intelligence among various human ethnic and racial groups. Sizable differences (ca. 10%-15%) in mean IQ have been detected among racial and ethnic groups in the United States. Generally, mean IQ and estimates of heritability are lower in economically disadvantaged groups. Discuss what these differences in IQ and heritability may mean. Base your discussion on a sound genetic basis.

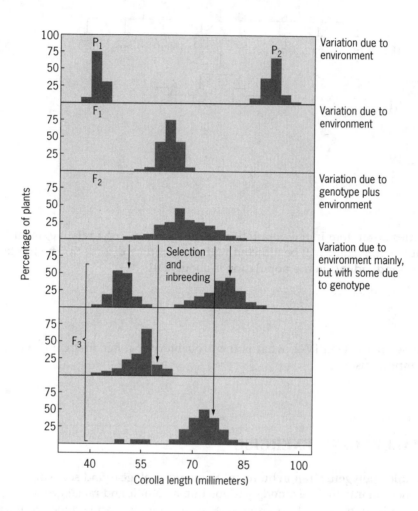

Figure 25.6   E. M. East studied the inheritance of corolla length in tobacco (*Nicotiana longiflora*). The figure shows the percentage frequencies with which individuals fall into various corolla-length classes. The P₁ and P₂ classes are pure lines; the F₁ is uniformly heterozygous.   Any variation in the P₁ and F₁ classes is due to environment. The F₂ is genetically heterogeneous. Variation in this generation is due to genetic and environmental factors. The means of the four F₃ populations are correlated with the corolla-length of the F₂ plants from which they were obtained by selfing.

SUMMARY OF KEY POINTS

A phenotype can exhibit a continuous pattern of variation and be controlled by a single gene with several alleles. The sum of the various genotypes produces a continuous pattern of variation. Experiments with wheat demonstrated that a continuously varying trait was determined by three independently assorting pairs of alleles showing incomplete dominance. Experiments with tobacco showed that a continuously varying trait was determined by both genetic and environmental factors. A threshold trait is one with a continuously distributed liability or risk in which individuals with a liability greater than a critical value (threshold) exhibit the phenotype.

A frequency distribution arranges measurements as a graph showing either the relative or absolute incidence of classes in a population. The mean is the sum of all measurements or values in a sample divided by the sample size. The mode is the single class in the statistical distribution having the greatest frequency. Variance and its square root, the standard deviation, are measures of variation in a population. The correlation coefficient is a measure of the extent to which variations in one variable are related to variations in another. Regression analysis assesses the relationship between two variables.

The phenotypic variance is equal to the variance due to genetic differences plus the variance due to environmental differences. Heritability is the proportion of phenotypic variation attributable to genetic differences. Broad-sense heritability includes dominant, additive, and epistatic alleles. Narrow-sense heritability includes only the additive effects of alleles. Narrow-sense heritability for a quantitative trait can be used to predict the outcome of a selective breeding program. Some of the variance in a quantitative trait may be caused by genotype-environmental interactions.

Inbreeding is the preferential mating between close relatives. It leads to an increase in homozygote frequency and a decrease in heterozygote frequency. Outbreeding is the preferential mating between unrelated individuals. Although inbreeding often leads to inbreeding depression, outbreeding can lead to hybrid vigor.

Quantitative trait loci (QTLs) are loci that influence quantitative traits. QTLs can be mapped using molecular markers distributed densely, but uniformly, over the genome.

Diabetes is a complex multifactorial disorder. Insulin-dependent diabetes (IDDM) is an autoimmune disease. Noninsulin-dependent diabetes (NIDDM) is due to several genes involved in glucose metabolism.

ANSWERS TO QUESTIONS AND PROBLEMS

1) correlation 2) continuous variation 3) discontinuous variation 4) broadsense heritability 5) narrow sense heritability 6) variance 7-14) see *Approaches to Problem Solving* section.

Ideas concerning the thought challenging exercise:

Unfortunately, the study of inheritance of human intelligence has not been free of racism and individual social and political agendas. Although it is generally agreed by geneticists and psychologists that intelligence is a quantitative and polygenic trait, there is no general agreement as to what the commonly used intelligence tests are measuring and the extent of educational and cultural bias in them. There is even greater disagreement concerning the role of the environment. The heritability component of "intelligence" is difficult to measure and the meaning of heritability is often not understood by the investigators. For example, some social scientists misuse the term heritability, thinking that high heritability means genetic superiority. If environment (education, economic, and social conditions) greatly influences IQ, then a lower mean IQ with a lower heritability would be expected in a group influenced by adverse environmental factors. Two groups could be genetically identical but have different mean IQs and different heritabilities simply due to differences in the environment.

APPROACHES TO PROBLEM SOLVING

7. In the cross *aa bb cc* (4 cm) X *AA BB CC* (10 cm), there are 6 cm differences in height between the parents. The 10 cm parent has six contributing alleles (capital letters)and the 4 cm plant has no alleles (lower case letters). Therefore, each contributing allele adds 1 cm to a residual phenotype (4 cm).

The F_1 progeny are all of the genotype *Aa Bb Cc*. Since they have three contributing alleles, they have a residual phenotype of 4 cm plus the effects of three contributing allele worth 1 cm each. The F_1 progeny therefore have 7 cm hypocotyls.

To determine the phenotypes and their frequencies in the F_2s, one could simply determine all the genotypes in regard to contributing alleles and their frequencies by methods presented in chapter 3 such as the a Punnett square or forked-line method. However, there is a much simpler way to obtain the F_2 phenotypic ratio. Consider the segregation of each gene pair independently. From the first gene pair *AA* X *aa*, the F_1 progeny are all *Aa*. The crossing of the F_1 progeny, *Aa* X *Aa*, leads to the following F_2 genotypic array: *AA* + 2*Aa* + *aa*. This is a binomial distribution represented by $(A + a)^2$. The F_2 genotypic array of the second gene pair is $(B +b)^2$, and that of the third gene pair is $(C + c)^2$. The F_2 genotypic array that represents the independent assortment of the three gene pairs is $(A + a)^2(B + c)^2(C + c)^2$.

Since all contributing alleles (*A, B, C*) have equal and additive effects, and the noncontributing alleles (*a, b, c*) do not contribute to the phenotype, all the contributing alleles may be designated by the same capital letter, e.g., H, and all noncontributing alleles may be designated by the corresponding lower case letter, h. By substituting these in place of *A a, B b*, and *C c*, the F_2 genotypic array may now be designated as follows:

$$(H + h)^2(H + h)^2(H + h)^2 = (H + h)^6.$$

The binomial $(H + h)^6$ is then expanded to give

$$1 H^6 + 6 H^5h^1 + 15 H^4h^2 + 20 H^3h^3 + 15 H^2h^4 + 6 H^1h^5 +1 h^6.$$

The coefficients indicate the proportions of the phenotypic classes. The exponents of H indicates the number of contributing alleles as indicate below.

$$1 H^6 + 6 H^5h^1 + 15 H^4h^2 + 20 H^3h^3 + 15 H^2h^4 + 6 H^1h^5 + 1 h^6$$

number of contributing alleles 6   5   4   3   2   1   0

By adding 1 cm to a residual phenotype (no. contributing alleles) of 4 cm, the following phenotypic ratio is obtained:

1/64 (10 cm): 6/64 (9 cm): 15/64 (8 cm): 20/64 (7 cm): 15/64 (6 cm): 6/64 (5 cm): 1/64 (4 cm).

The binomial distribution of the F_2 phenotypes is expected from the segregation of polygene loci.

The above approach can be utilized to depict phenotypic distributions from other crosses involving segregation of polygene loci.

8. To solve this problem, set it up like problem 7 and then look for the appropriate term in the expanded binomial representing the F_2s. In this case, there are four polygenes segregating in the F_1. The genotype of the F_1 can be initially presented as *Aa Bb Cc Dd*. The F2 genotypic array is $(A + a)^2(B + b)^2(C + c)^2(D + d)^2$. Substituting common letters for contributing and noncontributing alleles, e.g., X and x, the F_2 genotypic array may be represented by $(X + x)^8$. From the expanded binomial we look for the term that has four contributing alleles (the same as the F_1). This term is $70 X^4 x^4$. Therefore, 70/256 of the F_2 progeny would be expected to have the same phenotype as the F_1.

9. The key to solving this problem is the binomially distributed F_2 population. Of 4,080 F_2 plants only one produced seeds as small as the 40 mg parent and only one produced seed as large as the 60 mg parent plants. In the binomially distributed F_2 we expect $(1/4)^n$, where n is the number of segregating polygene loci, to have segregated to each parental extreme (homozygous for contributing or noncontributing alleles). Since $(1/4)^6 = 1/4096$, which is close to 1/4,080, an estimate of six segregating polygene pairs in the F_1 is obtained.

10. Inbreeding reduces the frequency of heterozygotes in a population. Self-pollination (self-fertilization) is the strictest form of inbreeding and reduces the frequency of heterozygotes by 1/2 each generation. This occurs because a heterozygous gene pair, e.g., *Aa*, upon selfing produces the following genotypes: 1/4 *AA* + 1/2 *Aa* +1/4 *aa*. The initial frequency of individuals heterozygous for the A and B alleles of lactic dehydrogenase is .48. It follows that after three generations of self pollination the frequency of heterozygotes should be $(.48)(0.5)(0.5)(0.5) = .06$.

11. A narrow-sense heritability of 0.1, determined by selection responses, indicates that the mean phenotype of the next generation should be moved 0.1 of the distance between the mean of the population and the mean of the plants selected. In this problem the selected population has 15% oil content, whereas the main population has 10% oil content. The mean oil content in the progeny of the selected plants should be $10\% + 0.1 (5\%) = 10.5\%$.

12. The data presented in the table are typical for the inheritance of a quantitative trait. The two parents differ in corolla length and the F_1 is intermediate. Although the parents are inbred and highly homozygous, there is still phenotypic variability due in part to the influence of the environment. The F_1 are heterozygous, but are all of the same genotype. The variability observed in the F_1 is again attributable to the environment and is of about the same magnitude as that present in each of the parental lines. The increase in variability and the binomial distribution observed in the F_2 is due to both genetic segregations of additive alleles and environmental components. Of 255 F_2 progeny, only one had a phenotype within the range of short corolla length parent and only one of the F_2 progeny had a corolla length within the range of the long corolla length parent. The number of genes segregating in the F_1 can be estimated by using the formula $(1/4)^n$. If $n = 4$, then we would expect about 1/256 to segregate to each parental extreme. Therefore, an estimate of four segregating gene pairs best fits the data.

13. (a) The inheritance in this cross indicates a quantitative trait. The F_1 is intermediate to the parents. The F_2 shows an increase in variability that is binomially distributed. To determine the number of polygene loci at which the parents carry different alleles, analyze the F_2 data to obtain an estimate of the number of polygene loci that segregated in the F_1. An estimate can be obtained from two parameters of the F_2 data. First, 2/510 or 1/255 F_2 progeny have the phenotype of each original parent. This is approximately $(1/4)^4$. Therefore, it is

estimated that four gene pairs segregated in the F_1. The parents must have carried different polygene alleles at four polygene loci.

(b) The genotype for the 20 mm parent can be designated as *aa bb cc dd*, and that of the 36 mm parent as *AA BB CC DD*. The genotype of the F_1 is *Aa Bb Cc Dd*.

(c) There is a 16 mm (36 mm - 20 mm) difference between the hair length of the parents. This is determined by a total of eight contributing alleles. Therefore, each contributing allele is responsible for adding 2 mm to hair length.

(d) The genotype of the F_1 is Aa Bb Cc Dd. The gametes produced by this F_1 can be represented as $(A + a)^1(B + b)^1(C + c)^1(D + d)^1$. An individual that is 24 mm has 2 contributing alleles (20 mm + 2 mm + 2 mm). If it is homozygous, both of these alleles must be of the same gene. The 24 mm rodent can be assigned the genotype *AA bb cc dd*. The gametes produced by this individual are of the genotype *A b c d*. Lets arbitrarily substitute H for contributing alleles and h for noncontributing alleles. This can be done because the contributing alleles are assumed to be equal and additive in effect. The gametes produced by the F_1 can now be represented by the binomial expression $(H +h)^1(H +h)^1(H +h)^1(H + h)^1 = (H + h)^4$. The phenotypes among progeny of this cross can be represented by the binomial $(H + h)^4 = 1H^4 + 4H^3h^1 + 6H^2h^2 + 4H^1h^3 + 1h^4$, plus one H allele obtained from the gametes of the homozygous 24 mm rodent all have.

$$1H^4 + 4H^3h^1 + 6H^2h^2 + 4H^1h^3 + 1h^4$$

Proportion	1/16	4/16	6/16	4/16	1/16
Contributing alleles	5	4	3	2	1
Hair length (mm)	30	28	26	24	22

14. (a) To solve this problem, first determine the probability that IV-1 is heterozygous. To do this, we must analyze the pedigree. When analyzing rare genetic disorders, assume that all members marrying into a pedigree are homozygous for the normal allele (in this case *J J*) unless there is evidence to the contrary. Since II-8 is homozygous for the jumping allele (*j j*), her parents, I-3 and I-4, were both heterozygous *J j*. Individual II-5 would have a 2/3 chance of being heterozygous. If heterozygous, he would have a 1/2 chance of passing the allele to his son, III-2, who in turn would have a 1/2 chance of passing the allele to his daughter, IV-1. The probability that IV-1 is heterozygous is (2/3)(1/2)(1/2) = 1/6 or 0.167. To produce a child with the jumping disorder, she would have to marry another individual who was carrying the allele. Since the disorder is very rare, we consider only the probability that she will marry a heterozygote for the jumping allele. The frequency of heterozygotes for the jumping allele in the population is 0.01. If she is heterozygous (P = 0.167) and marries another heterozygote (P = 0.01), then she will have a .25 chance of having a child (j j) with the jumping disorder. The total probability of IV-1 having a child afflicted with the jumping disorder if she marries at random is (.167)(0.01)(0.25) = 0.0004 or 1/2500.

(b) To solve this part of the problem, determine the probability that IV-1 and her cousin IV-2 are both heterozygous for the jumping allele. If II-5 is heterozygous (we have already taken the 2/3 probability into account when deterrmining the probability that IV-1 is heterozygous) he has a (1/2)(1/2) =1/4 chance of passing the recessive allele to his grandson IV-2. Therefore, the probability that they are both heterozygous is (1/6)(1/4) = 1/24. If IV-1 marries her cousin IV-2, and if they are both heterozygous, they would have have a 1/4 probability of having a child afflicted with the jumping disorder. Therefore, the total probability of having a child afflicted with the jumping disorder is (1/24)(1/4) = 1/96.

This problem illustrates how inbreeding can increase the frequency of rare alleles becoming homozygous. Related individuals share a proportion of common alleles.

26

The Genetic Control of Behavior

IMPORTANT CONCEPTS

A. Behavior is a coordinated neuromuscular response to changes or signals in the external and internal environment.
    1. This response requires the integration of sensory, neural, and hormonal factors.

B. Behavior is influenced by genetics and/or environment.
    1. Many gene products play a role directly and indirectly in the structure and function of the nervous system.
        a. The relationship between gene products and a specific behavior is very complex.
    2. In some cases genes may actually determine a behavioral trait.
        a. Behavior patterns with a simple genetic basis include nest-cleaning behavior in honeybees, circadian rhythms in *Drosophila* and mice, and barking characteristics in dogs.

C. Genetic imbalances in humans caused by aneuploidy, e.g., XXY, XXX, XYY, XO, and trisomy for chromosome number 21, result in abnormal aspects of mental development and hence abnormal behavior patterns and/or mental retardation.

D. Single gene mutations in humans may upset the normal development or functioning of neurons that result in abnormal behavioral patterns.

1. Examples of genetic diseases caused by mutant alleles having an adverse effect on the human nervous system are phenylketonuria (PKU), Lesch-Nyhan syndrome, and Huntington's disease.
   E. Studies of twins reared together and apart indicate that human behavior and personality are strongly influenced by the genotype.
   1. Complex traits such as human behavior and personality do not exhibit simple Mendelian inheritance, but rather are multifactorial.
   F. Much of human behavior is influenced by genotype/environment interactions and many human behavior conditions have been shown to be have a genetic component.
   1. Alcoholism is strongly influenced by the genotype.
   2. At least four different genes are associated with Alzheimer's disease.
   3. Intelligence is a complex trait that is influenced by genetic and environmental factors.
   4. Homosexual behavior may be influenced by the genotype.

IMPORTANT TERMS

In the space allotted, concisely define each term.

behavior:

circadian:

per gene:

type I alcoholism:

type II alcoholism:

dementia:

senile plaques:

amyloid protein:

neurofibrillary tangles:

amyloid precursor protein (APP):

early-onset familial Alzheimer's disease (FAD):

intelligence quotient (IQ):

IMPORTANT NAMES

In the space allotted, concisely state the major contribution of the individual or group.

N. Rothenbuhler:

J. P. Scott and J. L. Fuller:

Thomas Bouchard:

K. Blum:

J. Tiihonen:

A. Jensen:

S. LeVay:

J. Michael and R. Pillard:

D. Hamer:

TESTING YOUR KNOWLEDGE:

*In this section, answer the questions, fill in the blanks, and solve the problems in the space allotted. Problems noted with an * are solved in the Approaches to Problem Solving section at the end of the chapter.*

1. The coordinated neuromuscular response to changes or signals in the external and internal environment is called _____.

2. In *Drosophila*, the time the adult fly emerges from the pupal case, the fly's courtship pattern, and other circadian rhythms are influenced by the _____ gene.

3. The _____ gene in mice controls circadian rhythms.

4. A human genetic disease characterized by mental retardation and abnormal behavior patterns, including compulsive self-mutilation, is called _____ syndrome.

5. The _____ is derived by dividing the person's mental age by actual age and then multiplying by 100.

6. Name the disease that is characterized by a marked loss of neurons and the accumulation of thickened nerve cell processes surrounding an amyloid protein deposit.

*7. List the phenotypic and genotypic ratio in the progeny of a cross between a non-hygienic queen honeybee of the genotype $u^+u\ r^+r$ and a hygeinic drone of the genotype *uu rr*.

*8. A trait shows 97% concordance among monozygotic twin pairs and 45% concordance among dizygotic twin pairs. What do these figures reveal about the relative importance of genetics and environment in determining the trait?

*9. A trait shows 46% concordance among monozygotic twin pairs and 9% concordance among dizygotic twin pairs. What do these figures reveal about the relative importance of genetics and environment in determining the trait?

*10. A trait shows 95% concordance among monozygotic twin pairs and 92% concordance among dizygotic twin pairs. What do these figures reveal about the relative importance of genetics and environment in determining the trait?

THOUGHT CHALLENGING EXERCISE

Future studies of humans may indicate that additional types of abnormal behavior may have a strong genetic component. What affect might such findings have on the legal system of the United States?

KEY FIGURE

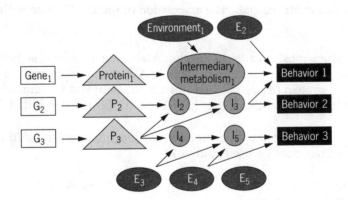

Figure 26.1 Behavior is influenced by the genotype (G) interacting with the environment (E). Genes code for proteins (P) whose function is influenced by multiple environmental factors.

SUMMARY OF KEY POINTS

A behavior is a coordinated neuromuscular response to environmental signals. Genes and the environment influence many behaviors.

Nest-cleaning behavior in bees is a behavior controlled by two independently assorting genes. Circadian rhythms are behaviors or functions that are synchronized with the 24-hour day/night cycle. *Drosophila's* circadian activity pattern is controlled by two genes, *per* and *tim*; *per* alleles lengthen or shorten the cycle or cause it to be arhythmic. They also affect other *Drosophila* behaviors. PER and TIM reach their highest levels during the day and form a complex that represses transcription of the *per* and *tim* genes at night. TIM is destroyed by light.

In carefully conducted breeding experiments using different dog breeds, it has been shown that many aspects of dog behavior are genetically controlled. The *Clock* gene in mice controls circadian rhythms and *fosB* controls nurturing behavior.

Various aneuploidic conditions have an effect on behavior. Trisomy 21 individuals are mentally retarded; XXY and XXX are abnormal in certain aspects of their mental functions; XYY males express some aggressive tendencies; and XO females tend to have hearing problems and problems with spatial perception.

PKU is a metabolic disorder caused by an enzyme deficiency that results in the accumulation of phenylalanine and phenylalanine-derived metabolites. Elevated levels of these substances cause central nervous system dysfunction. Lesch-Nyhan syndrome is caused by an enzymatic defect that disrupts purine metabolism and leads to mental retardation and self-mutilation behavior. Huntington's disease is a fatal neurodegenerative disease.

The Minnesota study of MZ twins reared apart presented strong evidence that personality and behavioral traits are heavily influenced by the genotype.

Alcoholism is strongly influenced by the genotype. One way the genotype may increase the risk for developing alcoholism is in the way it controls alcohol metabolism. Two genes involved with dopamine functions have been implicated in alcoholism. The association of the *A1* allele of the *DRD2* gene with alcoholism remains controversial. The association of the *DAT1* gene with alcoholism is a potentially important one.

People with Down syndrome almost invariably develop Alzheimer's disease, suggesting that the overexpression of a gene(s) on chromosome 21 is one cause of AD. A candidate gene on chromosome 21, encoding amyloid precursor protein (APP), has been identified. Amyloid protein, which accumulates in senile plaques, is cleaved from APP. Other genes have been implicated in familial AD.

Intelligence is a complex trait influenced by genetic and environmental factors. Although IQ scores are heritable, heritability estimates must be used cautiously.

Homosexual behavior is now thought to be influenced by the genotype. A candidate region on the tip of the X chromosome is currently being studied.

Much of human behavior is influenced by the genotype interacting with the environment. This interaction is complex.

ANSWERS TO QUESTIONS AND PROBLEMS

1) behavior 2) *per* 3) *clock* 4) Lesch-Nyhan 5) intelligence quotient or IQ 6) Alzheimer's disease

7-10) see *Approaches to Problem Solving* section.

Ideas on thought challenging exercise:

In the past, the legal system has shown some consideration for explanation (such as insanity) of an individual's criminal actions. If certain types of behavior are correlated with abnormal genes, it will be interesting to see how the courts will rule, i.e., will a person be accountable for their behavior even if a genetic condition may predispose them toward abnormal behavior, or will individuals (except under abnormally extreme mental conditions) be fully accountable for any criminal actions?

APPROACHES TO PROBLEM SOLVING

7.

Genotype	$u^+u\,r^+r$	$u^+u\,rr$	$uu\,r^+r$	$uu\,rr$
Phenotype	nonhygienic	no uncapping but removal	uncapping no removal	hygienic
Ratio	1	1	1	1

8. Dizygotic twins have 50% of their genes in common, whereas monozygotic twins are identical and have 100% of their genes in common. Therefore, a trait with a very strong genetic component and a negligible environmental component should show nearly 100% concordance in monozygotic twin pairs and 50% concordance in dizygotic twin pairs. The concordance becomes less and less as the environmental component increases. The 97% concordance among the monozygotic twin pairs and the 45% concordance among dizygotic twin pairs for the trait indicates that it is determined primarily by the genotype with little environmental influence on its expression.

9. The large difference between the concordance of the trait in monozygotic and dizygotic twin pairs suggest that the trait has a genetic component. The concordance in monozygotic twin pairs is much reduced, which suggests that the expression of the trait also has a major environmental component.

10. This trait shows very high concordance in both monozygotic and dizygotic twin pairs. This suggests that the trait is predominantly environmentally determined with little or no genetic component.

27

The Basic Principles of Population Genetics

THE EMERGENCE OF EVOLUTIONARY THEORY
    Early Evolutionary Theories
    The Voyage of the Beagle
    The Darwinian Revolution

THE THEORY OF POPULATION GENETICS
    The Hardy-Weinberg Principle of Genetic Equilibrium
    Applications of the Hardy-Weinberg Theorem

THE THEORY OF NATURAL SELECTION
    Two Forms of Natural Selection
    Selection in Natural Populations
    Modes of Natural Selection

CHANGES IN THE GENETIC STRUCTURES OF POPULATIONS
    Deviation From Random Mating: Inbreeding
    Genetic Drift: Random Changes in Small Populations
    Migration Changes Allele Frequencies
    Mutations: A Source of New Variation
    Are Mutations Truly Random?

A BALANCE BETWEEN MUTATION AND SELECTION

THE ROLE OF POPULATION GENETICS IN GENETIC COUNSELING

IMPORTANT CONCEPTS

  A. Modern evolutionary theory has its origin in the work of Charles Darwin who, in 1859, proposed a theory of evolution by natural selection.
    1. Darwin's theory of evolution was based upon several premises:
      a. Populations tend to increase in number until resources are limited.
      b. Limited resources result in competition.
      c. Individuals with favorable inherited variations can compete better and are more likely to survive and leave more progeny.
      d. Over time, favorable variants replace less favorable variants in the population.
  B. Population genetics, the discipline that studies changes in gene frequency in populations and permits the formulation of mechanisms that account for the processes of evolution, had its origins early in the 1900's.
    1. In 1908, G. H. Hardy and W. Weinberg independently developed the basic principle of population genetics known today as the Hardy-Weinberg theorem.
      a. The Hardy-Weinberg theorem states that an equilibrium is established for allelic and genotypic frequencies in large, randomly mating, sexually reproducing populations, and that this constancy is maintained generation after generation as long as no disruptive forces are acting on the population.
      b. The Hardy-Weinberg principles can be used to estimate allelic and genotypic frequencies in populations.
  C. Natural selection is the primary force that alters allele frequencies (disrupts Hardy-Weinberg equilibrium) and brings about evolutionary change.

1. The essence of natural selection is the increased production, by some genotypes relative to other genotypes, of reproductively successful progeny.
2. Fitness, the capacity of a phenotype to donate genes to the next generation, is an important component to natural selection.
   a. Selective pressure(s) is a measure of fitness and varies from 0 to 1.
   b. The stronger the selective pressure, the more rapid the rate of change in frequencies of alleles.
3. Two forms of selection are frequency-independent selection and frequency dependent selection.
   a. Frequency-independent selection is when an organism struggles directly with the physical constraints of the environment to survive and reproduce, and does not involve competition with other organisms.
   b. Frequency-dependent selection involves competition between organisms; the relative frequencies of the competing genotypes determine the fitness values, and these fitness values change.
4. Three modes of selection are directional selection, stabilizing selection, and disruptive selection.
   a. Directional selection favors an extreme phenotype, thus shifting the population mean in one or the other direction.
   b. Stabilizing selection favors an optimal phenotype and selects against both extremes, e.g., the heterozygote over the two homozygote classes.
   c. Disruptive selection favors the more extreme variants and selects against the intermediates, e.g., the two homozygote classes over the heterozygote.

D. Forces, in addition to selection, can cause a departure from Hardy-Weinberg equilibrium.
   1. Inbreeding, the mating between genetically related individuals, results in an increase in the frequency of homozygous genotypes, i.e., a change in genotypic frequencies, but does not cause a change in allelic frequencies.
   2. Genetic drift is a random fluctuation in allelic frequency from generation to generation resulting generally from small population size.
   3. The allelic frequency of a population may change due to the introduction of alleles from another population by migration.
   4. Mutation introduces alleles into a population and provides variation on which natural selection acts.
      a. Mutations may be harmful, beneficial, of neutral.
      b. If the mutant allele is deleterious, a mutation-selection equilibrium (for a recessive mutation, q = square root of u/s) is achieved when the rate of introduction of new alleles is balanced by the rate of elimination by selection.
      c. Mutations are preadaptive, not postadaptive.

E. Knowledge of gene frequencies in specific populations is extremely important in genetic counseling.

IMPORTANT TERMS

In the space allotted, concisely define each term.

special creation:

catastrophism:

essentialism:

population genetics:

Hardy-Weinberg principle:

fitness:

frequency-independent selection:

frequency-dependent selection:

heterozygote superiority:

balanced polymorphism:

industrial melanism:

stabilizing selection:

disruptive selection:

inbreeding:

consanguineous mating:

random genetic drift:

postadaptive mutation:

preadaptive mutation:

achondroplasia:

IMPORTANT NAMES

In the space allotted, concisely state the major contribution of the individual or group.

Charles Darwin:

Jean Baptiste Lamarck:

Charles Lyell:

Thomas Malthus:

Ernst Mayr:

William Castle:

G. H. Hardy and W. Weinberg:

E. B. Ford:

H. B. D. Kettlewell:

W. E. Kerr and Sewall Wright:

Bently Glass:

Hampton Carson:

John Cairns:

TESTING YOUR KNOWLEDGE:

*In this section, answer the questions, fill in the blanks, and solve the problems in the space allotted. Problems noted with an * are solved in the Approaches to Problem Solving section at the end of the chapter.*

1. The number of reproducing offspring donated to the next generation by a particular genotype, relative to that of another genotype, is called _____.

2. A type of selection in which an individual struggles directly with the physical constraints of the environment, and does not involve competition with other individuals, is called _____ _____ selection.

3. Undirected, spontaneous mutations that were present prior to selection are called _____ mutations.

4. Competition between individuals for limited resources results in _____ selection.

5. Selection that favors an extreme phenotype, thus shifting the population mean in one or the other direction is called _____ selection.

6. What term defines changes in allele frequencies, in small populations, due to chance fluctuations?

7. Selection that favors an optimal phenotype and selects against both extremes is called _____ selection.

*8. In a herd of 1000 cattle, the following phenotypes were observed: Red (RR), 100; roan (RR'), 260; white (R'R'), 640. The frequencies of the R and R' alleles are _____ and _____, respectively.

*9. The frequency of children homozygous for a certain recessive allele is 1/25000. Assuming Hardy-Weinberg equilibrium and that just two alleles of the gene occur in the population, what is the frequency of heterozygotes in the population?

10. The frequency of individuals with PKU is 1/12,000 in a southwestern United States population. What is the frequency of the PKU allele?

11. From blood samples of 100 humans, 35 have blood antigen M, 50 have blood antigens M and N, and, 15 have blood antigen N. What are the frequencies of the L^M and L^N allele in the sample?

12. A population in Hardy-Weinberg equilibrium has two alleles of a gene, A and a. If the frequency of the a allele is 0.3, the frequency of the AA individuals is expected to be _____.

*13. For a human population in Hardy-Weinberg equilibrium, a sex-linked dominant trait is found in 91% of the females. The frequency at which this trait is expected to occur in males is _____.

*14. Assuming Hardy-Weinberg equilibrium in a population with p = 0.8, what is the expected frequency of the heterozygous individuals after four generations of self-fertilization?

*15. If, in a population obeying Hardy-Weinberg equilibrium, there are ten times more homozygous dominant individuals (AA) as heterozygous individuals (Aa), what are the expected frequencies of the three genotypes, AA, Aa, aa?

*16. A phenotype due to a sex-linked dominant allele occurs in males in the frequency of 0.4. Assuming that the population is in Hardy-Weinberg equilibrium, what is the expected frequency of the dominant phenotype among females of the population?

*17. In a Hardy-Weinberg population with p = 0.7, what is the expected frequency of the heterozygote four generations later?

*18. If the heterozygous individuals of a population are selected against, the frequency of the least frequent allele of a gene would be expected:
(a) to increase in frequency.
(b) to decrease in frequency.
(c) to change at the same rate as the more common allele.
(d) to remain at the same frequency.
(e) none of the above

*19. What is the equilibrium frequency of a detrimental recessive allele if the mutation rate is 5×10^{-5} and the selection intensity is 0.9?

20. In a human population, a certain sex-linked recessive trait is found in females with a frequency of 0.81. Assuming Hardy-Weinberg equilibrium, what is the expected frequency of this trait among males?

THOUGHT CHALLENGING EXERCISE

Is it possible to eliminate or greatly reduce the frequency of individuals possessing relatively rare recessive traits by forbidding individuals possessing the recessive trait (aa) from reproducing, or in other words, carrying out complete selection against the homozygous recessive genotypes?

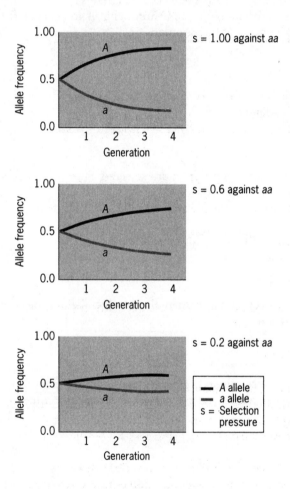

Figure 27.5   The effects of three different selective pressures on the frequencies of a pair of alleles (*A* and *a*) over four generations. At generation 0, *A* = *a* = 0.5.

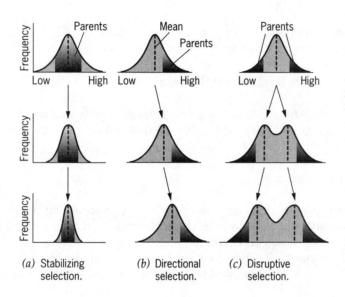

Figure 27.9 Three basic modes of selection and their effects on the mean (dashed lines) and variation of a normally distributed quantitative character.

SUMMARY OF KEY POINTS

Darwin's theory of evolution states that populations tend to increase exponentially but do not do so because resources are limited. Limited resources result in competition. Individuals with favorable inherited variations are more likely to survive and reproduce. Over time the genetic composition of the population changes, with favorable variants replacing less favorable variants.

The Hardy-Weinberg principle of genetic equilibrium describes the relationship between allele frequencies and genotypic frequencies under the assumption of random mating. This principle applies to loci with two alleles, loci with multiple alleles, and X-linked loci.

Fitness (w) is the relative reproductive ability of a genotype. Selection (s) refers to the intrinsic differences in the ability of genotypes to survive and reproduce. Quantitatively, $w = 1 - s$. The fitness of a phenotype may depend on its frequency in the population.

When the heterozygote is the most fit genotype, a balanced polymorphism results. Directional selection favors an extreme phenotype. Stabilizing selection favors the heterozygote or intermediate type over the extremes. Disruptive selection acts against the intermediate type.

Inbreeding, the mating between genetically related individuals, increases homozygosity and decreases heterozygosity. Genetic drift is the random fluctuation in allele frequency from generation to generation resulting from restricted population size.

One population may alter the genetic makeup of another population through migration. The net result of migration is to homogenize differences between populations.

Some mutations are harmful, some are beneficial, and some are neutral or nearly so. Experiments have shown that mutations are preadaptive, not postadaptive.

An equilibrium between mutations and selection will be achieved when the rate of introduction of new alleles is balanced by the rate of elimination by selection.

Knowledge of allele frequencies in populations or ethnic groups helps genetic counselors more accurately assess the risk for inheriting a genetic disorder.

ANSWERS TO QUESTIONS AND PROBLEMS

1) fitness 2) frequency-independent 3) preadaptive 4) frequency-dependent 5) directional
6) genetic drift 7) stabilizing 10) 0.009 11) 0.6, 0.4 12) 0.49 20) 0.9 8, 9,13-19) see *Approaches to Problem Solving* section below.

Ideas on thought challenging exercise:

Most single-gene detrimental genetic traits are caused by rare alleles; the efficacy of selection is very low because most of the recessive alleles occur in the heterozygous condition. Only those found in the homozygous state can be selected against. Therefore, selection against a recessive allele would be ineffective for all practical purposes. Furthermore, many recessive alleles may already be in an equilibrium between selection and mutation, and further selection could not eliminate the allele from the population.

APPROACHES TO PROBLEM SOLVING

8. To solve this problem the number of alleles of each type can be summed and divided by 2000. Since each individual is diploid, the total number of alleles of 1000 individuals is 2000.

$$\text{Frequency of } R \text{ is } p = \frac{2(100)R'R' + 260 \, R'R}{2000} = 0.23. \qquad \text{Since } p + q = 1, q = 0.77.$$

9. The frequency of homozygous recessive individuals is $q^2 = 1/25000 = 0.00004$. Therefore q is the square root of $0.00004 \cong 0.006$. $p = 1 - q = 0.994$. The frequency of heterozygotes in the population is $2pq \cong 0.012$.

13. Sex-linked genes in equilibrium are distributed p^2 (*AA*) + 2pq (*Aa*) + q^2 (*aa*) among females. Since males are hemizygous, all genotypes are expressed and are distributed p (*A*) + q (*a*). Therefore, allelic frequencies can be estimated directly by looking at the frequencies of the male phenotypes, A and a. In this problem, the frequency of the dominant phenotype in females is $0.91 = p^2 + 2pq$. It follows that $q^2 = 1 - (p^2 + 2pq) = 0.09$. Then q = 0.3 and p = 0.7. The frequency of this sex-linked dominant trait should occur in males at a frequency of p = 0.7.

14. Self-fertilization reduces the frequency of the heterozygote by 0.5 each generation. The initial frequency of heterozygotes in the population is $2pq = 2 \times 0.8 \times 0.2 = 0.32$. The frequency of the heterozygote after four generations of self-fertilization is expected to be (0.32)(0.5)(0.5)(0.5)(0.5) = 0.02.

15. There are ten times as many homozygous dominant individuals (*AA*) than heterozygous individuals (*Aa*) in the population. This can be mathematically expressed as $p^2 = 10 (2pq)$. Substituting (1 - p) for q leads to the equation $p^2 = 10 \times 2(p)(1-p)$. This is solved for p = 0.952. The value for q = 1 -p = 0.048. The expected frequencies of the genotypes are p^2 (*AA*) ≅ 0.906, 2pq (*Aa*) ≅ 0.091, and q^2 (*aa*) ≅ 0.002.

16. The sex-linked dominant allele occurs in males at a frequency of p = 0.4. It occurs in females at a frequency of $p^2 + 2pq = 1 - q^2$. Therefore, calculate the frequency of the recessive phenotype ($q^2 = 0.36$). The frequency of the dominant phenotype in females is 0.64.

17. The population is in equilibrium with the frequency of heterozygotes being 2pq = 0.42. If no forces act to change the genotypic frequencies, the frequency of heterozygotes should still be 0.42 after four generations.

18. Each heterozygote removed from the population takes away one allele of each type. This will have a proportionally greater affect on the least frequent allele. Therefore, the frequency of the least frequent allele will decrease with selection against the heterozygotes of the population.

19. This problem can be solved by using the formula $q = \sqrt{(u/s)}$, where u (5×10^{-5}) is the mutation rate and s (0.9) is the selection intensity.

$$q = \sqrt{(0.00005 / 0.9)} = 0.007$$

28

Genetics and Speciation

GENETIC VARIABILITY IN NATURAL POPULATIONS
    Phenotypic Variation
    Polymorphism of Chromosome Structure
    The Classical Versus Balanced Models of Variation
    Genetic Variation at the Molecular Level

THE SPECIES CONCEPT

REPRODUCTIVE ISOLATION: THE KEY TO THE SPECIES CONCEPT

MODES OF SPECIATION
    The Allopatric Mode of Speciation
    Sympatric Speciation: A Controversial Mode
    Parapatric Speciation
    Quantum Speciation

PUNCTUATED EQUILIBRIUM: THE RATE OF EVOLUTIONARY CHANGE IS NOT ALWAYS GRADUAL

PRIMATE EVOLUTION AS SEEN IN THE CHROMOSOMES

HUMANS AND CHIMPANZEES: EVOLUTION AT TWO LEVELS

IMPORTANT CONCEPTS

A. Evolution acts on genetic and phenotypic variation in a population.
   1. Some phenotypic variation in a population is discontinuous and some continuous.
   2. Genetic variation in a population can be considered according to the classical and balance views.
      a. According to the classic view, individuals are homozygous for the wild-type alleles at most loci, and mutant alleles are rare.
      b. In the balanced view, there is no single wild-type allele at most gene loci. Several alleles of a gene in the population are more or less functionally equivalent, and heterozygosity is the norm.
   3. Members of a population may exhibit considerable variation in chromosome structure.
   4. Genetic variation can be analyzed at the molecular level by studying proteins and nucleic acids.
      a. Studies of molecular variation support the balance view of genetic variation.
B. The two major species concepts are the phenetic species and the biological species.
   1. The phenetic species concept is based on phenotypic similarities.
   2. The more widely accepted biological species concept is based on reproductive isolation.
      a. Numerous prezygotic and postzygotic mechanisms keep species reproductively isolated.
C. There are several mechanisms that lead to the formation of new species.
   1. The classic allopatric model of speciation requires the physical separation of two populations in order to achieve reproductive isolation.
   2. In the more controversial sympatric mode of speciation, geographic separation of two populations is not necessary in order to achieve reproductive isolation.
   3. Parapatric speciation is a variant of sympatric speciation that often involves chromosomal repatterning in a small subpopulation.

4. Quantum speciation is relatively rapid and occurs at the margins of a population.
5. Speciation by polyploidy occurs predominantly in plants when doubling of entire chromosome sets leads to reproductive isolation.
6. The punctuated equilibrium theory suggests that there are long periods during which there is little change, followed by punctuated bursts of speciation events.
   a. The gradual and steady change in the genetic composition of populations over time, leading eventually to reproductive isolation and the formation of new species, stands in contrast to punctuated equilibrium.
   b. There is considerable evidence in the fossil record consistent with the punctuated equilibrium view, though it remains controversial.

D. The evolutionary history of the primates can be traced in their chromosomes.
   1. The chromosomes of humans, chimpanzees, gorillas, and orangutans can be related to each other by patterns of inversions and other structural changes.

IMPORTANT TERMS

In the space allotted, concisely define each term.

phyletic speciation:

branching speciation:

classic model:

balance model:

Phenetic (phenotypic) species concept:

biological species concept:

sibling species:

prezygotic isolating mechanisms:

postzygotic isolating mechanisms:

allopatric speciation:

subspecies(race):

character displacement:

sympatric speciation:

parapatric speciation:

quantum speciation:

punctuated equilibrium:

IMPORTANT NAMES

In the space allotted, concisely state the major contribution of the individual or group.

Richard Lewontin:

Harry Harris:

Charles Darwin:

Theodosius Dobzhansky:

L. E. Hurd and R. M. Eisenberg:

C. A. Tauber and M. J. Tauber:

George Gaylord Simpson:

Verne Grant:

Stephen J. Gould and Niles Eldridge:

P. G. Williamson:

Jeremy Jackson and Alan Cheetham:

M-C. King and Alan Wilson:

TESTING YOUR KNOWLEDGE:

*In this section, answer the questions, fill in the blanks, and solve the problems in the space allotted. Problems noted with an * are solved in the Approaches to Problem Solving section at the end of the chapter.*

1. The _____ view of population variation considers a wild-type allele to exist at most loci and mutations to be rare and generally detrimental.

2. The _____ species concept characterizes species in terms of phenotypic features.

3. The key event in the formation of species is _____.

4. The _____ model of variation has no single wild-type alleles but, rather, several more or less equivalent alleles of a gene exist in the gene pool in various combinations to confer a more general fitness on the population.

5. The individual who detected, using gel electrophoresis of proteins, an unexpectedly large amount of genetic variability in *Drosophila willistonii* was _____.

6. The _____ species concept defines species as a group of interbreeding populations reproductively isolated from other such groups.

7. List the six different categories of prezygotic isolating mechanisms proposed by Theodosius Dobzhansky.

8. The process in which there is selection for accentuated differences in species' characters under conditions of interspecific competition is called _____.

9. A process by which reproductive isolation occurs among groups of individuals within a continuous interbreeding population is called _____.

10. List the three different categories of postzygotic isolating mechanisms proposed by Theodosius Dobzhansky.

11. The study of green lacewing flies by Tauber and Tauber suggests that _____ speciation may have occurred.

12. Evolution of subspecies of the Old World mole rat, *Spalax ehrenbergi*, is an example of _____ speciation.

13. The evolution of *Clarkia lingulata* from *C. biloba*, involving rapid chromosomal rearrangements, is the best documented example of _____ speciation.

14. The detailed morphological analysis of fossilized bryozoans by Jackson and Cheetham indicating that species remained virtually unchanged for millions of years interrupted by bursts of speciation events is one of the strongest studies supporting the theory of _____.

15. The mode of speciation that may occur when two or more races of a species become geographically isolated is called _____speciation.

*16. A systemist is studying a plant species (species A) widely dispersed on well-developed soils in the Sacramento River valley. Upon exploring the surrounding hills, two closely related species were found, one (species B) growing on serpentine outcrops and the second (species C) on very dry gravely sites. Although there was no indication that hybridization occurred naturally, both species B and C could be crossed with species A (grown under greenhouse conditions) to produce partially fertile hybrids. Chromosome counts were obtained; root tips of species A had 18 chromosomes, species B had 18 chromosomes, and species C had 36 chromosomes. The hybrid between A and B had five bivalents and two quadrivalents at metaphase I of meiosis. The hybrid of species A and C had 9 bivalents and 9 univalents at metaphase I of meiosis. What do these data tell us about the evolution of species B and C?

THOUGHT CHALLENGING EXERCISE

The statement is often made that modern medicine, by allowing individuals of almost all genotypes to live and reproduce, is leading to an accumulation of detrimental alleles and the degeneration of the human gene pool. Based upon your understanding of population and evolutionary genetics, is there any scientific basis for such a viewpoint?

Allopatric Speciation Model

Parental species

1. A single population or group of similar populations in a homogeneous environment.

2. The environment becomes partly diversified in physical or biotic factors, or new populations are built up from migrants into new environments. The populations become genetically diversified, giving rise to races with different ecological requirements but which nevertheless can still exchange genes at their boundaries. No reproductive isolating mechanisms have developed.

3. Further differentiation and migration produce geographic isolation of some races and subspecies.

4. Some subspecies acquire enough genetic differences causing them to be reproductively isolated from the original population and from each other.

5. Further changes in the environment permit some of the newly evolved species to re-enter the area still occupied by the original population. Because of past differentiation, the two sympatric species exploit the environment in different ways and are prevented from merging by the barriers of reproductive isolation. Natural selection against the formation of hybrids promotes reinforcement of the isolating mechanisms and further differentiation in the ways the two species exploit their environment.

Species A   Species B

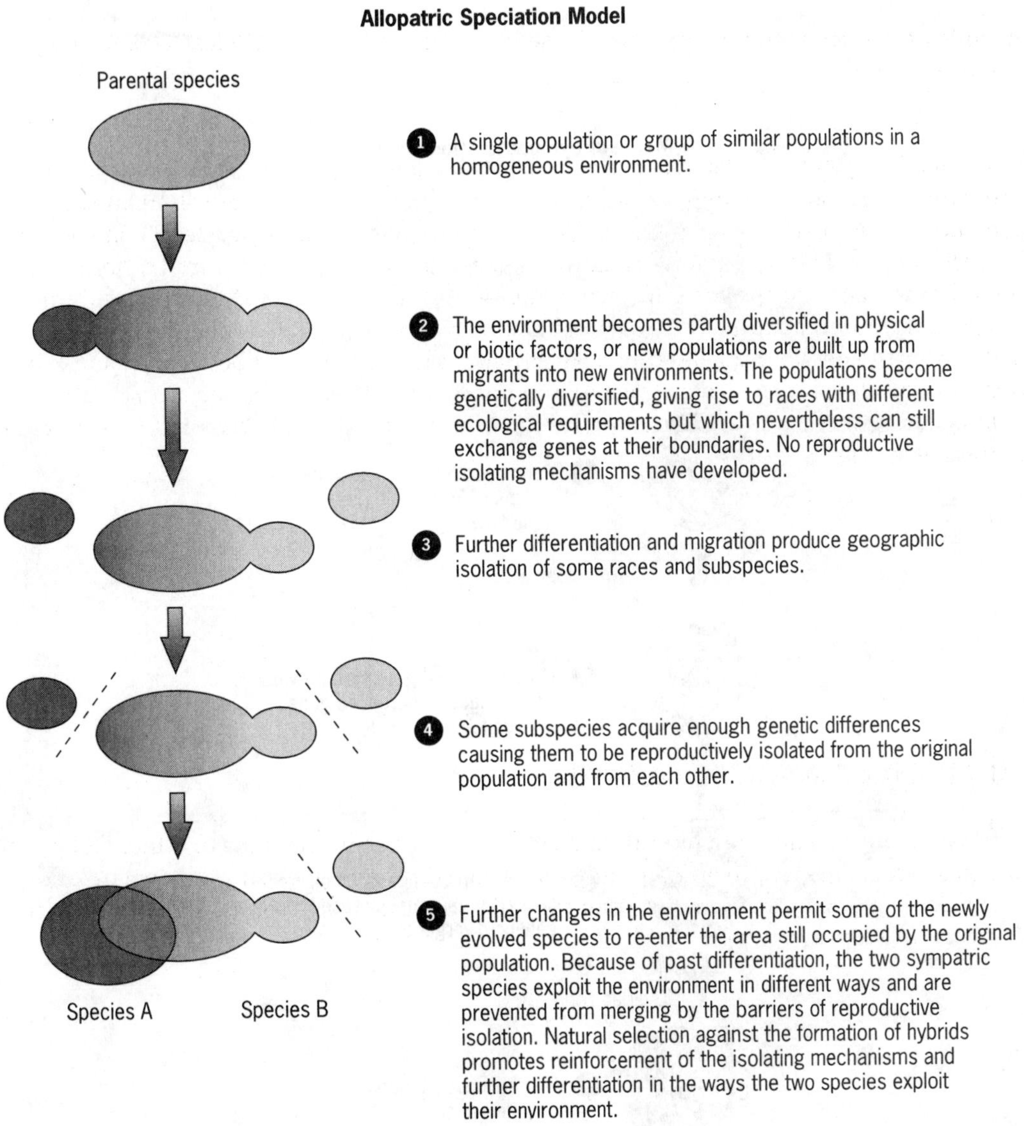

Figure 28.11 A schematic model of allopatric speciation.

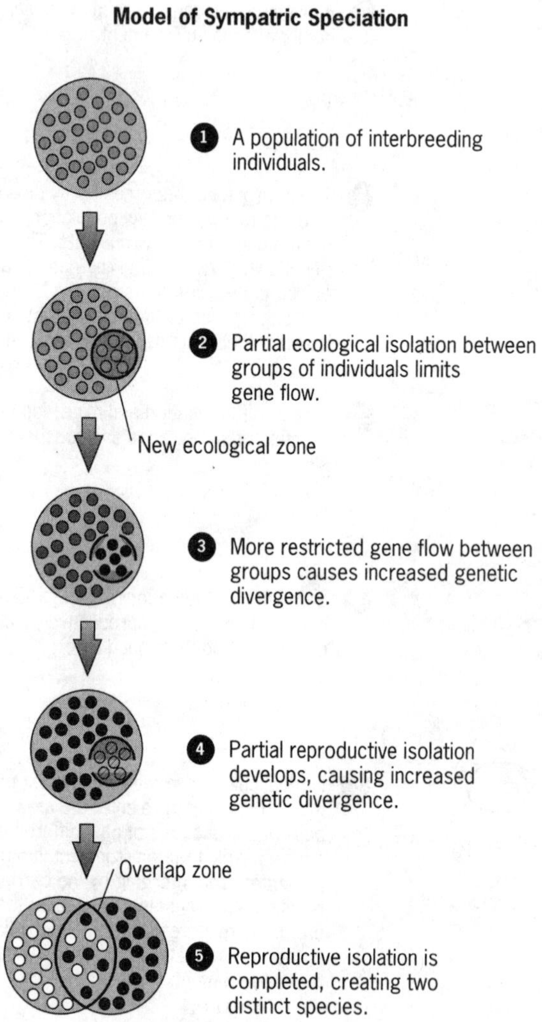

Model of Sympatric Speciation

1 A population of interbreeding individuals.

2 Partial ecological isolation between groups of individuals limits gene flow.

New ecological zone

3 More restricted gene flow between groups causes increased genetic divergence.

4 Partial reproductive isolation develops, causing increased genetic divergence.

Overlap zone

5 Reproductive isolation is completed, creating two distinct species.

Figure 28.13   A schematic model of sympatric speciation.

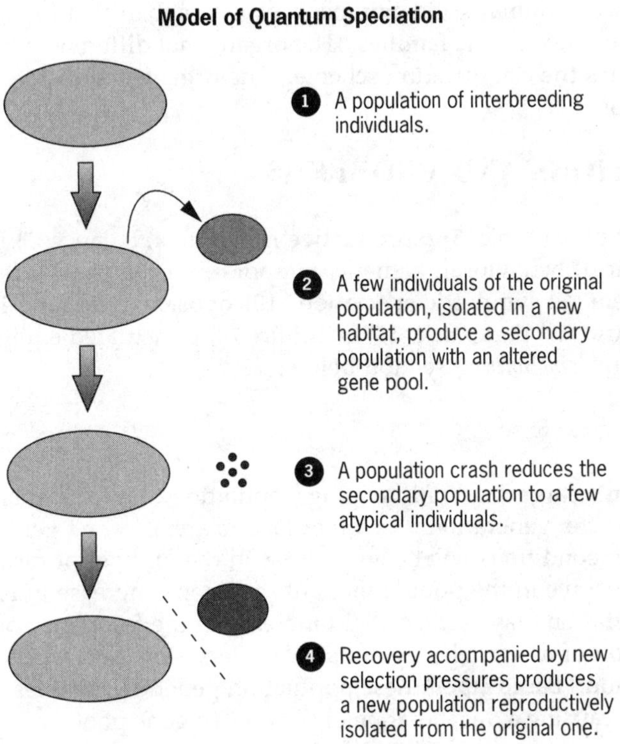

Model of Quantum Speciation

① A population of interbreeding individuals.

② A few individuals of the original population, isolated in a new habitat, produce a secondary population with an altered gene pool.

③ A population crash reduces the secondary population to a few atypical individuals.

④ Recovery accompanied by new selection pressures produces a new population reproductively isolated from the original one.

Figure 28.15   A schematic model of quantum speciation.

SUMMARY OF KEY POINTS

Members of a population have considerable genetic variation. Some populations are polymorphic for chromosome rearrangements which can be associated with effects on fitness. Many populations are polymorphic for electrophoretic protein variants and for DNA sequence variants. This variation is probably maintained by a combination of recurrent mutation and balancing selection.

The phenotypic species concept groups organisms according to phenotypic similarities. The biological species concept is based on reproductive isolation.

Prezygotic isolating barriers to reproduction prevent the union of male and female gametes. In the event of a successful fertilization between two species, postzygotic isolating barriers to reproduction assure that the hybrid will not successfully reproduce.

Allopatric speciation is a gradual process of speciation via geographical isolation. Sympatric speciation occurs without geographic separation. Parapatric speciation is a rapid process that does not require geographic isolation. Quantum speciation is the budding off of a new species from a semi-isolated peripheral population of the ancestral species.

Punctuated equilibrium refers to patterns of speciation in which populations exhibit little change for millions of years, punctuated by periods of rapid speciation.

A comparison of the banding patterns of human, chimpanzee, gorilla, and orangutan chromosomes shows clear evolutionary relationships. The evolution of the primates is characterized by chromosome rearrangements, including inversions, and translocations.

The molecular distance between humans and chimpanzees is less than that for sibling species, yet the two species belong to different taxonomic families. The organismal differences between the two species are quite large, which justifies the classification scheme. The organismal differences may be due to differences in gene regulation.

ANSWERS TO QUESTIONS AND PROBLEMS

1) classic **2)** phenetic or phenotypic **3)** reproductive isolation **4)** balance **5)** Richard Lewontin **6)** biological **7)** ecological, behavioral, gametic, temporal, mechanical, different pollinators **8)** character displacement **9)** sympatric speciation **10)** hybrid inviability, hybrid sterility, hybrid breakdown **11)** sympatric **12)** parapatric **13)** quantum **14)** punctuated equilibrium **15)** allopatric **16)** see *Approaches to Problem Solving* section below.

Ideas on thought challenging exercise:

If selection is relaxed within a large randomly mating population, the frequencies of alleles will remain the same generation after generation if no other factors are affecting gene frequency. The first generation after lax selection conditions will be an increase in the number of formerly lethal or sublethal individuals that survive in the population, but this won't increase in future generations. Over many generations, mutation may cause a slight increase in the frequency of detrimental alleles. However, the rate of addition of new alleles by mutation is very slow and will reach an equilibrium with the rate of back mutation. Based upon these population genetic principles, it can be clearly stated that modern medicine is not causing a degeneration of the human gene pool.

APPROACHES TO PROBLEM SOLVING

16. Species B grows on a very restricted and specialized habitat so we may suspect that it evolved from species A. The two quadrivalents in the hybrid of A × B indicates that the genome of species B differs from that of species A by two translocations. Therefore, the postulated evolution of species B from species A involved restructuring of the chromosomes in the form of two reciprocal translocations. Species B is tetraploid as is indicative by its chromosome number. The homology of nine chromosomes of species C with the nine chromosomes of species A (9 bivalents) in the hybrid indicates that the chromosomes of species A are homologous to the nine chromosomes of species C. It appears that A hybridized with another species to form the allotetraploid species C.

PART II

Complete Answers and Solutions to Text Questions and Problems

ANSWERS
TO
QUESTIONS AND PROBLEMS

CHAPTER 2

2.1 (a) 0

(b) 0

(c) 0

(d) 0

(e) 0

(f) +

(g) +

2.2 (a) 23

(b) 23

(c) 23

(d) 23

(e) 46

(f) 46

(g) 23

(h) 46

2.3 (a) 200

(b) 50

2.4 Parents: $M//m$ female $\times$ $m//m$ male

Gametes: $\frac{1}{2}M$ and $\frac{1}{2}m$; all m

Offspring: $\frac{1}{2}M//m$ and $\frac{1}{2}m//m$; $\frac{1}{2}$ of the progeny will have myopia.

2-5 See Figure 2.13 (oogenesis). At the start of oogenesis, the woman has a pair of chromosomes, one carrying M and one carrying m; at the completion of oogenesis, a mature egg carrying the M allele, when fertilized by an m carrying sperm, produces a myopic child ($M//m$).

2.6 The chromosome mechanism is similar in plants and animals. Division of the cytoplasmic part of the flexible animal cell is accomplished by constriction (cytokinesis), whereas the rigid plant cell forms a partition or cell plate.

2.7 Meiosis includes a pairing (synapsis) of homologous maternal and paternal chromosomes. In the cell division that follows, the chromosomes that have previously paired separate. This results in a reduction of chromosome number from $2n$ (diploid) to n (haploid).

2.8 (a) Many plants have male and female parts on the same plant or in the same flower. Unlike animals, plants have a gametophyte stage that consists (in higher plants) of a few cell divisions.

(b) The chromosome mechanism is essentially the same during the formation of gametes in plants and animals.

2.9 An egg nucleus and two polar nuclei are developed in the ovule. Two haploid nuclei are introduced into the ovule by the pollen tube. One nucleus fuses with the egg nucleus to produce the $2n$ (diploid) zygote and the other with the two polar nuclei to produce the $3n$ (triploid) endosperm nucleus. The zygote develops into the embryo, and the endosperm forms the nutrient material that supports the developing embryo.

2.10 The father is $A//a\ h//h$ and the mother is $a//a\ H//h$; There are two classes of sperm: $A/\ h/$ and $a/\ h/$; there are two classes of eggs: $a/\ H/$ and $a/\ h/$. An $a/\ H/$ egg fertilized by an $A/\ h/$ sperm will produce an $A//a\ H//h$ zygote.

2.11 See Figure 2.13 (oogenesis). The primary oocyte is $a//a\ H//h$; following meiosis I and II, $\frac{1}{2}$ of the mature eggs will carry a and H chromosomes.

2.12 See Figure 2.14 (spermatogenesis). The primary spermatocyte is $A//a\ h//h$; following meiosis I and II, half of the mature sperm will carry the A and h chromosomes.

2.13 The man is $A//a\ B//b$.

2.14 Bisexual organisms. Asexual reproduction provides for no genetic variation, except that of rare mutations. Self-fertilization tends toward homozygosity or pure lines. Bisexual reproduction in higher organisms is associated with great hereditary variation through segregation, independent assortment, and crossing over between maternal and paternal chromosomes.

2.15 (a) Early primary oocyte
(b) prophase of meiosis I
(c) suspended prophase I
(d) the first meiotic division is completed just before ovulation of each egg.

2.16 For each pair of chromosomes, there are 2 different classes of gamete possible. Therefore, the number of different classes of gametes is $2 \times 2 \times 2 \times 2 = 2^4 = 16$.

2.17 Four.

2.18 Fourteen chromosomes in the diploid state means there are 7 pairs. Using the same logic as 2.16, there would be 2^7, or 128 different classes of gametes produced (not counting variation generated by crossing over).

2.19 Haploid cells differentiate into gametes that fuse to produce a diploid zygote. The diploid zygote then undergoes meiosis to produce four haploid cells, each of which divides mitotically to produce the haploid individual. A haploid individual cannot undergo meiosis directly because its chromosomes do not exist as homologous pairs.

2.20 Meiosis maintains constancy of chromosome number from generation to generation; and through segregation, independent assortment, and crossing over, meiosis produces new gene combinations that generate genetic variability in a species. Increased genetic variability produces increased phenotypic variability, which is crucial to the evolutionary success of any species.

CHAPTER 3

3.1 **(a)** All tall

(b) $\frac{3}{4}$ tall, $\frac{1}{4}$ dwarf

(c) all tall

(d) $\frac{1}{2}$ tall, $\frac{1}{2}$ dwarf

3.2 Round (WW) × wrinkled (ww) → F_1 round (Ww); F_1 self-fertilized → F_2: $\frac{3}{4}$ round (2 WW: 1 Ww), $\frac{1}{4}$ wrinkled (ww). The expected results in the F_2 are 5493 round, 1831 wrinkled. To compare the observed and expected results, compute χ^2 with one degree of freedom:

$$\frac{(5474-5493)^2}{5493} + \frac{(1850-1831)^2}{1831} = 0.307,$$ which is not significant at the 5% level. Thus, the results are consistent with the Principle of Segregation.

3.3 The data suggest that coat color is controlled by a single gene with two alleles, C (gray) and c (albino), and that C is dominant over c. On this hypothesis, the crosses are: Gray (CC) × albino (cc) → F_1 gray (Cc); $F_1 \times F_1$ → $\frac{3}{4}$ gray (2 CC: 1 Cc), $\frac{1}{4}$ albino (cc). The expected results in the F_2 are 203 gray, 67 albino. To compare the observed and expected results, compute χ^2 with one degree of freedom: $\frac{(198-203)^2}{203} + \frac{(67-72)^2}{72} = 0.470$, which is not significant at the 5% level. Thus, the results are consistent with the hypothesis.

3.4 **(a)** Woman's genotype Pp, father's genotype Pp, mother's genotype pp

(b) $\frac{1}{2}$

3.5 **(a)** Checkered, red (CC BB) × plain, brown ($cc\ bb$) → F_1 all checkered, red ($Cc\ Bb$)

(b) F_2 progeny: $\frac{9}{16}$ checkered, red (C- B-), $\frac{3}{16}$ plain, red (cc B-), $\frac{3}{16}$ checkered, brown (C- bb), $\frac{1}{16}$ plain, brown ($cc\ bb$)

3.6 **(a)** colored, normal (CC Vv) × white, normal ($cc\ Vv$)

(b) colored, normal ($Cc\ Vv$) × colored, normal ($Cc\ Vv$)

(c) colored, normal ($Cc\ Vv$) × white, waltzing ($cc\ vv$)

3.7 Among the F_2 progeny with long, black fur, the genotypic ratio is 1 $BB\ RR$: 2 $BB\ Rr$: 2 $Bb\ RR$: 4$Bb\ Rr$; thus $\frac{1}{9}$ of the rabbits with long, black fur are homozygous for both genes.

3.8 **(a)** all red

(b) $\frac{1}{2}$ red, $\frac{1}{2}$ roan

(c) all roan

(d) $\frac{1}{4}$ red, $\frac{1}{2}$ roan, $\frac{1}{4}$ white

(e) $\frac{1}{2}$ roan, $\frac{1}{2}$ white

(f) all white

3.9 Half the children from $Aa \times aa$ matings would be albino. In a family of three children, the chance that one will be normal and two albino is $3 \times \left(\frac{1}{2}\right)^{1} \times \left(\frac{1}{2}\right)^{2} = \frac{3}{8}$.

3.10 **(a)** $\left(\frac{3}{4}\right)^{4} = \frac{81}{256}$

(b) $4 \times \left(\frac{3}{4}\right)^{3} \times \left(\frac{1}{4}\right)^{1} = \frac{108}{256}$

(c) $6 \times \left(\frac{3}{4}\right)^{2} \times \left(\frac{1}{4}\right)^{2} = \frac{54}{256}$

(d) $4 \times \left(\frac{3}{4}\right)^{1} \times \left(\frac{1}{4}\right)^{3} = \frac{12}{256}$

3.11 Man ($Cc\,ff$) × woman ($cc\,Ff$).

(a) $cc\,ff$, $\quad \frac{1}{2} \times \frac{1}{2} = \frac{1}{4}$

(b) $Cc\,ff$, $\quad \frac{1}{2} \times \frac{1}{2} = \frac{1}{4}$

(c) $cc\,Ff$, $\quad \frac{1}{2} \times \frac{1}{2} = \frac{1}{4}$

(d) $Cc\,Ff$, $\quad \frac{1}{2} \times \frac{1}{2} = \frac{1}{4}$

3.12 **(a)** dominant

(b) recessive

(c) recessive

(d) uncertain

3.13

	F_1 gametes	F_2 genotypes	F_2 phenotypes
(a)	2	3	2
(b)	$2 \times 2 = 4$	$3 \times 3 = 9$	$2 \times 2 = 4$
(c)	$2 \times 2 \times 2 = 8$	$3 \times 3 \times 3 = 27$	$2 \times 2 \times 2 = 8$
(d)	2^{n}	3^{n}	2^{n}, where n is the number of genes

3.14 On the hypothesis, the expected number in each class is 27.5; χ^2 with three degrees of freedom is calculated as $\dfrac{(31-27.5)^2}{27.5} + \dfrac{(26-27.5)^2}{27.5} + \dfrac{(27-27.5)^2}{27.5} + \dfrac{(26-27.5)^2}{27.5} = 0.618$, which is not significant at the 5% level. Thus, the results are consistent with the hypothesis of two independently assorting genes, each segregating two alleles.

3.15 $\left(\dfrac{1}{2}\right)^3 = \dfrac{1}{8}$

3.16 (a) $6 \times \left(\dfrac{1}{2}\right)^2 \times \left(\dfrac{1}{2}\right)^2 = \dfrac{6}{16}$

(b) $\left(\dfrac{1}{2}\right)^4 = \dfrac{1}{16}$

(c) 2 boys, 2 girls

(d) $1 - $ probability that all four are boys $= 1 - \left(\dfrac{1}{2}\right)^4 = \dfrac{15}{16}$

3.17 $\dfrac{20}{64} + \dfrac{10}{64} + \dfrac{5}{64} + \dfrac{1}{64} = \dfrac{36}{34}$

3.18 (a) zero

(b) $\dfrac{1}{2}$

3.19 (a) $\dfrac{1}{2} \times \dfrac{1}{4} = \dfrac{1}{8}$

(b) $\dfrac{1}{2} \times \dfrac{1}{2} \times \dfrac{1}{4} = \dfrac{1}{16}$

(c) $\dfrac{2}{3} \times \dfrac{2}{3} \times \dfrac{1}{4} = \dfrac{1}{6}$

(d) $\dfrac{2}{3} \times \dfrac{1}{2} \times \dfrac{1}{2} \times \dfrac{1}{4} = \dfrac{1}{24}$

3.20 (a) recessive

(b) dominant

3.21 For III-1 × III-2, the chance of an affected child is $\dfrac{1}{2}$. For IV-2 × IV-3, the chance is zero.

3.22 $\dfrac{1}{2}$

3.23 **(a)** $\left(\dfrac{1}{4}\right)^3 = \dfrac{1}{64}$

(b) $\left(\dfrac{1}{2}\right)^3 = \dfrac{1}{8}$

(c) $3 \times \left(\dfrac{1}{2}\right)^1 \times \left(\dfrac{1}{2}\right)^2 = \dfrac{3}{8}$

(d) 1 – probability that the offspring is not homozygous for the recessive allele of any gene

$= 1 - \left(\dfrac{3}{4}\right)^3 = \dfrac{37}{64}$

(e) $\left(\dfrac{3}{4}\right)^3 = \dfrac{27}{64}$

CHAPTER 4

4.1 M and MN

4.2 **(a)** all wild-type

(b) $\dfrac{3}{4}$ wild-type, $\dfrac{1}{4}$ albino

(c) $\dfrac{3}{4}$ wild-type, $\dfrac{1}{4}$ chinchilla

(d) $\dfrac{1}{2}$ chinchilla, $\dfrac{1}{2}$ albino

(e) $\dfrac{3}{4}$ wild-type, $\dfrac{1}{4}$ himalayan

(f) $\dfrac{1}{2}$ himalayan, $\dfrac{1}{2}$ albino

4.3

	Parents	Offspring
(a)	yellow × yellow	2 yellow: 1 light belly
(b)	yellow × light belly	2 yellow: 1 light belly: 1 black and tan
(c)	black and tan × yellow	2 yellow: 1 black and tan: 1 black
(d)	light belly × light belly	all light belly
(e)	light belly × yellow	1 yellow: 1 light belly
(f)	agouti × black and tan	1 agouti: 1 black and tan
(g)	black and tan × black	1 black and tan: 1 black
(h)	yellow × agouti	1 yellow: 1 light belly
(i)	yellow × yellow	2 yellow: 1 light belly

4.4 (a) S^1S^3, S^2S^3

(b) S^1S^3, S^1S^4, S^2S^3, S^2S^4

(c) none

(d) S^3S^5, S^3S^6, S^4S^5, S^4S^6

4.5 (a) all AB

(b) 1 A : 1 B

(c) 1 A : 1 B : 1 AB : 1 O

(d) 1 A : 1 O

4.6 No. The woman must be I^OI^O; if her mate is I^AI^B, they could not have an I^OI^O child.

4.7 No. The woman is I^AI^B. One man could be either I^AI^A or I^AI^O; the other could be either I^BI^B or I^BI^O. Given the uncertainty in the genotype of each man, either could be the father of the child.

4.8 The woman is I^OI^O L^ML^M; the man is I^AI^B L^ML^N; the blood types of the children will be A and M, A and N, B and M, and B and N, all equally likely.

4.9 Cross homozygous *waltzing* with homozygous *tango*. If the mutations are alleles, all the offspring will have an uncoordinated gait; if they are not alleles, all the offspring will be wild-type. If the two mutations are alleles, they could be denoted with the symbols v (*waltzing*) and v^t (*tango*).

4.10 The individuals III-4 and III-5 must be homozygous for recessive mutations in different genes, that is, one is *aa BB* and the other is *AA bb*; none of their children is deaf because all of them are heterozygous for both genes (*Aa Bb*).

4.11 $\frac{9}{16}$ dark red, $\frac{7}{16}$ brownish-purple

4.12 No. The test for allelism cannot be performed with dominant mutations.

4.13 The allele for yellow fur is homozygous lethal.

4.14 Because the mother is *Bb*, the chance that her daughter is *Bb* is $\frac{1}{2}$. The chance that the daughter will have a child who becomes bald as an adult is P(daughter is *Bb*) × P(daughter transmits *B* to child) × P(child is male) = $\frac{1}{2}$ × $\frac{1}{2}$ × $\frac{1}{2}$ = $\frac{1}{8}$.

4.15 Dominant. The condition appears in every generation, and nearly every affected individual has an affected parent. The exception, IV-2, had a father who carried the ataxia allele but did not manifest the trait—an example of incomplete penetrance.

4.16 (a) $\frac{3}{4}$ walnut, $\frac{1}{4}$ rose

(b) $\frac{1}{2}$ walnut, $\frac{1}{2}$ pea

(c) $\frac{3}{8}$ walnut, $\frac{3}{8}$ rose, $\frac{1}{8}$ pea, $\frac{1}{8}$ single

(d) $\frac{1}{2}$ rose, $\frac{1}{2}$ single

4.17 $Rr\ pp \times Rr\ Pp$

4.18 $\frac{12}{16}$ white, $\frac{3}{16}$ yellow, $\frac{1}{16}$ green

4.19 $\frac{13}{16}$ white, $\frac{3}{16}$ colored

4.20 $\frac{9}{16}$ dark red (wild-type), $\frac{3}{16}$ brownish-purple, $\frac{3}{16}$ bright red, $\frac{1}{16}$ white

4.21 9 black: 3 gray: 52 white

4.22 9 black: 39 gray: 16 white

CHAPTER 5

5.1 The male-determining sperm carries a Y chromosome; the female-determining sperm carries an X chromosome.

5.2 Cross the singed male to wild-type females, and then intercross the offspring. If the singed bristle phenotype is due to an X-linked mutation, approximately half the F_2 males, but none of the F_2 females, will show it.

5.3 All the daughters will be green, and all the sons will be rosy.

5.4 The cross is go/go $+/+$ female x $+/Y$ bw/bw male → F$_1$: $go/+$ $bw/+$ females (wild-type eyes and body) and go/Y $bw/+$ males (golden body, wild-type eyes). An intercross of the F$_1$ offspring yields the following F$_2$ phenotypes in both sexes:

Body	Eyes	Genotype	Proportion
golden	brown	go/go or Y bw/bw	$\dfrac{1}{2} \times \dfrac{1}{4} = \dfrac{1}{8}$
golden	wild-type	go/go or Y $+/bw$ or $+$	$\dfrac{1}{2} \times \dfrac{3}{4} = \dfrac{3}{8}$
wild-type	brown	$+/go$ or Y bw/bw	$\dfrac{1}{2} \times \dfrac{1}{4} = \dfrac{1}{8}$
wild-type	wild-type	$+/go$ or Y $+/bw$ or $+$	$\dfrac{1}{2} \times \dfrac{3}{4} = \dfrac{3}{8}$

5.5 XX is female, XY is male, XXX is female (but barely viable), XO is male (but sterile).

5.6

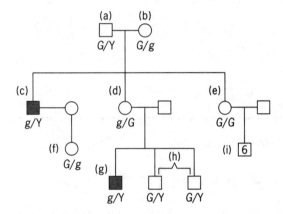

5.7 No. Defective color vision is caused by an X-linked mutation. The son's X chromosome came from his mother, not his father.

5.8 The risk for the child is P(woman transmits mutant allele) × P(child is male) = $\dfrac{1}{2} \times \dfrac{1}{2} = \dfrac{1}{4}$.

5.9 The risk for the child is P(mother is C/c) × P(mother transmits c) × P(child is male)
$= \dfrac{1}{2} \times \dfrac{1}{2} \times \dfrac{1}{2} = \dfrac{1}{8}$; if the couple has already had a child with color blindness,

P(mother is C/c) = 1, and the risk for each subsequent child is $\dfrac{1}{4}$.

5.10 Each of the rare vermilion daughters must have resulted from the union of an $X(v) X(v)$ egg with a Y-bearing sperm. The diplo-X eggs must have originated through nondisjunction of the X chromosomes during oogenesis in the mother. However, we cannot determine if the nondisjunction occurred in the first or the second meiotic division.

5.11 Each of the rare white-eyed daughters must have resulted from the union of an $X(w) X (w)$ egg with a Y-bearing sperm. The rare diplo-X eggs must have originated through nondisjunction of the X chromosomes during the second meiotic division in the mother.

5.12 Female

5.13 Three-fourths will be phenotypically female (genotypically tfm/Tfm, Tfm/Tfm, or tfm/Y). Among the females, $\frac{2}{3}$ (tfm/Tfm and Tfm/Tfm) will be fertile; the tfm/Y females will be sterile.

5.14 Male

5.15 **(a)** Female
   (b) intersex
   (c) intersex
   (d) male
   (e) female
   (f) male

5.16 **(a)** $\frac{1}{2} X^{bb} X^{bb}$ females, $\frac{1}{2} X^{bb} Y^{+}$ wild-type males

   (b) $\frac{1}{2} X^{+} X^{bb}$ wild-type females, $\frac{1}{2} X^{bb} Y^{bb}$ bobbed males

   (c) $\frac{1}{4} X^{+} X^{+}$ wild-type females, $\frac{1}{4} X^{+} X^{bb}$ wild-type females, $\frac{1}{4} X^{+} Y^{bb}$ wild-type males, $\frac{1}{4} X^{bb} Y^{bb}$ bobbed males

   (d) $\frac{1}{4} X^{+} X^{bb}$ wild-type females, $\frac{1}{4} X^{bb} X^{bb}$ bobbed females, $\frac{1}{4} X^{+} Y^{+}$ wild-type males, $\frac{1}{4} X^{bb} Y^{+}$ wild-type males

5.17 Yes. The gene for feather patterning is on the Z chromosome. If we denote the allele for barred feathers as B and the allele for nonbarred feathers as b, the crosses are: B/B (barred) male × b/W (nonbarred) female → F_1: B/b (barred) males and B/W (barred) females. Intercrossing the F_1 produces B/B (barred) males, B/b (barred) males, B/W (barred) females, and b/W (nonbarred) females, all in equal proportions.

5.18 *Drosophila* does not achieve dosage compensation by inactivating one of the X chromosomes in females.

5.19 **(a)** zero

(b) one

(c) one

(d) two

(e) three

(f) zero

5.20   Metaphase   Anaphase

CHAPTER 6

6.1   Use one of the banding techniques.

6.2   Both terms imply an increase in the number of copies of each chromosome; with polyploidy, the multiple copies separate from each other, whereas with polyteny, they do not.

6.3   46, XX, 22q– or 46, XY, 22q–, depending on the sex chromosome constitution

6.4   Union of a diploid gamete and a haploid gamete would produce a triploid; union of two haploid gametes followed by duplication of all the chromosomes would produce a tetraploid.

6.5   In allotetraploids, each member of the different sets of chromosomes can pair with a homologous partner during prophase I and then disjoin during anaphase I. In triploids, disjunction is irregular because homologous chromosomes associate during prophase I by forming either bivalents and univalents or by forming trivalents.

6.6   The fertile plant is an allotetraploid with 7 pairs of chromosomes from species A and 9 pairs of chromosomes from species B; the total number of chromosomes is $(2 \times 7) + (2 \times 9) = 32$.

6.7   The F_1 hybrid had 5 chromosomes from species X and 7 from species Y, for a total of 12. When this hybrid was backcrossed to species Y, the few progeny that were produced had $5 + 7 = 12$ chromosomes from the hybrid and 7 from species Y, for a total of 19. This hybrid was therefore a triploid. Upon self-fertilization, a few F_2 plants were formed, each with 24 chromosomes. Presumably, the chromosomes in these plants consisted of $2 \times 5 = 10$ from species X and $2 \times 7 = 14$ from species Y. These vigorous and fertile F_2 plants were therefore allotetraploids.

6.8   XX is female, XY is male, XO is female (but sterile), XXX is female, XXY is male (but sterile), XYY is male.

6.9 $\frac{1}{2}$

6.10 The fly is a gynandromorph, that is, a sexual mosaic. The yellow tissue is X(y)/O and the gray tissue is X(y)/X(+). This mosaicism must have arisen through loss of the X chromosome that carried the wild-type allele, presumably during one of the early embryonic cleavage divisions.

6.11 Approximately half the progeny should be disomic *ey*/+ and half should be monosomic *ey*/O. The disomic progeny will be wild-type, and the monosomic progeny will be eyeless.

6.12 Nondisjunction must have occurred in the mother. The color blind woman with Turner syndrome was produced by the union of an X-bearing sperm, which carried the mutant allele for color blindness, and a nullo-X egg.

6.13 XYY men would produce more children with sex chromosome abnormalities because their three sex chromosomes will disjoin irregularly during meiosis. This irregular disjunction will produce a variety of aneuploid gametes, including the XY, YY, XYY, and nullo sex chromosome constitutions.

6.14 The chromosomes are large, polytene, and banded by differential condensation of the chromatin. Homologous chromosomes are also paired, permitting the detection of chromosome aberrations such as deletions, duplications and inversions.

6.15 The animal is heterozygous for an inversion:

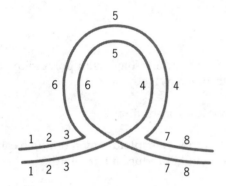

6.16 (a) Deletion:

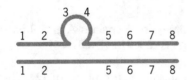

(b) Duplication:

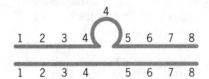

6.16 (continued)

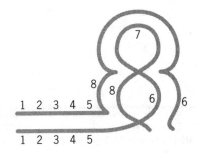

6.17

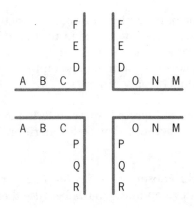

6.18 The exceptional male, whose genotype is $bw/+ st/+$, is heterozygous for a translocation between chromosomes 2 and 3. It is not possible to determine whether the translocation is between the two mutant chromosomes or between the two wild-type chromosomes, that is, whether it is T(bw; st) or T(+; +); however it clearly is not between a mutant chromosome and a wild-type chromosome, that is, T(bw; +) or T(+; st). If it were, the progeny would be either brown or scarlet, not either wild-type or white.

6.19 The boy carries a translocation between chromosome 21 and another chromosome, say No. 14. He also carries a normal chromosome 21 and a normal chromosome 14. The boy's sister carries the translocation, one normal chromosome 14, and two normal copies of chromosome 21.

6.20 A variegated position effect originates when a euchromatic gene is transposed via a chromosome rearrangement into or near heterochromatin, which then suppresses the gene's function.

6.21 All the daughters will be yellow-bodied, and all the sons will be white-eyed.

6.22 Zygotes produced by this couple will be either trisomic or monosomic for chromosome 21. Thus, 100% of their viable children will develop Down syndrome.

6.23   The three populations are related by a series of inversions:

P1 1 2 3̣4 5 6 7 8 9̣10

P2 1 2 3̣9 8̣7 6 5̣4̣10

P3 1 2 3 9 8̣5 6 7̣4 10

6.24   The mother is heterozygous for a reciprocal translocation between the long arms of the large and small chromosomes; a piece from the long arm of the large chromosome has been broken off and attached to the long arm of the short chromosome. The child has inherited the rearranged large chromosome and the normal small chromosome from the mother. Thus, because the rearranged large chromosome is deficient for some of its genes, the child is hyperploid.

6.25   The phenotype in the female offspring is mosaic because one of the X chromosomes is inactivated in each of their cells. If the translocated X is inactivated, the autosome attached to it could also be partially inactivated by a spreading of the inactivation process across the translocation breakpoint. This spreading could therefore inactivate the color-determining gene on the translocated autosome and cause patches of tissue to be phenotypically mutant.

6.26   The sons will have bright red eyes because they will inherit the Y chromosome with the bw^+ allele from their father. The daughters will have white eyes because they will inherit an X chromosome from their father.

6.27   XX zygotes will develop into males because one of their X chromosomes carries the *TDF* gene that was translocated from the Y chromosome. XY zygotes will develop into females because their Y chromosome has lost the *TDF* gene.

CHAPTER 7

7.1   The class represented by 351 offspring indicates that at least two of the three genes are linked.

7.2   No. The genes *a* and *d* could be very far apart on the same chromosome--so far apart that they recombine freely, that is, 50% of the time.

7.3   A two-strand double crossover must have occurred.

7.4 $\left(\dfrac{1}{19}\right)^2 = 0.0027.$

7.5 $\dfrac{7!}{7^7} = 0.0061$

7.6   20%

7.7 **(a)** Cross: $a^+ b^+/a^+ b^+ \times a\,b/a\,b$
Gametes: $a^+ b^+$ from one parent, $a\,b$ from the other
F_1: $a^+ b^+/a\,b$

(b) 40% $a^+ b^+$, 40% $a\,b$, 10% $a^+ b$, 10% $a\,b^+$

(c) F_2 from testcross: 40% $a^+ b^+/a\,b$, 40% $a\,b/a\,b$, 10% $a^+ b/a\,b$, 10% $a\,b^+/a\,b$

(d) Coupling linkage phase

(e) F_2 from intercross:

	Sperm			
	40% $a^+ b^+$	40% $a\,b$	10% $a^+ b$	10% $a\,b^+$
40% $a^+ b^+$	16% $a^+ b/a^+ b^+$	16% $a^+ b/a\,b$	4% $a^+ b/a^+ b$	4% $a^+ b/a\,b^+$
Eggs 40% $a\,b$	16% $a\,b/a^+ b^+$	16% $a\,b/a\,b$	4% $a\,b/a^+ b$	4% $a\,b/a\,b^+$
10% $a^+ b$	4% $a^+ b/a^+ b^+$	4% $a^+ b/a\,b$	1% $a^+ b/a^+ b$	1% $a^+ b/a\,b^+$
10% $a\,b^+$	4% $a\,b^+/a^+ b^+$	4% $a\,b^+/a\,b$	1% $a\,b^+/a^+ b$	1% $a\,b^+/a\,b^+$

Summary of phenotypes:

a^+ and b^+	66%
a^+ and b	9%
a and b^+	9%
a and b	16%

7.8 **(a)** Cross: $a^+ b/a^+ b \times a\,b^+/a\,b^+$
Gametes: $a^+ b$ from one parent, $a\,b^+$ from the other
F_1: $a^+ b/a\,b^+$

(b) 40% $a^+ b$, 40% $a\,b^+$, 10% $a^+ b^+$, 10% $a\,b$

(c) F_2 from testcross: 40% $a^+ b/a\,b$, 40% $a\,b^+/a\,b$, 10% $a^+ b^+/a\,b$, 10% $a\,b/a\,b$

(d) Repulsion linkage phase

7.8 **(e)** F_2 from intercross:

		Sperm			
		40% $a^+ b$	40% $a\ b^+$	10% $a^+ b^+$	10% $a\ b$
	40% $a^+ b$	16% $a^+ b/a^+ b$	16% $a^+ b/a\ b^+$	4% $a^+ b/a^+ b^+$	4% $a^+ b/a\ b$
Eggs	40% $a\ b^+$	16% $a\ b^+/a^+ b$	16% $a\ b^+/a\ b^+$	4% $a\ b^+/a^+ b^+$	4% $a\ b^+/a\ b$
	10% $a^+ b^+$	4% $a^+ b^+/a^+ b$	4% $a^+ b^+/a\ b^+$	1% $a^+ b^+/a^+ b^+$	1% $a^+ b^+/a\ b$
	10% $a\ b$	4% $a\ b/a^+ b$	4% $a\ b/a\ b^+$	1% $a\ b/a^+ b^+$	1% $a\ b/a\ b$

Summary of phenotypes:

a^+ and b^+	51%
a^+ and b	24%
a and b^+	24%
a and b	1%

7.9 Coupling heterozygotes $a^+ b^+/a\ b$ would produce the following gametes: 30% $a^+ b^+$, 30% $a\ b$, 20% $a^+ b$, 20% $a\ b^+$; repulsion heterozygotes $a^+ b/a\ b^+$ would produce the following gametes: 30% $a^+ b$, 30% $a\ b^+$, 20% $a^+ b^+$, 20% $a\ b$. In each case, the frequencies of the testcross progeny would correspond to the frequencies of the gametes.

7.10 No. The leaf color and tassel seed traits assort independently.

7.11 Yes. Recombination frequency $= \dfrac{24 + 26}{126 + 24 + 26 + 124} = 0.167.$

Cross:

$\dfrac{b\ vg}{b^+ vg^+}$ female $\times$ $\dfrac{b\ vg}{b\ vg}$ male

$\dfrac{b\ vg}{b^+ vg^+}$	$\dfrac{b\ vg}{b\ vg}$	$\dfrac{b\ vg}{b^+ vg}$	$\dfrac{b\ vg}{b\ vg^+}$
126	124	24	26

7.12 Yes. Recombination frequency = $\dfrac{23 + 26}{23 + 127 + 124 + 26}$ = 0.163.

Cross:

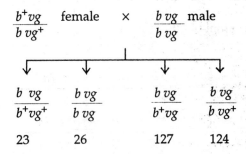

$\dfrac{b^+ vg}{b\ vg^+}$ female $\times$ $\dfrac{b\ vg}{b\ vg}$ male

$\dfrac{b\ vg}{b^+vg^+}$	$\dfrac{b\ vg}{b\ vg}$	$\dfrac{b\ vg}{b^+vg}$	$\dfrac{b\ vg}{b\ vg^+}$
23	26	127	124

7.13 Yes. Recombination frequency is estimated by the frequency of black offspring among the colored offspring: $\dfrac{34}{66 + 34}$ = 0.34. Cross:

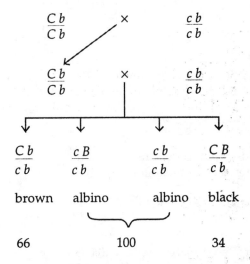

$\dfrac{C\ b}{C\ b}$ $\times$ $\dfrac{c\ b}{c\ b}$

$\dfrac{C\ b}{C\ b}$ $\times$ $\dfrac{c\ b}{c\ b}$

$\dfrac{C\ b}{c\ b}$	$\dfrac{c\ B}{c\ b}$	$\dfrac{c\ b}{c\ b}$	$\dfrac{C\ B}{c\ b}$
brown	albino	albino	black
66	100		34

7.14 Plant I has the genotype $D\ P/d\ p$, and when crossed to a $d\ p/d\ p$ plant, produces four classes of progeny:

$\dfrac{D\ P}{d\ p}$	$\dfrac{d\ p}{d\ p}$	$\dfrac{D\ p}{d\ p}$	$\dfrac{d\ p}{d\ p}$
81	79	22	17

Plant II has the genotype $D\ p/d\ P$, and when crossed to a $d\ p/d\ p$ plant, produces four classes of progeny:

$\dfrac{D\ p}{d\ p}$	$\dfrac{d\ P}{d\ p}$	$\dfrac{D\ P}{d\ p}$	$\dfrac{d\ p}{d\ p}$
21	18	5	4

7.14 (continued)

If the two plants are crossed ($D\,P/d\,p \times D\,p/d\,P$), the phenotypes of the offspring can be predicted from the following table.

Gametes from plant I

		$D\,P$ 0.40	$d\,p$ 0.40	$D\,p$ 0.10	$d\,P$ 0.10
Gametes	$D\,p$ 0.40	$D\,p/D\,P$ 0.16	$D\,p/d\,p$ 0.16	$D\,p/D\,p$ 0.04	$D\,p/d\,P$ 0.04
from	$d\,P$ 0.40	$d\,P/D\,P$ 0.16	$d\,P/d\,p$ 0.16	$d\,P/D\,p$ 0.04	$d\,P/d\,P$ 0.04
plant II	$D\,P$ 0.10	$D\,P/D\,Pa^+$ 0.04	$D\,P/d\,p$ 0.04	$D\,P/D\,p$ 0.01	$D\,P/d\,P$ 0.01
	$d\,p$ 0.10	$d\,p/D\,P$ 0.04	$d\,p/d\,p$ 0.04	$d\,p/D\,p$ 0.01	$d\,p/d\,P$ 0.01

Summary of phenotypes
tall, spherical   0.54
tall, pear   0.21
dwarf, spherical   0.21
dwarf, pear   0.04

7.15 Because the two chromosomes assort independently, the genetic makeup of the gametes (and, therefore, of the backcross progeny) can be obtained from the following table.

Chromosome 3 in gametes

		$c\,d$ 0.425	c^+d^+ 0.425	$c\,d^+$ 0.075	c^+d 0.075
Chromo-	$a\,b$	$a\,b\,c\,d$	$a\,b\,c^+d^+$	$a\,b\,c\,d^+$	$a\,d\,c^+d$
some 2	0.40	0.17	0.17	0.03	0.03
in	a^+b^+	$a^+b^+c\,d$	$a^+b^+c^+d^+$	$a^+b^+c\,d^+$	$a^+b^+c^+d$
gametes	0.40	0.17	0.17	0.03	0.03
	a^+b	$a^+b\,c\,d$	$a^+b\,c^+d^+$	$a^+b\,c\,d^+$	$a^+b\,c^+d$
	0.10	0.0425	0.0425	0.0075	0.0075
	$a\,b^+$	$a\,b^+c\,d$	$a\,b^+c^+d^+$	$a\,b^+c\,d^+$	$a\,b^+c^+d$
	0.10	0.0425	0.0425	0.0075	0.0075

7.16 **(a)** The F$_1$ females, which are $sr\ e^+/sr^+\ e$, produce four types of gametes: 46% $sr\ e^+$, 46% $sr^+\ e$, 4% $sr\ e$, 4% $sr^+\ e^+$.

(b) The F$_1$ males, which have the same genotype as the F$_1$ females, produce two types of gametes: 50% $sr\ e^+$, 50% $sr^+\ e$; remember, there is no recombination in *Drosophila* males.

(c) 46% striped, gray; 46% unstriped, ebony; 4% striped, ebony; 4% unstriped, gray.

(d) The offspring from the intercross can be obtained from the following table.

		Sperm	
		$sr\ e^+$ 0.50	$sr^+\ e$ 0.50
Eggs	$sr\ e^+$ 0.46	$sr\ e^+/sr\ e^+$ 0.23	$sr\ e^+/sr^+\ e$ 0.23
	$sr^+\ e$ 0.46	$sr^+\ e/sr\ e^+$ 0.23	$sr^+\ e/sr^+\ e$ 0.23
	$sr\ e$ 0.04	$sr\ e/sr\ e^+$ 0.002	$sr\ e/sr^+\ e$ 0.002
	$sr^+\ e^+$ 0.04	$sr^+\ e^+/sr\ e^+$ 0.002	$sr^+\ e^+/sr^+\ e$ 0.002

Summary of phenotypes

striped, gray	0.25
unstriped, gray	0.50
striped, ebony	0
unstriped, ebony	0.25

7.17 **(a)** The F$_1$ females, which are $cn\ vg^+/cn^+\ vg$, produce four types of gametes: 45% $cn\ vg^+$, 45% $cn^+\ vg$, 5% $cn^+\ vg^+$, 5% $cn\ vg$.

(b) 45% cinnabar eyes, normal wings; 45% reddish-brown eyes, vestigial wings; 5% reddish-brown eyes, normal wings; 5% cinnabar eyes, vestigial wings.

7.18 In the enumeration below, classes 1 and 2 are parental types, classes 3 and 4 result from a single crossover between *st* and *ss*, classes 5 and 6 result from a single crossover between *ss* and *e*, and classes 7 and 8 result from a double crossover, with one of the exchanges between *st* and *ss* and the other between *ss* and *e*.

7.18 (continued)

Class	Phenotypes	(a) Frequency with no interference	(b) Frequency with complete interference
1	scarlet, spineless	0.3784	0.37
2	ebony	0.3784	0.37
3	scarlet, ebony	0.0616	0.07
4	spineless	0.0616	0.07
5	scarlet, spineless, ebony	0.0516	0.06
6	wild-type	0.0516	0.06
7	scarlet	0.0084	0
8	spineless, ebony	0.0084	0

7.19 In the enumeration below, classes 1 and 2 are parental types, classes 3 and 4 result from a single crossover between *Pl* and *Sm*, classes 5 and 6 result from a single crossover between *Sm* and *Py*, and classes 7 and 8 result from a double crossover, with one of the exchanges between *Pl* and *Sm* and the other between *Sm* and *Py*.

Class	Phenotypes	(a) Frequency with no interference	(b) Frequency with complete interference
1	Purple, salmon, pigmy	0.405	0.40
2	green, yellow, normal	0.405	0.40
3	purple, yellow, normal	0.045	0.05
4	green, salmon, pigmy	0.045	0.05
5	purple, salmon, normal	0.045	0.05
6	green yellow pigmy	0.045	0.05
7	purple, yellow, pigmy	0.005	0
8	green, salmon, normal	0.005	0

7.20 In the enumeration below, classes 1 and 2 are parental types, classes 3 and 4 result from crossing over between Tu and J2, and classes 5 and 6 result from crossing over between J2 and Gl3; only the chromosome from the triply heterozygous F_1 plant is shown. Because interference is complete, there are no double crossover progeny.

Class	Genotype	Frequency
1	*tu j2 gl3*	0.445
2	*Tu J2 Gl3*	0.445
3	*tu J2 Gl3*	0.025
4	*Tu j2 gl3*	0.025
5	*tu j2 Gl3*	0.030
6	*Tu J2 gl3*	0.030

7.21 The double crossover classes, which are the two that were not observed, establish that the gene order is y—w—ec. Thus, the F_1 females had the genotype $y\ w\ ec/+\ +\ +$. The distance between y and w is estimated by the frequency of recombination between these two genes: $\dfrac{8+7}{1000} = 0.015$; similarly, the distance between w and ec is $\dfrac{18+23}{1000} = 0.041$. Thus, the genetic map for this segment of the X chromosome is y—1.5 cM—w—4.5 cM—ec.

7.22 The yellow, bar and vermilion classes, with a total of 50 progeny, result from double crossovers. Thus, the order of the genes is y--v--B, and the F_1 females had the genotype $y\ v\ B/+\ +\ +$. The distance between y and v is the average number of crossovers between them: $\dfrac{244+50}{1000} = 29.4$ cM; likewise, the distance between v and B is $\dfrac{160+50}{1000} = 21.0$ cM. Thus, the genetic map is y--—29.4 cM—v--21.0 cM—B.

7.23 (a) Two of the classes (the parental types) vastly outnumber the other six classes (recombinant types)

(b) $st\ +\ +/+\ ss\ e$

(c) st—ss—e

(d) $\dfrac{(145 + 122) \times 1 + (18) \times 2}{1000} = 30.3$ cM

(d) $\dfrac{122 + 18}{1000} = 14.0$ cM

(e) $\dfrac{0.018}{0.163 \times 0.140} = 0.789$

(f) $st\ ++/+\ ss\ e$ females $\times$ $st\ ss\ e\ /\ st\ ss\ e$ males $\rightarrow$ 2 parental classes and 6 recombinant classes.

7.24 The female will produce four kinds of gametes: 30% $w\ +$, 30% $+\ dor$, 20% $w\ dor$, and 20% $+\ +$; thus, 80% of the progeny will be mutant (either white or deep orange), and 50% will be pigmented (either red or deep orange).

7.25 Ignore the female progeny and base the map on the male progeny. The parental types are $++\ z$ and $x\ y\ +$. The two missing classes ($+\ y\ +$ and $x\ +\ z$) must represent double crossovers; thus, the gene order is y—x—z. The distance between y and x is $\dfrac{32 + 27}{1000} = 5.9$ cM and that between x and z is $\dfrac{31 + 39}{1000} = 7.0$ cM. Thus, the map is y—5.9 cM—x—7.0 cM—z. The coefficient of coincidence is zero.

7.26 The pedigree, with genotypes, is

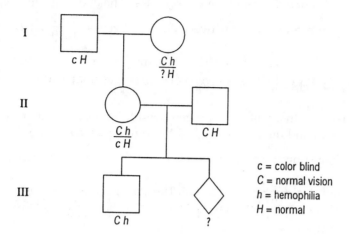

For the second child, III-2, the probability that it will have hemophilia is $\frac{1}{2}$ x $\frac{1}{2}$ = $\frac{1}{4}$; the probability that it will have color blindness is $\frac{1}{2}$ x $\frac{1}{2}$ = $\frac{1}{4}$; the probability that it will have both conditions is $\frac{0.1}{2}$ x $\frac{1}{2}$ = 0.025;

the probability that it will have neither condition is $\frac{0.1}{2}$ x $\frac{1}{2}$ = 0.025.

7.27 $\left(\dfrac{P}{2}\right)^2$

7.28 5

7.29 From the parental classes, $+ + c$ and $a\,b\,+$, the heterozygous females must have had the genotype $+ + c / a\,b\,+$. The missing classes, $+\,b\,+$ and $a + c$, which would represent double crossovers, establish that the gene order is b—a—c. The distance between b and a is $\dfrac{96 + 110}{1000}$ = 20.6 cM and that between a and c is $\dfrac{65 + 75}{1000}$ = 14.0 cM.

Thus, the genetic map is b—20.6 cM—a—14.0 cM—c.

7.30 2.4 chiasmata

7.31 5.4 chiasmata

7.32 100 cM

7.33 M2 carries an inversion that suppresses recombination in the chromosome.

7.34 A two-strand double crossover within the inversion; the exchange points of the double crossover must lie between the genetic markers and the inversion breakpoints.

CHAPTER 8

8.1 PD >> NPD, so the genes are linked; the distance between the genes is estimated as

$$\frac{\frac{1}{2} \times 23 + 3}{48} = 30 \text{ cM.}$$

8.2 The *pr* gene is on chromosome 3.

8.3 The distance is half the frequency of second division segregation asci: $\frac{1}{2} \times \frac{84}{200} = 21$ cM.

8.4 Distance between *a* and *b* is $\dfrac{\frac{1}{2} \times (220 + 14)}{2000} = 5.85$ cM; distance between *a* and centromere is $\frac{1}{2} \times \frac{220}{2000} = 5.5$ cM; distance between *b* and centromere is $\frac{1}{2} \times \frac{14}{2000} = 0.35$ cM. Thus, the genetic map is *a*—5.5 cM—centromere—0.35 cM—*b*.

8.5 The *arg* and *thi* loci are unlinked; however, the *arg* and *leu* loci are linked. The distance between *arg* and *leu* is $\dfrac{\frac{1}{2} \times 44}{200} = 11$ cM; the distance between *arg* and its centromere is $\frac{1}{2} \times \frac{1}{300} = 0.17$ cM; thus the genetic map for the chromosome that carries *arg* and *leu* is centromere—0.17 cM—*arg*—11 cM—*leu*. The *thi* gene is very tightly linked to its centromere.

8.6 (a) 2 crossovers

   (b) one exchange between *x* and *y*, the other between *y* and *z*

   (c) 4-strand double crossover.

8.7 The exceptional females result from crossing over between the genes and the centromere. The *y* locus is farther away from the centromere than the *sn* locus.

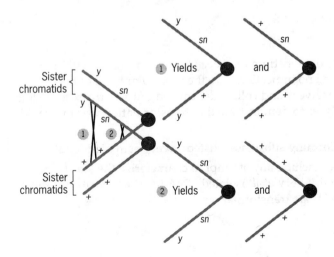

8.8 Band 7

8.9 CH/ch (c = color blind, C = normal vision; h = hemophilia, H = normal)

8.10 Ch/cH

8.11 40% Ab, 40% aB, 10% AB, 10% ab

8.12 ABc/abC

8.13 Chromosome 18

8.14 $G6PD$, $HPRT$, and PGK are in Xq distal to the translocation breakpoint; NP is in 14q proximal to the translocation breakpoint.

8.15 ACP is in 2p distal to the translocation breakpoint.

8.16 The fragile site and the HPA gene are separated by 8 cM.

8.17 **I)** A—10 cM—B—6 cM—C
     II) B—10 cM—A—16 cM—C

8.18 I

8.19 Chromosome 10

8.20 **(a)** $(0.25)^6 = 0.000244$
     (b) $(0.5)^6 = 0.015625$
     (c) 1.806
     (d) The sum of the lod scores for complete linkage is greater than 3; thus, the two genes are tightly linked.

CHAPTER 9

9.1 **(a)** Griffith's *in vivo* experiments demonstrated the occurrence of transformation in pneumococcus. They provided no indication as to the molecular basis of the transformation phenomenon. Avery and colleagues carried out *in vitro* experiments, employing biochemical analyses to demonstrate that transformation was mediated by DNA.

  (b) Griffith showed that a transforming substance existed; Avery et al. defined it as DNA

  (c) Griffith's experiments did not include any attempt to characterize the substance responsible for transformation. Avery *et al.* isolated DNA in "pure" form and demonstrated that it could mediate transformation.

9.2 **(a)** No effect

(b) no effect

(c) DNase will destroy the capacity of the extract to transform type IIR cells to Type IIIS by degrading the DNA in the extract. Protease and RNase will degrade the proteins and RNA, respectively, in the extract. They will have no effect, since the proteins and RNA are not involved in transformation.

9.3 Purified DNA from Type III cells was shown to be sufficient to transform Type II cells. This occurred in the absence of any dead Type III cells.

9.4 About $\frac{1}{2}$ protein, $\frac{1}{2}$ DNA. A single long molecule of DNA is enclosed within a complex "coat" composed of many proteins.

9.5 DNA contains phosphorus (normally ^{31}P) but no sulfur; it can be labeled with ^{32}P. Proteins contain sulfur (normally ^{32}S) but usually no phosphorus; they can be labeled with ^{35}S.

9.6 **(a)** The objective was to determine whether the genetic material was DNA or protein.

(b) By labeling phosphorus, a constituent of DNA, and sulfur, a constituent of protein, in a virus, it was possible to demonstrate that only the labeled phosphorus was introduced into the host cell during the viral reproductive cycle. The DNA was enough to produce new phages.

(c) Therefore DNA, not protein, is the genetic material.

9.7 **(a)** The ladderlike pattern was known from X-ray diffraction studies. Chemical analyses had shown that a 1:1 relationship existed between the organic bases adenine and thymine and between cytosine and guanine. Physical data concerning the length of each spiral and the stacking of bases were also available.

(b) Watson and Crick developed the model of a double helix, with the rigid strands of sugar and phosphorus forming spirals around an axis, and hydrogen bonds connecting the complementary bases in nucleotide pairs.

9.8 **(a)** A multistranded, spiral structure was suggested by the X-ray diffraction patterns. A double-stranded helix with specific base-pairing nicely fits the 1:1 stoichiometry observed for A:T and G:C in DNA.

(b) Use of the known hydrogen-bonding potential of the bases provided a means of holding the two complementary strands in a stable configuration in such a double helix.

9.9 **(a)** 400,000

(b) 20,000

(c) 400,000

(d) 68,000 nm

9.10 3'-C A G T A C T G-5'.

9.11 **(a)** DNA has one atom less of oxygen than RNA in the sugar part of the molecule. In DNA, thymine replaces the uracil that is present in RNA. (In certain bacteriophages, DNA-containing uracil is present.) DNA is most frequently double-stranded, but bacteriophages such as φX174 contain single-stranded DNA. RNA is most frequently single-stranded. Some viruses, such as the Reoviruses, however, contain double-stranded RNA chromosomes.

9.12 13%.

9.13 No. TMV RNA is single-stranded. Thus the base-pair stoichiometry of DNA does not apply.

9.14 **(a)** False
(b) false
(c) true
(d) true
(e) true
(f) true
(g) false
(h) true
(I) true
(j) false
(k) true
(l) false
(m) true.

9.15 **(1)** The nucleosome level; the core containing an octamer of histones plus 146 nucleotide pairs of DNA arranged as $1\frac{3}{4}$ turns of a supercoil (see Figure 9.25), yielding an approximately 10 nm diameter spherical body; or juxtaposed, a roughly 10 nm diameter fiber.
(2) The 30 nm fiber observed in condensed mitotic and meiotic chromosomes; it appears to be formed by coiling or folding the 10-nm nucleosome fiber.
(3) The highly condensed mitotic and meiotic chromosomes (for example, metaphase chromosomes); the tight folding or coiling maintained by a "scaffold" composed of nonhistone chromosomal proteins (see Figure 9.28).

9.16 In the diploid nucleus of *D. melanogaster*, 10^6 nucleosomes would be present; these would contain 2×10^6 molecules of each histone, H2a, H2b, H3, and H4.

9.17 The satellite DNA fragments would renature much more rapidly than the main-band DNA fragments. In *D. virilus* satellite DNAs, all three have repeating heptanucleotide-pair sequences. Thus essentially every 40 nucleotide-long (average) single-stranded fragment from one strand will have a sequence complementary (in part) with every single-stranded fragment from the complementary strand. Many of the nucleotide-pair sequences in main-band DNA will be unique sequences (present only once in the genome).

9.18 It indicates that highly repetitive DNA sequences do not contain structural genes specifying RNA and polypeptide gene products.

9.19 **(a)** **(1)** euchromatin
(2) euchromatin
(3) heterochromatin
(b) **(1)** yes
(2) no

9.19 **(c)** **(1)** Most of the single-copy DNA sequences are believed to be structural genes encoding proteins: structural proteins and the vast repertoire of enzymes employed by living organisms

(2) Essentially nothing is known regarding the functions of the highly repetitive DNA sequences—your hypotheses are probably as valid as anyone else's.

(d) Some moderately repetitive DNA sequences specify products such as ribosomal RNA molecules that are required by cells in large quantities. Others are believed to be binding sites for proteins that regulate gene expression and replication of the multiple replicons in the giant DNA molecules of eukaryotic chromosomes. Some moderately repetitive sequences probably play structural roles in chromosomes, especially during the condensations of mitosis and meiosis. Others undoubtedly carry out functions that are still unknown.

9.20 Interphase. Chromosomes are for the most part metabolically inactive (very little transcription) during the various stages of condensation in mitosis and meiosis.

9.21 **(a)** **(1)** Centromeres function as spindle-fiber attachment sites on chromosomes; they are required for the separation of homologous chromosomes to opposite poles of the spindle during anaphase I of meiosis and for the separation of sister chromatids during anaphase of mitosis and anaphase II of meiosis.

(2) Telomeres provide at least three important functions: (i) prevention of exonucleolytic degradation of the ends of the linear DNA molecules in eukaryotic chromosomes, (ii) prevention of the fusion of ends of DNA molecules of different chromosomes, and (iii) provision of a mechanism for replication of the distal tips of linear DNA molecules in eukaryotic chromosomes.

(b) Yes. Most telomeres studied to date contain DNA sequence repeat units (for example, TTAGGG in human chromosomes), and, at least in some species, telomeres terminate with single-stranded 3' overhangs that form "hairpin" structures. The bases in these hairpins exhibit unique patterns of methylation that presumably contribute to the structure and stability of telomeres.

(c) Telomerase adds the terminal DNA sequences or telomeres to the linear chromosomes in eukaryotes.

(d) The broken ends resulting from irradiation will not contain telomeres; as a result, the free ends of the DNA molecules are apparently subject to the activities of enzymes such as exonucleases, ligases, and the like, which modify the ends. They can regain stability by fusing to broken ends of other DNA molecules that contain terminal telomere sequences.

9.22 Viscoelastometry is a procedure used to measure the viscosity of molecules in solution. In addition, viscoelastometric methods can be used to estimate the sizes of the largest DNA molecules present in aqueous solutions. By using viscoelastometry, scientists have obtained evidence which indicates that all of the DNA present in chromosomes of eukaryotes exists as giant, "chromosome-size" DNA molecules (one huge DNA molecule per chromosome). These data eliminated early models of chromosome structure with multiple DNA molecules joined end-to-end by protein or RNA "linkers."

9.23 **(a)** Two. The axial region of a "lampbrush" chromosome contains the two chromatids of one homologous chromosome (postreplication).

(b) One. Each lateral loop of a "lampbrush" chromosome is a segment of a single chromatid.

9.24 Nonhistone chromosomal proteins. The "scaffold" structures of metaphase chromosomes can be observed by light microscopy after removal of the histones by differential extraction procedures.

9.25 **(a)** Histones have been highly conserved throughout the evolution of eukaryotes. A major function of histones is to package DNA into nucleosomes and chromatin fibers. Since DNA is composed of the same four nucleotides and has the same basic structure in all eukaryotes, one might expect that the proteins that play a structural role in packaging this DNA would be similarly conserved.

(b) The nonhistone chromosomal proteins exhibit the greater heterogeneity in chromatin from different tissues and cell types of an organism. The histone composition is largely the same in all cell types within a given species—consistent with the role of histones in packaging DNA into nucleosomes. The nonhistone chromosomal proteins include proteins that regulate gene expression. Because different sets of genes are transcribed in different cell types, one would expect heterogeneity in some of the nonhistone chromosomal proteins of different tissues.

9.26 The rate of DNA renaturation is proportional to the square of the concentration of single strands (C^2), because renaturation requires a collision between and a reassociation of two complementary molecules (single strands) that are present in the reaction mixture in *identical concentrations*. These single strands were originally present in the same double helix. Thus, the reaction depends on the concentrations of both strands (with both concentrations equal, therefore, $C \times C = C^2$), whereas the rates of most other bimolecular reactions are a function of the concentrations (often different, therefore, $C_1 \times C_2$) of both reacting substances. The fact that DNA renaturation rates are proportional to C^2 is thus a direct consequence of the double-helix structure of DNA with each double helix composed of two complementary single strands.

9.27 The 6 percent of the human DNA that is already renatured at $t = 0$ in standard DNA renaturation experiments results from the presence of single strands that themselves contain complementary sequences with opposite chemical polarity (one sequence reading 5' to 3' complementary to another sequence reading 3' to 5'). Single strands containing such complementary sequences form double-stranded "hairpin" or "foldback" structures. Such reactions are concentration independent, because collisions between two molecules are not required for the renaturation events to occur. Thus they occur very fast (too fast to be measured in the standard renaturation experiments) with unimolecular reaction kinetics. Some DNA sequences that exhibit unimolecular renaturation kinetics are called palindromes; such DNAs contain sequences that are the same when read in opposite direction starting from a central point of symmetry (Chapter 19).

9.28 **(a)** One μg of human DNA will contain, on average, 3.04×10^5 copies of the genome. Using an average molecular weight per nucleotide pair of 660, the "molecular" weight of the entire human genome is 1.98×10^{12} ($3 \times 10^9 \times 660$). Thus, 1.98×10^{12} grams (1 "mole" = number of grams equivalent to the "molecular" weight) of human DNA will contain, on average, 6.02×10^{23} molecules [Avogadro's number = number of molecules (here, copies of the genome) present in one "mole" of a substance]. One gram will contain, on average,

$$(3.04 \times 10^{11})\left(\frac{6.02 \times 10^{23}}{1.98 \times 10^{12}}\right) \text{ copies of the genome; thus, 1 μg will contain, on average,}$$

3.04×10^5 copies of the human genome.

(b) One copy of the human genome weighs approximately

$$(3.3 \times 10^{-12} \text{ g})\left(\frac{1.98 \times 10^{12} \text{ g per "mole"}}{6.02 \times 10^{23} \text{ molecules per "mole"}}\right) \text{ or } 3.3 \times 10^{-6} \text{ μg.}$$

(c) By analogous calculations, 1 μg of *Arabidopsis thaliana* DNA contains, on average, 1.18×10^7 copies of the genome.

9.28 **(d)** Similarly, one copy of the *A. thaliana* genome weighs approximately 8.4×10^{-8} μg.

(e) In carrying out molecular analyses of the structures of genomes, geneticists frequently need to know how many copies of a genome are present, on average, in a given quantity of DNA.

CHAPTER 10

10.1 **(a)** **(i)** One-half of the DNA molecules with ^{15}N in both strands and $\frac{1}{2}$ with ^{14}N in both strands

(ii) all DNA molecules with one strand containing ^{15}N and the complementary strand containing ^{14}N

(iii) all DNA molecules with both strands containing roughly equal amounts of ^{15}N and ^{14}N.

(b) **(i)** $\frac{1}{4}$ of the DNA molecules with ^{15}N in both strands and $\frac{3}{4}$ with ^{14}N in both strands

(ii) $\frac{1}{2}$ of the DNA molecules with one strand containing ^{15}N and the complementary strand containing ^{14}N and the other $\frac{1}{2}$ with ^{14}N in both strands

(iii) all DNA molecules with both strands containing about $\frac{1}{4}$ ^{15}N and $\frac{3}{4}$ ^{14}N.

See Figure 10.2.

10.2 One-half of the DNA molecules fully heavy (^{15}N in both strands); the other half of the molecules "hybrid" (^{15}N in one strand, ^{14}N in the complementary strand).

10.3 **(a)** Both $3' \rightarrow 5'$ and $5' \rightarrow 3'$ exonuclease activities.

(b) The $3' \rightarrow 5'$ exonuclease "proofreads" the nascent DNA strand during its synthesis. If a mismatched base pair occurs at the 3'-OH end of the primer, the $3' \rightarrow 5'$ exonuclease removes the incorrect terminal nucleotide before polymerization proceeds again. The $5' \rightarrow 3'$ exonuclease is responsible for the removal of RNA primers during DNA replication and functions in pathways involved in the repair of damaged DNA (see Chapter 13).

(c) Yes, both exonuclease activities appear to be very important. Without the $3' \rightarrow 5'$ proofreading activity during replication, an intolerable mutation frequency would occur. The $5' \rightarrow 3'$ exonuclease activity is essential to the survival of the cell. Conditional mutations that alter the $5' \rightarrow 3'$ exonuclease activity of DNA polymerase I are lethal to the cell under conditions where the exonuclease is nonfunctional.

10.4 **(a)**

3' Ⓟ-TGCGAATTAGCGACAT-Ⓟ 5'
5' Ⓟ-ATCGGTACGACGCTTAATCGCTGTA-OH 3';

note that DNA synthesis will *not* occur on the left end since the 3'-terminus of the potential primer strand is blocked with a phosphate group—all DNA polymerases require a free 3'-OH terminus.

(b) The first step will be the removal of the mismatched C (exiting as dCMP) from the 3'-OH primer terminus by the 3' → 5' exonuclease ("proofreading") activity.

10.5 If nascent DNA is labeled by exposure to ^{3}H-thymidine for very short periods of time, continuous replication predicts that the label would be incorporated into chromosome-sized DNA molecules, whereas discontinuous replication predicts that the label would first appear in small pieces of nascent DNA (prior to covalent joining, catalyzed by polynucleotide ligase).

10.6

PROTEINS	FUNCTIONS
1. DNA polymerase III	1. (a) Catalyzes polymerization (covalent extension) of new DNA chains. (b) The 3' → 5' exonuclease activity "proofreads" the product, removing any mismatched base-pairs at the 3' end of the primer strand.
2. DNA polymerase I	2. Removes the RNA primers (5' → 3' exonuclease activity) and replaces them with new DNA strands (5' → 3' polymerase activity).
3. DNA ligase	3. Catalyzes covalent joining of "Okazaki fragments."
4. Primase (*dnaG* protein)	4. Catalyzes RNA primer synthesis.
5. DNA gyrase	5. Catalyzes the formation of negative supercoils; facilitates unwinding?
6. DNA helicase	6. Catalyzes unwinding.
7. DNA single-strand binding protein	7. Maintains an "extended" single-stranded template; aids unwinding?
8. *dnaB* protein	8. Required for initiation of replication.
9. Proteins i, n, n', *dnaC* protein	9. "Prepriming"—required prior to initiation of primer synthesis.
10. *dnaI, dnaJ*, etc., proteins	10. Required; but functions unknown.

10.7

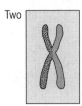 Two    Plus two   For both the large and small chromosomes

10.8

 Two    Plus two   For both the large and small chromosomes.

10.9 That DNA replication was unidirectional rather than bidirectional. As the intracellular pools of radioactive ^{3}H-thymidine are gradually diluted after transfer to nonradioactive medium, less and less ^{3}H-thymidine will be incorporated into DNA at each replicating fork. This will produce autoradiograms with tails of decreasing grain density at each growing point. Since such tails appear at only one end of each track, replication must be unidirectional. Bidirectional replication would produce such tails at both ends of an autoradiographic track (see Figure 10.29).

10.10 Interphase. Chromosomes are for the most part metabolically inactive (very little transcription) during the various stages of condensation in mitosis and meiosis.

10.11 DNA polymerases α, β, δ, and ε are located in the nuclei of cells; polymerase γ is located in mitochondria and chloroplasts. Current evidence suggests that polymerases α and δ are both required for the replication of nuclear DNA. Polymerase δ is thought to catalyze the continuous synthesis of the leading strand, and polymerase α is believed to catalyze discontinuous synthesis of the lagging strand because it contains the primase activity required for the repeated initiation of "Okazaki fragments." Polymerase γ presumably catalyzes replication of organellar chromosomes. Polymerases β and ε function in DNA repair pathways like DNA polymerase I of *E. coli*.

10.12 (a) Given bidirectional replication of a single replicon, each replication fork must traverse 2×10^6 nucleotide pairs in *E. coli* and 3×10^7 nucleotide pairs in the largest *Drosophila* chromosome. If the rates were the same in both species, it would take 15 times $\dfrac{3 \times 10^7}{2 \times 10^6}$ as long to replicate the *Drosophila* chromosome or 10 hours (40 minutes $\times$ 15 = 600 minutes).

(b) If replication forks in *E. coli* move 20 times as fast as replication forks in *Drosophila* (100,000 nucleotide pairs per minute/ 5,000 nucleotide pairs per minute), the largest *Drosophila* chromosome would require 8.3 days (10 hours $\times$ 20 = 200 hours) to complete one round of replication.

(c) Each *Drosophila* chromosome must contain many replicons in order to complete replication in less than 10 minutes.

10.13 Sucrose velocity density gradient centrifugation is the standard technique for separating DNA molecules in this size range. Pulsed-field gel electrophoresis (Chapter 9) could also be used.

10.14 (a) and (b)

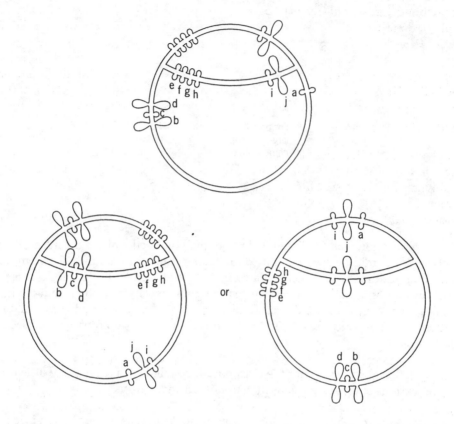

or

10.15 (a) DNA gyrase

(b) primase

(c) the 5' → 3' exonuclease activity of DNA polymerase I

(d) the 5' → 3' polymerase activity of DNA polymerase III

(e) the 3' → 5' exonuclease activity of DNA polymerase III.

10.16 In eukaryotes, the rate of DNA synthesis at each replication fork is about 2,500 to 3,000 nucleotide pairs per minute. Large eukaryotic chromosomes often contain 10^7 to 10^8 nucleotide pairs. A single replication fork could not replicate the giant DNA in one of these large chromosomes fast enough to permit the observed cell generation times.

10.17 The 5' → 3' exonuclease activity of DNA polymerase I is essential to the survival of the bacterium, whereas the 5' → 3' polymerase activity of the enzyme is not essential.

10.18 No. *E. coli* strains carrying *polA* mutations that eliminate the 3' → 5' exonuclease activity of DNA polymerase I will exhibit unusually high mutation rates.

10.19 Because A:T base pairs are held together by only two hydrogen bonds instead of the three hydrogen bonds present in G:C base pairs, the two strands of A:T-rich regions of double helices are separated more easily, providing the single-stranded template regions required for DNA replication.

10.20 Rolling-circle replication begins when an endonuclease cleaves one strand of a circular DNA double helix. This cleavage produces a free 3'-OH on one end of the cut strand, allowing it to function as a primer. The discontinuous synthesis of the lagging strand requires the *de novo* initiation of each Okazaki fragment, which requires DNA primase activity.

10.21 DNA polymerase III does not have a 5' → 3' exonuclease activity that acts on double-stranded nucleic acids. Thus it cannot excise RNA primer strands from replicating DNA molecules. DNA polymerase I is present in cells at much higher concentrations and functions as a monomer. Thus DNA polymerase I is able to catalyze the removal of RNA primers from the vast number of Okazaki fragments formed during the discontinuous replication of the lagging strand.

10.22 DNA helicase unwinds the DNA double helix, and single-strand DNA-binding protein coats the unwound strands, keeping them in an extended state. DNA gyrase catalyzes the formation of negative supercoiling in E. coli DNA, and this negative supercoiling behind the replication forks is thought to drive the unwinding process because superhelical tension is reduced by unwinding the complementary strands.

10.23 DNA polymerase I is a single polypeptide of molecular weight 109,000, whereas DNA polymerase III is a complex multimeric protein. The DNA polymerase holoenzyme has a molecular mass of about 900,000 daltons and is composed of at least 20 different polypeptides. The *dnaN* gene product, the β subunit of DNA polymerase III, forms a dimeric clamp that encircles the DNA molecule and prevents the enzyme from dissociating from the template DNA during replication.

10.24 DnaA protein initiates the formation of the replication bubble by binding to the 9-bp repeats of *OriC*. DnaA protein is known to be required for the initiation process because bacteria with temperature-sensitive mutations in the *dnaA* gene cannot initiate DNA replication at restrictive temperatures.

10.25 The primosome is a protein complex that initiates the synthesis of Okazaki fragments during lagging strand synthesis. The major components of the *E. coli* DNA primosome are DNA primase and DNA helicase. Geneticists have been able to show that both DNA primase and DNA helicase are required for DNA replication by demonstrating that mutations in the genes encoding these enzymes result in the arrest of DNA synthesis in mutant cells under conditions where the altered proteins are inactive.

10.26 Nucleosomes and replisomes are both large macromolecular structures, and the packaging of eukaryotic DNA into nucleosomes raises the question of how a replisome can move past a nucleosome and replicate the DNA in the nucleosome in the process. The most obvious solution to this problem would be to completely or partially disassemble the nucleosome to allow the replisome to pass. The nucleosome would then reassemble after the replisome had passed. One popular model has the nucleosome disassembling into two half-nucleosomes, allowing the replisome to move past it (see Figure 10.32b).

10.27 Grow *E. coli* cells for a few seconds in medium containing ^{3}H-thymidine, isolate total DNA from these cells, and determine the sizes of the radioactive DNA molecules by sucrose velocity density gradient centrifugation. If replication is continuous on one strand and discontinuous on the other strand, 50 percent of the radioactivity will be present in Okazaki fragments that are 1000 to 2000 nucleotides long and the other 50 percent will be present in large (chromosome-size) DNA molecules. If replication is discontinuous on both strands, all of the radioactivity will be present in Okazaki fragments.

10.28 The product of the first gene is required for DNA chain extension, whereas the product of the second gene is only required for the initiation of DNA synthesis.

10.29 (1)   DNA replication usually occurs continuously in rapidly growing prokaryotic cells but is restricted to the S phase of the cell cycle in eukaryotes.

(2)   Most eukaryotic chromosomes contain multiple origins of replication, whereas most prokaryotic chromosomes contain a single origin of replication.

(3)   Prokaryotes utilize two catalytic complexes that contain the same DNA polymerase to replicate the leading and lagging strands, whereas eukaryotes utilize two distinct DNA polymerases for leading and lagging strand synthesis.

(4)   Replication of eukaryotic chromosomes requires the partial disassembly and reassembly of nucleosomes as replisomes move along parental DNA molecules. In prokaryotes, replication probably involves a similar partial disassembly/reassembly of nucleosome-like structures.

(5)   Most prokaryotic chromosomes are circular and thus have no ends. Most eukaryotic chromosomes are linear and have unique termini called telomeres that are added to replicating DNA molecules by a unique, RNA-containing enzyme called telomerase.

10.30 (a)   2,000 to 4,000 Okazaki fragments.

(b)   300,000 to 600,000 Okazaki fragments.

10.31 The chromosomes of haploid yeast cells that carry the *est1* mutation become shorter during each cell division. Eventually, chromosome instability results from the complete loss of telomeres, and cell death occurs because of the deletion of essential genes near the ends of chromosomes.

CHAPTER 11

11.1 **(a)** RNA contains the sugar ribose, which has an hydroxyl (OH) group on the 2-carbon; DNA contains the sugar 2-deoxyribose, with only hydrogens on the 2-carbon. RNA usually contains the base uracil at positions where thymine is present in DNA. However, some DNAs contain uracil, and some RNAs contain thymine. DNA exists most frequently as a double helix (double-stranded molecule); RNA exists more frequently as a single-stranded molecule; but some DNAs are single-stranded and some RNAs are double-stranded.

(b) The main function of DNA is to store genetic information and to transmit that information from cell to cell and from generation to generation. RNA stores and transmits genetic information in some viruses that contain no DNA. In cells with both DNA and RNA,

   (1) mRNA acts as in intermediary in protein synthesis, carrying the information from DNA in the chromosomes to the ribosomes (sites at which proteins are synthesized)

   (2) tRNAs carry amino acids to the ribosomes and function in codon recognition during the synthesis of polypeptides

   (3) rRNA molecules are essential components of the ribosomes.

(c) DNA is located primarily in the chromosomes (with some in cytoplasmic organelles, such as mitochondria and chloroplasts), whereas RNA is located throughout cells.

11.2 3'—ACGUCUGU—5'

11.3 3'—GACTA—5'

11.4 The genetic information of cells is stored in DNA, which is located predominantly in the chromosomes. The gene products (polypeptides) are synthesized primarily in the cytoplasm on ribosomes. Some intermediate must therefore carry the genetic information from the chromosomes to the ribosomes. RNA molecules (mRNAs) were shown to perform this function by means of RNA pulse-labeling and pulse-chase experiments combined with autoradiography (see Fig. 11.6). The enzyme RNA polymerase was subsequently shown to catalyze the synthesis of mRNA using chromosomal DNA as a template. Finally, the mRNA molecules synthesized by RNA polymerase were shown to faithfully direct the synthesis of specific polypeptides when used in *in vitro* protein synthesis systems.

11.5 Protein synthesis occurs on ribosomes. In eukaryotes, most of the ribosomes are located in the cytoplasm and are attached to the extensive membranous network of endoplasmic reticulum. Some protein synthesis also occurs in cytoplasmic organelles such as chloroplasts and mitochondria.

11.6 Both prokaryotic and eukaryotic organisms contain messenger RNAs, transfer RNAs, and ribosomal RNAs. In addition, eukaryotes contain small nuclear RNAs. Messenger RNA molecules carry genetic information from the chromosomes (where the information is stored) to the ribosomes in the cytoplasm (where the information is expressed during protein synthesis). The linear sequence of triplet codons in an mRNA molecule specifies the linear sequence of amino acids in the polypeptides produced during translation of that mRNA.

11.6 (continued)

Transfer RNA molecules are small (about 80 nucleotides long) molecules that carry amino acids to the ribosomes and provide the codon-recognition specificity during translation. Ribosomal RNA molecules provide part of the structure and function of ribosomes; they represent an important part of the machinery required for the synthesis of polypeptides. Small nuclear RNAs are structural components of spliceosomes, which excise noncoding intron sequences from nuclear gene transcripts.

11.7 The entire nucleotide-pair sequences—including the introns—of the genes are transcribed by RNA polymerase to produce primary transcripts that still contain the intron sequences. The intron sequences are then spliced out of the primary transcripts to produce the mature, functional RNA molecules. In the case of protein-encoding nuclear genes of higher eukaryotes, the introns are spliced out by complex macromolecular structures called spliceosomes (see Figure 11.31).

11.8 "Self-splicing" of RNA precursors demonstrates that RNA molecules can also contain catalytic sites; this property is not restricted to proteins.

11.9 Spliceosomes excise intron sequences from nuclear gene transcripts to produce the mature mRNA molecules that are translated on ribosomes in the cytoplasm. Spliceosomes are complex macromolecular structures composed of snRNA and protein molecules (see Figure 11.32).

11.10 The introns of protein-encoding nuclear genes of higher eukaryotes almost invariably begin (5') with GT and end (3') with AG. In addition, the 3' subterminal A in the "TACTAAC box" is completely conserved; this A is involved in bond formation during intron excision.

11.11 (a) Sequence 5. It contains the conserved intron sequences: a 5' GU, a 3' AG, and a UACUAAC internal sequence providing a potential bonding site for intron excision. Sequence 4 has a 5' GU and a 3' AG, but contains no internal A for the bonding site during intron excision.

(b) 5'—UAGUCUCAA—3'; the putative intron from the 5' GU through the 3' AG has been removed.

11.12 This is a wide-open question at present! There is much speculation, but little hard evidence. One popular hypothesis is that introns enhance exon shuffling by increasing recombination events between sequences encoding adjacent domains of a polypeptide. Also, in one yeast mitochondrial gene, the introns contain open reading frames that encode "maturases" that splice out these introns—a neat negative feedback control. Other introns may be merely relics of evolution.

11.13

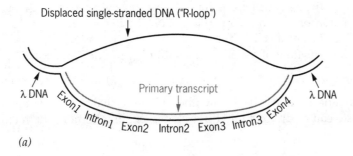

Displaced single-stranded DNA ("R-loop")

λ DNA   λ DNA

Primary transcript

Exon1 Intron1 Exon2 Intron2 Exon3 Intron3 Exon4

(a)

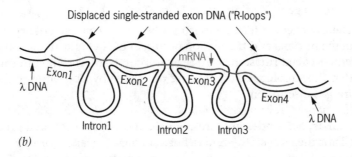

Displaced single-stranded exon DNA ("R-loops")

mRNA

λ DNA

Exon1   Exon2   Exon3

Exon4

λ DNA

Intron1   Intron2   Intron3

(b)

11.14 If there is a promoter located upstream from this DNA segment, the nucleotide sequence of this portion of the RNA transcript will be 5'-UACGAUGACGAUAAGCGACAUAGC-3'. If there is no upstream promoter, this segment of DNA will not be transcribed.

11.15 Assuming that there is a -35 sequence upstream from the consensus -10 sequence in this segment of the DNA molecule, the nucleotide sequence of the transcript will be 5'- ACCCGACAUAGCUACGAUGACGAUAAGCGACAUAGC-3'.

11.16 Given the consensus -35 and -10 sequences in this segment of DNA and the fact that transcripts almost always start with a purine, the predicted nucleotide sequence of the transcript is 5'-ACCCGACAUAGCUACGAUGACGAUA-3'.

11.17 Assuming that there is a CAAT box located upstream from the TATA box shown in this segment of DNA, the nucleotide sequence of the transcript will be 5'-ACCCGACAUAGCUACGAUGACGAUA-3'.

11.18 Given the vast amount of information that can be stored on a small computer chip by using a binary code, it is clear that large quantities of genetic information can be stored in the genomes of organisms by using the four-letter alphabet of the genetic code.

11.19 Although in theory it would be possible to produce six different enzyme activities by alternative splicing of one gene transcript or by combining the polypeptide products of two or three genes into different combinations with distinct enzyme activities, in reality these possibilities are unlikely. In living organisms, six enzymes are usually specified by at least six genes—with each gene encoding a single polypeptide. However, many enzymes are composed of two or more distinct polypeptides. Thus the synthesis of six enzymes may require more than six genes.

11.20 According to the the central dogma, genetic information is stored in DNA and is transferred from DNA to RNA to protein during gene expression. RNA tumor viruses store their genetic information in RNA, and that information is copied into DNA by the enzyme reverse transcriptase after a virus infects a host cell. Thus the discovery of RNA tumor viruses or retroviruses—retro for backwards flow of genetic information—provided an exception to the central dogma.

11.21 DNA, RNA, and protein synthesis all involve the synthesis of long chains of repeating subunits. All three processes can be divided into three stages: chain initiation, chain elongation, and chain termination.

11.22 The two stages of gene expression are:
- (1) transcription—the transfer of genetic information from DNA to RNA
- (2) translation—the transfer of genetic information from RNA to protein. In eukaryotes, transcription occurs in the nucleus and translation occurs in the cytoplasm on complex macromolecular structures called ribosomes. In prokaryotes, transcription and translation are often coupled with mRNA molecules often being translated by ribosomes while still being synthesized during transcription.

11.23 The primary transcripts of eukaryotes undergo more extensive post-transcriptional processing than those of prokaryotes. Thus the largest differences between mRNAs and primary transcripts occur in eukaryotes. Transcript processing is usually restricted to the excision of terminal sequences in prokaryotes. In contrast, eukaryotic transcripts are usually modified by:
- (1) the excision of intron sequences
- (2) the addition of 7-methyl guanosine caps to the 5' termini
- (3) the addition of poly(A) tails to the 3' termini. In addition, the sequences of some eukaryotic transcripts are modified by RNA editing processes.

11.24 The four types of RNA molecules that are involved in gene expression are mRNAs, rRNAs, tRNAs, and snRNAs. mRNA molecules carry genetic information from genes to the sites of protein synthesis and specify the amino acid sequences of polypeptides. rRNAs are major structural components of the ribosomes and provide functions required for translation. tRNA molecules are the adapters that provide amino acid-codon specificity during translation; each tRNA is activated by a specific amino acid and contains an anticodon sequence that is complementary or partially complementary to one, two, or three codons in mRNAs. snRNAs are structural components of the spliceosomes that excise introns from gene transcripts in eukaryotes. snRNAs perform their splicing functions in the nucleus. mRNAs carry information from the nucleus to the cytoplasm, so they function in both compartments of the cell. However, their most prominant function is to direct the synthesis of polypeptides during translation, which occurs in the cytoplasm. rRNAs and tRNAs perform their functions during translation in the cytoplasm.

11.25 The gene that contains a single exon will be transcribed in the least time. The rate of RNA chain extension is the same—about 30 nucleotides per second—for both intron and exon sequences. Thus the synthesis of the long intron sequence will take about 23 minutes. On the other hand, the time required to translate the two mRNAs will be the same because intron sequences are spliced out of transcripts prior to their translation.

11.26 In eukaryotes, the genetic information is stored in DNA in the nucleus, whereas proteins are synthesized on ribosomes in the cytoplasm. How could the genes, which are separated from the sites of protein synthesis by a double membrane—the nuclear envelope, direct the synthesis of polypeptides without some kind of intermediary to carry the specifications for the polypeptides from the nucleus to the cytoplasm? Researchers first used labeled RNA and protein precursors and autoradiography (see Figure 11.6) to demonstrate that RNA synthesis and protein synthesis occurred in the nucleus and the cytoplasm, respectively.

11.27 A simple pulse- and pulse/chase-labeling experiment will demonstrate that RNA is synthesized in the nucleus and is subsequently transported to the cytoplasm. This experiment has two parts:

(1) Pulse label eukaryotic culture cells by growing them in $[^3H]$uridine for a few minutes, and localize the incorporated radioactivity by autoradiography.

(2) Repeat the experiment, but this time add a large excess of nonradioactive uridine to the medium in which the cells are growing after the labeling period, and allow the cells to grow in the nonradioactive medium for about an hour. Then localize the incorporated radioactivity by autoradiography. The expected results are shown in Figure 11.6

11.28 The first evidence for the involvement of unstable RNA intermediaries in protein synthesis was Volkin and Astrachan's demonstration of a burst of unstable RNA synthesis in bacteriophage T2-infected *E. coli*. They showed that the RNAs synthesized after T2 infection had half-lives of a few minutes and the nucleotide composition of phage T2 DNA. Subsequently, Spiegelman and colleagues showed that the unstable RNA molecules synthesized in phage T4-infected cells could form RNA-DNA duplexes with denatured T4 DNA, but not with denatured *E. coli* DNA. Their results demonstrated that the unstable RNAs were transcribed from T4 DNA templates, not from *E. coli* DNA templates. Shortly thereafter, Brenner, Jacob, and Meselson demonstrated that phage T4 proteins are synthesized on ribosomes present in *E. coli* cells prior to their infection by T4 phage. Together, the results of these experiments firmly established the concept of unstable RNA intermediaries or messenger RNAs, as we now call them.

11.29 During DNA synthesis,

(1) both strands of DNA are used as templates for the synthesis of complementary nascent strands

(2) the precursors are deoxribonucleoside triphosphates

(3) new chains are initiated with RNA primers

(4) the parental DNA strands are completely separated

(5) chain extension occurs at a rate of about 500 nucleotides per second (prokaryotes)

During RNA synthesis,

(1) only one strand of DNA is used as a template for the synthesis of a complementary nascent strand

(2) the precursors are ribonucleoside triphosphates

(3) new chains are initiated *de novo*,

(4) synthesis occurs within a localized region of strand separation

(5) chain extension occurs at a rate of 40 to 50 nucleotides per second (prokaryotes).

11.30 RNA-DNA duplexes will be formed when the template strand is used, but not when the nontemplate strand is used. Only one strand—the template strand—of most genes is transcribed. Thus RNA will contain nucleotide sequences complementary to the template strand, but not to the nontemplate strand.

11.31 The first preparation of RNA polymerase is probably lacking the sigma subunit and, as a result, initiates the synthesis of RNA chains at random sites along both strands of the *argH* DNA. The second preparation probably contains the sigma subunit and initiates RNA chains only at the site used *in vivo*, which is governed by the position of the -10 and -35 sequences of the promoter.

11.32 In eukaryotes, transcription occurs in the nucleus and translation occurs in the cytoplasm. Because these processes occur in different compartments of the cell, they cannot be coupled as they are in prokaryotes.

11.33 The simplest procedure for determining which of the three RNA polymerases catalyzes the transcription of the gene is to measure the sensitivity of its transcription to α-amanitin. Since the gene is expressed in cells growing in culture, you can simply add α-amanitin to the culture medium. If the gene is transcribed by RNA polymerase I, α-amanitin will have no effect on its transcription. If the gene is transcribed by RNA polymerase II, its transcription will be completely blocked by the presence of α-amanitin. If the gene is transcribed by RNA polymerase III, the rate of transcription will be reduced, but not blocked, by α-amanitin.

11.34 TATA and CAAT boxes. The TATA and CAAT boxes are usually centered at positions -30 and -80, respectively, relative to the startpoint (+1) of transcription. The TATA box is responsible for positioning the transcription startpoint; it is the binding site for the first basal transcription factor that interacts with the promoter. The CAAT box enhances the efficiency of transcriptional initiation.

11.35 (1) Intron sequences are spliced out of gene transcripts to provide contiguous coding sequences for translation.

     (2) The 7-methyl guanosine caps added to the 5' termini of most eukaryotic mRNAs help protect them from degradation by nucleases and are recognized by proteins involved in the initiation of translation.

     (3) The poly(A) tails at the 3' termini of mRNAs play an important role in their transport from the nucleus to the cytoplasm and enhance their stability.

11.36 RNA editing sometimes leads to the synthesis of two or more distinct polypeptides from a single mRNA. Guide RNAs are partially complementary to the pre-mRNAs that are edited and serve as templates for the editing process. During editing, uracils are inserted in pre-mRNAs opposite adenines in the guide RNAs.

11.37 The introns of tRNA precursors, *Tetrahymena* rRNA precursors, and nuclear pre-mRNAs are excised by completely different mechanisms.

     (1) Introns in tRNA precursors are excised by cleavage and joining events catalyzed by splicing nucleases and ligases, respectively.

     (2) Introns in *Tetrahymena* rRNA precursors are excised autocatalytically.

     (3) Introns of nuclear pre-mRNAs are excised by spliceosomes. snRNAs are involved in nuclear pre-mRNA splicing as structural components of spliceosomes. In addition, snRNA U1 is required for the cleavage events at the 5' termini of introns; U1 is thought to base-pair with a partially complementary consensus sequence at this position in pre-mRNAs.

11.38 Some individuals with systemic lupus erythematosus produce antibodies that react with proteins in snRNPs. These antibodies have been used to immunoprecipitate snRNPs, facilitating their purification and subsequent characterization.

11.39 This zygote will probably be lethal because the gene product is essential and the elimination of the 5' splice site will almost certainly result in the production of a nonfunctional gene product.

11.40 RNA-DNA duplexes would be expected when RNA isolated from nuclei is used in the renaturation experiment, but not when cytoplasmic RNA is used. The introns of genes are transcribed along with the exons. Thus the primary transcripts present in nuclei will contain both exon and intron sequences. However, intron sequences are excised by nuclear spliceosomes before the transcripts are exported to the cytoplasm. Thus cytoplasmic RNAs does not contain intron sequences and will not form RNA-DNA duplexes when incubated with single-stranded intron sequences.

CHAPTER 12

12.1 Proteins are long chainlike molecules made up of amino acids linked together by peptide bonds. Proteins are composed of carbon, hydrogen, nitrogen, oxygen, and usually sulfur. They provide the enzymatic capacity and much of the structure of living organisms. DNA is composed of phosphate, the pentose sugar 2-deoxyribose, and four nitrogen-containing organic bases (adenine, cytosine, guanine, and thymine). DNA stores and transmits the genetic information in most living organisms. Protein synthesis is of particular interest to geneticists because proteins are the primary gene products—the key intermediates through which genes control the phenotypes of living organisms.

12.2 Protein synthesis occurs on ribosomes. In eukaryotes, most of the ribosomes are located in the cytoplasm and are attached to the extensive membranous network of endoplasmic reticulum. Some protein synthesis also occurs in cytoplasmic organelles such as chloroplasts and mitochondria.

12.3 Ribosomes are from 10 to 20 nm in diameter. They are located primarily in the cytoplasm of cells. In bacteria, they are largely free in the cytoplasm. In eukaryotes, many of the ribosomes are attached to the endoplasmic reticulum. Ribosomes are complex structures composed of over 50 different polypeptides and three to five different RNA molecules.

12.4 **(a)** The nucleus, specifically the nucleoli.

(b) The cytoplasm.

12.5 Messenger RNA molecules carry genetic information from the chromosomes (where the information is stored) to the ribosomes in the cytoplasm (where the information is expressed during protein synthesis). The linear sequence of triplet codons in an mRNA molecule specifies the linear sequence of amino acids in the polypeptide(s) produced during translation of that mRNA. Transfer RNA molecules are small (about 80 nucleotides long) molecules that carry amino acids to the ribosomes and provide the codon-recognition specificity during translation. Ribosomal RNA molecules provide part of the structure and function of ribosomes; they represent an important part of the machinery required for the synthesis of polypeptides.

12.6 **(a)** Polysomes are formed when two or more ribosomes are simultaneously translating the same mRNA molecule. Ribosomes are usually spaced about 90 nucleotides apart on an mRNA molecule. Thus, polysome size is determined by mRNA size.

(b) A ribosome, which contains rRNA molecules, can participate in the synthesis of any polypeptide specified by the ribosome-associated mRNA. In that sense, rRNA is *nonspecific*. Messenger RNAs and tRNAs, in contrast, are *specific*, in directing the synthesis of a particular polypeptide or set of polypeptides (mRNA) or in attaching to a particular amino acid (tRNA).

(c) Transfer RNA molecules are much smaller (about 80 nucleotides) than DNA or mRNA molecules. They are single-stranded molecules but have complex secondary structures because of the base pairing between different segments of the molecules.

12.7 A specific aminoacyl-tRNA synthetase catalyzes the formation of an amino acid-AMP complex from the appropriate amino acid and ATP (with the release of pyrophosphate). The same enzyme then catalyzes the formation of the aminoacyl-tRNA complex, with the release of AMP. The amino acid-AMP and aminoacyl-tRNA linkages are both high-energy phosphate bonds.

12.8 Synthetic RNA molecules (polyuridylic acid molecules) containing only the base uracil were prepared. When these synthetic molecules were used to activate *in vitro* protein synthesis systems, small polypeptides containing only the amino acid phenylalanine (polyphenylalanine molecules) were synthesized. Codons composed only of uracil were therefore shown to specify phenylalanine. Similar experiments were carried out using synthetic RNA molecules with different base compositions. Later, *in vitro* systems activated with synthetic RNA molecules with known repeating base sequences were developed. Ultimately, *in vitro* systems in which specific aminoacyl-tRNAs where shown to bind to ribosomes activated with specific mini-mRNAs, which were trinucleotides of known base sequence, were developed and used in codon identification.

12.9 **(a)** The genetic code is degenerate in that all but 2 of the 20 amino acids are specified by two or more codons. Some amino acids are specified by six different codons. The degeneracy occurs largely at the third or 3′ base of the codons. "Partial degeneracy" occurs where the third base of the codon may be either of the two purines or either of the two pyrimidines and the codon still specifies the same amino acid. "Complete degeneracy" occurs where the third base of the codon may be any one of the four bases and the codon still specifies the same amino acid.

(b) The code is ordered in the sense that related codons (codons that differ by a single base change) specify chemically similar amino acids. For example, the codons CUU, AUU, and GUU specify the sructurally related amino acids, leucine, isoleucine, and valine, respectively.

(c) The code appears to be almost completely universal. Known exceptions to universality include strains carrying suppressor mutations that alter the reading of certain codons (with low efficiencies in most cases) and the use of UGA as a tryptophan codon in yeast and human mitochondria.

12.10 Blueprints transcribed into building instructions and translated into structures composed of boards, bricks, and mortar by skilled craftsmen and craftswomen may be likened to DNA, mRNA, and tRNA functions in the assembly of amino acids into polypeptides by ribosomes and other required factors.

12.11 **(a)** Met → Val. This substitution occurs as a result of a transition. All other amino acid substitutions listed would require transversions.

12.12 His → Arg results from a transition; His -> Pro would require a transversion (not induced by 5-bromouracil).

12.13 (a) By a complex reaction involving mRNA, ribosomes, initiation factors (IF-1, IF-2, and IF-3), GTP, the initiator codon AUG, and a special initiator tRNA (tRNA$_f^{Met}$). It also appears to involve a base-pairing interaction between a base sequence near the 3'-end of the 16S rRNA and a base sequence in the "leader sequence" of the mRNA.

(b) By recognition of one or more of the chain-termination codons (UAG, UAA, and UGA) by the appropriate protein release factor (RF-1 or RF-2).

12.14 At least 813 nucleotides.

12.15 Crick's wobble hypothesis explains how the anticodon of a given tRNA can base-pair with two or three different mRNA codons. Crick proposed that the base-pairing between the 5' base of the anticodon in tRNA and the 3' base of the codon in mRNA was less stringent than normal and thus allowed some "wobble" at this site. As a result, a single tRNA often recognizes two or three of the related codons specifying a given amino acid (see Table 12.3).

12.16 (a) Inosine.

(b) Two.

12.17 (a) Singlet and doublet codes provide a maximum of 4 and $(4)^2$ or 16 codons, respectively. Thus neither code would be able to specify all 20 amino acids.

(b) 20

(c) $(20)^{146}$

12.18 Translation occurs by very similar mechanisms in prokaryotes and eukaryotes; however, there are some differences.

(1) In prokaryotes, the initiation of translation involves base pairing between a conserved sequence (AGGAGG) — the Shine-Dalgarno box — in mRNA and a complementary sequence near the 5' end of the 16S rRNA. In eukaryotes, the initiation complex forms at the 5' end of the transcript when a cap-binding protein interacts with the 7-methyl guanosine on mRNA. The complex then scans the mRNA processively and initiates translation (with a few exceptions) at the AUG closest to the 5' terminus.

(2) In prokaryotes, the amino group of the initiator methionyl-tRNA$_f^{Met}$ is formylated; in eukaryotes, the amino group of methionyl-tRNA$_i^{Met}$ is not formylated.

(3) In prokaryotes, two soluble protein release factors (RFs) are required for chain termination. RF-1 terminates polypeptides in response to UAA and UAG codons; RF-2 terminates chains in response to UAA and UGA codons. In eukaryotes, one release factor responds to all three termination codons.

12.19 (a) Attachment of an amino acid to the correct tRNA.

(b) ecognition of termination codons UAA and UAG and release of the nascent polypeptide from the tRNA in the P site of the ribosome.

(c) Formation of a peptide bond between the amino group of the aminoacyl-tRNA in the A site and the carboxyl group of the growing polypeptide on the tRNA in the P site.

(d) Formation of the initiation complex required for translation; all steps leading up to peptide bond formation.

(e) Translocation of the peptidyl-tRNA from the A site on the ribosome to the P site.

12.20 Assuming 0.34 nm per nucleotide pair in B-DNA, a gene 68 nm long would contain 200 nucleotide pairs. Given the triplet code, this gene would contain $\frac{200}{3}$ = 66.7 triplets, one of which must specifiy chain termination. Disregarding the partial triplet, this gene could encode a maximum of 65 amino acids.

12.21 **(a)** A nonsense mutation changes a codon specifying an amino acid to a chain-termination codon, whereas a missense mutation changes a codon specifying one amino acid to a codon specifying a different amino acid.

(b) Missense mutations are more frequent.

(c) Of the 64 codons, only three specify chain termination. Thus the number of possible missense mutations is much larger than the number of possible nonsense mutations. Moreover, nonsense mutations almost always produce nonfunctional gene products. As a result, nonsense mutations in essential genes are usually lethal in the homozygous state.

12.22 426 nucleotides—3 × 121 = 423 specifying amino acids plus three (one codon) specifying chain termination.

12.23 **(a)** The incoming aminoacyl-tRNA enters the *A* site of the ribosome, whereas the nascent polypeptide-tRNA occupies the *P* site.

(b) In order for peptide bond formation to occur, the amino group of an aminoacyl-tRNA must be placed in juxtaposition to the carboxyl group of a peptidyl-tRNA. For this to occur, ribosomes must contain binding sites for at least two tRNAs.

12.24 **(a)** Related codons often specify the same or very similar amino acids. As a result, single base-pair substitutions frequently result in the synthesis of identical proteins (degeneracy) or proteins with amino acid substitutions involving very similar amino acids.

(b) Leucine and valine have very similar structures and chemical properties; both have nonpolar side groups and fold into essentially the same three-dimensional structures when present in polypeptides. Thus, substitutions of leucine for valine or valine for leucine seldom alter the function of a protein.

12.25 **(a)** The Shine–Dalgarno sequence is a conserved polypurine tract, consensus AGGAGG, that is located about seven nucleotides upstream from the AUG initiation codon in mRNAs of prokaryotes. It is complementary to, and is believed to base-pair with, a sequence near the 5' terminus of the 16S ribosomal RNA.

(b) Prokaryotic mRNAs with the Shine–Dalgarno sequence deleted are either not translated or are translated inefficiently.

12.26 **(a)** Ribosomes and spliceosomes both play essential roles in gene expression, and both are complex macromolecular structures composed of RNA and protein molecules.

(b) Ribosomes are located in the cytoplasm; spliceosomes in the nucleus. Ribosomes are larger and more complex than splicesomes.

12.27 Incorporation of alanine into polypeptide chains.

12.28 Met-Ser-Ile-Cys-Leu-Phe-Gln-Ser-Leu-Ala-Ala-Gln-Asp-Arg-Pro-Gly

12.29 NH_2-Met-Ala-Ile-Cys-Leu-Phe-Gln-Ser-Leu-Ala-Ala-Gln-Asp-Arg-Pro-Gly-COOH.

12.30 *Amber* (UAG). This is the only nonsense codon that is related to tryptophan, serine, tyrosine, leucine, glutamic acid, glutamine, and lysine codons by a single base-pair substitution in each case.

CHAPTER 13

13.1 **(a)** transition

(b) transition

(c) transversion

(d) transversion

(e) frameshift

(f) transition.

13.2 **(a)** *ClB* method

(b) attached-X method (see Chapter 6, pp. 121-123).

13.3 Bacteria treated with a mutagen or expected to carry mutations may be introduced into media with particular drugs in appropriate concentrations. Colonies that appear have originated from cells carrying preexisting mutations for resistance. This may be verified by the replica-plating technique (see Figure 13.3). The frequency of mutations of wild-type (drug-sensitive) cells to drug resistance can be measured in the presence or absence of the drug.

13.4 Probably not. A human is larger than a bacterium, with more cells and a longer life span. If mutation frequencies are calculated in terms of cell generations, the rates for human cells and bacterial cells are similar.

13.5 A dominant mutation presumably occurred in the woman in whom the condition was first known.

13.6 The sex-linked gene is carried by mothers, and the disease is expressed in half of their sons. Such a disease is difficult to follow in pedigree studies because of the recessive nature of the gene, the tendency for the expression to skip generations in a family line, and the loss of the males who carry the gene. One explanation for the sporadic occurrence and tendency for the gene to persist is that, by mutation, new defective genes are constantly being added to the load already present in the population.

13.7 Plants can be propagated vegetatively, but no such methods are available for widespread use in animals.

13.8 The sheep with short legs could be mated to unrelated animals with long legs. If the trait is expressed in the first generation, it could be presumed to be inherited and to depend on a dominant gene. On the other hand, if it does not appear in the first generation, F_1 sheep could be crossed back to the short-legged parent. If the trait is expressed in one-half of the backcross progeny, it might be presumed to be inherited as a simple recessive. If two short-legged sheep of different sex could be obtained, they could be mated repeatedly to test the hypothesis of dominance. In the event that the trait is not transmitted to the progeny that result from these matings, it might be considered to be environmental or dependent on some complex genetic mechanism that could not be identified by the simple test used in the experiments.

13.9 Enzymes may discriminate among the different nucleotides that are being incorporated. Mutator enzymes may utilize a higher proportion of incorrect nucleotides, whereas antimutator enzymes may select fewer incorrect bases in DNA replication. In the case of the phage T4 DNA polymerase, the relative efficiencies of polymerization and proofreading by the polymerase's $3' \rightarrow 5'$ exonuclease activity play key roles in determining the mutation rate.

13.10 If both mutators and antimutators operate in the same living system, an optimum mutation rate for a particular organism in a given environment may result from natural selection.

13.11 *Dt* is a mutator gene that induces somatic mutations in developing kernels.

13.12 (a) Yes

(b) A block would result in the accumulation of phenylalanine and a decrease in the amount of tyrosine, which would be expected to result in several different phenotypic expressions.

13.13 These hemoglobins can be distinguished by mobility of molecules in an electric field (electrophoretic mobility) and by the amino acid sequences of their β polypeptides.

13.14

AMINO ACID	mRNA	DNA
Glumatic acid	$-$GAA$\rightarrow$	$-$GAA$\rightarrow$
		$\leftarrow$ C T T $-$ ◄— Transcribed strand
		↓ Mutation
Valine	$-$GUA$\rightarrow$	$-$GTA$\rightarrow$
		$\leftarrow$ C AT $-$
		↓ Mutation
Lysine	$-$AAA$\rightarrow$	$-$AAA$\rightarrow$
		$\leftarrow$ T T T $-$

13.15 The label "molecular disease" became common in speaking of sickle-cell anemia because its molecular basis (the substitution of a valine residue for the glutamic acid residue at amino acid position number 6 in the b chain) was recognized quite early during the emergence of the science of molecular biology. Actually, most if not all inherited diseases probably have very similar molecular bases. We just don't know what the molecular defects are in most instances.

13.16 Mutations: transitions, transversions, and fraeshifts.

13.17 Irradiate the nonresistant strain and plate the irradiated organisms on a medium containing streptomycin. Those that survive and produce colonies are resistant. They could then be replicated to a medium without streptomycin. Those that survive would be of the first type; those that can live with streptomycin but not without it would be the second type.

13.18 3%; 4%; 6%.

13.19 Each quantum of energy from the X rays that is absorbed in a cell has a certain probability of hitting and breaking a chromosome. Hence, the greater the number of quanta of energy or dosage, the more likely breaks are to occur. The rate at which this dosage is delivered does not change the probability of each quantum inducing a break.

13.20 The person receiving a total of 100 r would be expected to have twice as many mutations as the one receiving 50 r.

13.21 During the replicating process, untraviolet light produces mispairing alterations mostly in pyrimidines (for example, cytosine to thymine transitions). Thymine may be altered to cytosine (or a modified pyrimidine with the base-pairing potential of cytosine), which pairs with guanine. A reverse mutation may occur when cytosine is changed to thymine (or a derivative of cytosine with the hydrogen-bonding potential of thymine), which pairs with adenine. The T-A base pair may thus be changed to a C-G, and the reverse mutation may occur from C-G to T-A.

13.22 Nitrous acid brings about a substitution of an OH group for an NH_2 group in those bases (A, C, and G) having NH_2 side groups. In so doing, adenine is converted to hypoxanthine, which base-pairs with cytosine, and cytosine is converted to uracil, which base-pairs with adenine. The net effects are GC $\Leftrightarrow$ AT base-pair substitutions (see Figure 13.18).

13.23 Transitions.

13.24 Nitrous acid acts as a mutagen on either replicating or nonreplicating DNA and produces transitions from A to G or C to T, whereas 5-bromouracil does not affect nonreplicating DNA but acts during the replication process causing GC $\Leftrightarrow$ AT transitions. 5-Bromouracil must be incorporated into DNA during the replication process in order to induce mispairing of bases and thus mutations.

13.25 Mutations induced by acridine dyes are primarily insertions or deletions of single base-pairs. Such mutations alter the reading frame (the in-phase triplets specifying mRNA codons) for that portion of the gene distal (relative to the direction of transcription and translation) to the mutation (see Figure 13.15b). This would be expected to totally change the amino acid sequences of polypeptides distal to the mutation site and produce inactive polypeptides. In addition, such frameshift mutations frequently produce in-frame termination codons that result in truncated proteins.

13.26 Proline and serine.

13.27 No. Leucine → proline would occur more frequently. Leu (CUA) $\xrightarrow{5\text{-BU}}$ Pro (CGA) occurs by a single base-pair transition, whereas Leu (CUA) $\xrightarrow{5\text{-BU}}$ Ser (UCA) requires two base-pair transitions. Recall that 5-bromouracil (5-BU) induces only transitions (see Figure 13.17).

13.28 No. 5-Bromouracil is mutagenic only to replicating nucleic acids.

13.29 Yes:

DNA: ←GGX—   ←GGX—

            CGX′→ $\xrightarrow{HNO_2}$ —UCX′→

              ↓   ↓

mRNA: GG X   A G X

              ↓   ↓

Polypeptide: Gly   Ser or Arg

                           (depending on X)

or

DNA: ←GGX—   ←GGX—

            —CGX′→ $\xrightarrow{HNO_2}$ —CUX′→

              ↓   ↓

mRNA: GGX   G A X

              ↓   ↓

Polypeptide: Gly   Asp or Glu

                           (depending on X)

or

DNA: ←GGX—   ←GGX—

            —CGX′→ $\xrightarrow{HNO_2}$ —UUX′→

              ↓   ↓

mRNA: GGX   A A X

              ↓   ↓

Polypeptide: Gly   Asn or Lys

                       (depending on X)

Note: The X at the third position in each codon in mRNA and in each triplet of base pairs in DNA refers to the fact that there is complete degeneracy at the third base in the glycine codon. Any base may be present in the codon, and it will still specify glycine.

13.30 No. The glycine codon is GGX, where X can be any one of the four bases. Because of this complete degeneracy at the third position of the glycine codon, changing X to any other base will have no effect (that is, the codon will still specify glycine). Nitrous acid deaminates guanine (G) to xanthine, but xanthine still base-pairs with cytosine. Thus guanine is not a target for mutagenesis by nitrous acid.

13.31 Tyr → Cys substitutions; Tyr to Cys requires a transition, which is induced by nitrous acid. Tyr to Ser would require a transversion, and nitrous acid is not expected to induce transversions.

13.32 (b) Met → Thr. 5-Bromouracil induces transitions, not transversions. All other changes listed require transversions.

13.33 5′-AUGCCCUUUGGG**GAAAGG**UUUCCCUAA-3′

CHAPTER 14

14.1 Prior to 1940, the gene was considered a "bead-on-a-string," not subdivisible by recombination or mutation. Today, the gene is considered to be the unit of genetic material that codes for one polypeptide. The unit of structure, not subdivisible by recombination or mutation, is known to be the single nucleotide pair.

14.2 The recombination observed between lz^S and lz^g, two functionally allelic mutations at the *lozenge* locus of *Drosophila*.

14.3 The *cis-trans* test, which defines the unit of genetic material specifying the amino acid sequence of one polypeptide.

14.4 They provide powerful selective sieves for identifying rare recombinants. This is accomplished by using the restrictive environmental conditions to select wild-type recombinant progency from crosses between pairs of conditional lethal mutants.

14.5 Two genes; mutations 1, 2, 3, 4, 5, 6, and 8 are in one gene; mutation 7 is in a second gene.

14.6 Four genes; mutations 1 and 2 in one gene; mutations 3 and 4 in a second gene; mutations 5 and 6 in a third gene; mutations 7 and 8 in a fourth gene.

14.7 The size of the gene (assuming that all nucleotide pairs in the gene are capable of undergoing base-pair substitutions, as seems highly probable). Dominant lethal alleles and recessive lethal alleles in haploids will (under normal conditions) exist only transiently, of course.

14.8 Homoalleles are structurally and functionally allelic; they are not separable by recombination. Heteroalleles are functionally allelic (based on *cis-trans* tests), but are structurally nonallelic (based on recombination tests). Heteroalleles thus result from mutations occuring at different sites within a gene.

14.9 (1) Cross the two white-flowered varieties. The F_1 plants will be *trans*-heterozygotes. If the F_1 plants have white flowers, the two varieties probably carry mutations in the same gene, causing white flowers.

(2) Cross white-flowered varieties with red-flowered varieties and self-pollinate or intercross the F_1 plants. If alleles of a single gene are involved, monohybrid F_2 ratios should be observed in all cases.

14.10 (a) Five genes

(b) Mutations 1, 3, and 5 are in one gene; mutations 7 and 8 are in a second gene; mutations 2, 4, and 6 identify genes 3, 4, and 5, respectively.

14.11 A maximum of three mutant homoalleles in addition to the wild-type base pair at any one site.

14.12 (a) True

(b) False

(c) True

(d) True

(e) True.

14.13 The observed complementation between ry^2 and ry^{42} is *intragenic* complementation. Xanthine dehydrogenase is a dimeric protein, and dimers that contain one polypeptide encoded by the ry^2 allele and one polypeptide encoded by the ry^{42} allele are partially active. Presumably, the wild-type segment of the ry^2 polypeptide somehow stabilizes the mutant segment of the ry^{42} polypeptide, and vice versa, yielding a dimer with enzymatic activity.

14.14 *Am* mutations result in UAG chain-termination codons within the coding sequence of the mRNA product of a gene; they thus produce truncated polypeptide gene-products. Since all *am* mutant alleles of a gene will produce polypeptides lacking the COOH-terminus, they would not be expected to exhibit intragenic complementation except in very rare cases. In contrast, most *ts* mutations are caused by missence mutations that change the amino acid sequence of the polypeptide gene-product making it more heat-labile. However, most *ts* mutant alleles produce a complete, although altered, gene-product. As a result, *ts* mutant alleles often exhibit intragenic complementation when the active form of the protein gene-product is a homomultimer. For this reason, *cis-trans* tests carried out with *am* mutants, not *ts* mutants, have been used whenever possible to operationally define the genes of phage T4.

14.15 (a) The seven *sus* mutants are located in three different genes.

(b) Mutant strains 1, 3, and 7 contain mutations in one gene; 4, 5, and 6 carry mutations in a second gene; 2 has a mutation in a third gene.

14.16 It depends on how you define alleles. If every variation in nucleotide sequence is considered to be a different allele, even if the gene product and the phenotype of the organism carrying the mutation are unchanged, then the number of alleles will be directly related to gene size. However, if the nucleotide sequence change must produce an altered gene product or phenotype before it is considered a distinct allele, then there will be a positive correlation, but not a direct relationship, between the number of alleles of a gene and its size in nucleotide pairs. The relationship is more likely to occur in prokaryotes where most genes lack introns. In eukaryotic genes, nucleotide sequence changes within introns usually are neutral; that is, they do not affect the activity of the gene product or the phenotype of the organism. Thus, in the case of eukaryotic genes with introns, there may be no correlation between gene size and number of alleles producing altered phenotypes.

14.17 *White* and *eosin* are located in the same gene; *carnation* is located in a different gene.

14.18 (a) Four genes

(b) Mutations 1, 3, and 7 are in one gene; mutations 4 and 6 are in a second gene; mutation 2 is in a third gene; and mutation 5 is in a fourth gene.

14.19 One gene; all seven mutations are in the same gene.

14.20 No. Because the two mutations map to different chromosomes, they could not be located within the same gene, at least, based on our current concept of the gene. However, if two transcripts are spliced in *trans* (see Figure 18.5), it is possible for two parts of a gene—a nucleotide sequence encoding one polypeptide — to be located on two different chromosomes. Remember that our concept of the gene has evolved considerably since it was introduced by Mendel in 1865, and it will undoubtedly continue to evolve in the future.

14.21 Several enzymes were shown to contain two or more different polypeptides, and these polypeptides were sometimes controlled by genes that mapped to different chromosomes. Thus the mutations clearly were not in the same gene.

14.22 *Neurospora* has many advantages over humans as an experimental organism. The most important advantages are the ability to grow organisms under carefully controlled conditions, to enhance mutation frequency by treatment with mutagenic agents, and to perform controlled crosses for genetic analysis.

14.23 **(a)** Homoalleles
   (b) heteroalleles.

14.24 One. The *trpA*58 and *trpA*78 mutations alter the same codon. If they alter the 5' and 3' nucleotide pairs of the triplet, they will be separated by the middle nucleotide pair of the triplet specifying one mRNA codon.

14.25 All four mutations are in the same gene. However, ry^{42} and ry^{406} exhibit intragenic complementation with each other, whereas ry^5 and ry^{41} do not.

14.26 They are usually referred to as gene segments because they are joined together by somatic recombination to produce a sequence of nucleotide pairs that encodes a single polypeptide.

14.27 Several different polypeptides can be produced from a single gene by alternate pathways of transcript splicing. In the case of the tropomyosins, the exons of the transcripts are known to be spliced together in different combinations to produce overlapping, but distinct polypeptides.

14.28 No. These mutations are located far apart on the X chromosome. They are separated by millions of nucleotide pairs and many other genes, and it is virtually impossible for mutations located that far apart to be part of the same gene. For these mutations to be located in the same gene, all of the intervening genes would have to be part of a huge intron or some type of *trans* splicing (see Figure 18.5) would have to occur.

14.29 Mutations 1, 2, and 4 do not complement one another and thus appear to be located in the same gene, as do mutations 3 and 5. The anomaly is that mutation 7 does not complement mutations 3, 5, and 6, even though mutation 6 does complement mutations 3 and 5.
   (a) There are four simple explanations of the seemingly anomalous complementation behavior of mutation 7.
      (1) It is a deletion spanning all or parts of two genes.
      (2) It is a double mutation with defects in two genes.
      (3) It is a polar mutation in the promoter-proximal gene of a multigenic transcription unit.
      (4) It exhibits rare intergenic noncomplementation with either mutations 3 and 5 or mutation 6 because it is present in a gene that encodes a product that interacts with the product of the other gene.

14.29 (b) Three simple genetic operations will distinguish between these four possibilities.

(1) Reversion. Plate a large number of mutant 7 phage on *E. coli* strain Z and look for wild-type revertants.

(2) Backcross mutant 7 to wild-type phage and test the mutant progeny for the ability to complement mutations 3 and 6.

(3) Introduce F's carrying tRNA nonsense suppressor genes into *E. coli* strain Z and determine whether any of them suppress the *loz7* mutation.

(c) If mutation 7 is a deletion, it will not revert, and, if it is a double mutation, the reversion rate will probably be below the level of detection in your experiment. On the other hand, if it is a polar nonsense mutation or a noncomplementing missense mutation, *loz*+ revertants will be obtained. If mutation 7 is a deletion, no new genotypes will be produced in the backcross to wild-type. However, if it is a double mutation, some recombinant single-mutant progeny will be produced in the backcross to wild-type, and these single mutations will complement either mutation 3 or mutation 6. If mutation 7 is a polar nonsense mutation, it should be suppressed by one or more of the tRNA suppressor genes introduced into *E. coli* strain Z. If mutation 7

(1) reverts to *loz*+

(2) does not yield any single-mutant progeny in the backcross to wild-type, and

(3) is not suppressed by any of the suppressor tRNA genes, then rare intergenic noncomplementation is probably responsible for its unusual behavior.

CHAPTER 15

15.1 Because wild-type recombinants are rarest in cross (a), this must be the double crossover class, and *m* must be the middle gene. In crosses (b) and (c), wild-type recombinants can be generated by single exchanges.

15.2 Key points that should be addressed in this review question include a contrast between meiotic crossing over and phage recombination and the populational aspect of phage recombination. Breakage and reunion would be a key similarity with respect to mechanism.

15.3 The chromosome theory of inheritance is the theory that genes are located on chromosomes. The gene is DNA, and Hershey and Chase's studies on bacterial viruses (Chapter 9) demonstrated that DNA was the genetic material. Studies on the mechanism of recombination have correlated genetic recombination with the physical exchange of DNA molecule segments.

15.4 Regardless of the position taken, the discussion should include a critical look at the remark relating to the fact that there is a wide variety of viral types with various replication schemes, host cells, physical structures, genomes, etc. Yet in spite of the differences, viruses all require host cells for reproduction and use host cell resources. At one level, there are striking differences, but at another level, there are similarities.

15.5 The evidence suggests that the genes are linked and that the order is $a - b - c$. The rarest classes ($a^+ b\, c^+$ and $a\, b^+ c$) when compared to the noncrossover classes (abc and $a^+ b^+ c^+$) establish b as the middle gene. The distance between a and b is $0.05 + 0.05 + 0.02 + 0.02 = 0.14$, or 14 map units; the distance between b and c is $0.08 + 0.09 + 0.02 + 0.02 = 0.21$, or 21 map units. The coefficient of coincidence (observed double cross-overs/expected double crossovers) is $0.040/0.036 = 1.11$. There are slightly more double exchanges than expected, probably reflecting the multiple exchanges occurring in the population of DNA molecules.

15.6 These data show that recombination occurred between all three phage strains in each cross to produce recombinants carrying alleles from all three strains. This illustrates the multiple rounds of exchange that phage experience.

15.7 (a) The data suggest that x and h are 18 map units apart; y and h are 9 map units apart; and z and h are 2 map units apart.

(b) The possible maps for these four genes are:

   1. x - - - - -9- - - - -y- - - - 7- - -z- -2- -h

   2. x - - - - - -16- - - - - -z- -2- -h- - - - -9- - - - -y

   3. x - - - - -9- - - - -y- - - - -9- - - - -h- -2- -z

   4. x - - - - - - -18- - - - - - -h- -2- -z- - - -7- - - -y

(c) With 13 map units between y and z, orders 2 and/or 3 are the most likely.

(d) A way to resolve these data is to construct a circular genetic map.

15.8 We expect to find 9 double crossovers ($0.012 \times 0.009 \times 80{,}000$) and we observe 76. The coefficient of coincidence $= 76/9 = 8.44$; the interference is $1 - 8.44 = -7.44$ (negative interference).

15.9 One possibility is that $rIIB_2$ and $rIIB_3$ are point mutations 0.9 map unit apart; since $rIIB_1$ does not recombine with either of them, it may be a deletion that spans the B_2 and B_3 sites. Additional tests would need to be done to confirm that B_1 is a deletion, such as reverse mutation analysis.

15.10 The most reliable way to determine if an rII mutation is a deletion is to do a reverse mutation analysis. Deletions do not revert, but point mutations do.

15.11 Because the two mutants fail to grow on K cells, they do not complement each other, so they are different mutants of the same gene. The few wild-type progeny that appear after a cross on B cells indicate that recombination has occurred between the mutant sites, producing wild-type progeny. (The double mutant recombinants do not grow on K cells.)

15.12 When referring to a virus, we usually use the term heteroduplex DNA molecule to refer to the heterozygous state in which one DNA strand has the sequence of one allele and the other strand has the sequence of the other allele. In a general way, this is comparable to a Drosophila heterozygote carrying a different allele on each member of a homologous pair of chromosomes.

15.13 Linear chromosomes that are terminally redundant and circularly permuted produce a circular genetic map.

15.14 Lambda is terminally redundant for 12 base pairs, but it is not circularly permuted. When the circular chromosome produces a linear form, it breaks at a specific point (cos) so that genes on each side of the break are always separated and do not appear next to each other. All linear forms of the chromosome have the same gene sequence from left to right.

15.15 The new mutant maps in the A6c-A6d region.

15.16 No because the ends of the fragments are not redundant.

CHAPTER 16

16.1 Recombination has occurred between the two strains, producing wild-type (prototrophic) bacteria.

16.2 Perform two experiments:

(1) determine whether the process is sensitive to Dnase

(2) determine whether cell-cell contact is required for the process to take place. The cell-cell contact requirement can be assessed by a U-tube experiment. If the process is sensitive to DNase, it is similar to transformation. If cell-cell contact is required, it is similar to conjugation. If it is neither sensitive to DNase nor requires cell-cell contact, it is similar to transduction.

16.3

Recombination Process	Agent Mediating DNA Transfer	Size of DNA units Transferred	State of Donor DNA in Recombinant Cell
Transformation	Active uptake of free DNA	Small (about 1/200 to 1/100 of a chromosome)	Single-stranded; integrated
Transduction	Bacteriophage	Small (about 1/00 to 1/50 of a chromosome)	Double-stranded; integrated into host chromosome (except in abortive transduction)
Sexduction	F factor	Variable	Initially added to the host cell as separate plasmid; may undergo recombination with host chromosome to yield stable transductant

16.4 (a) In F$^-$ cells, no F factor is present; F$^+$ cells have an autonomous F factor; Hfr cells have an integrated F factor.

(b) F$^+$ and Hfr cells have F (sex) pili; F$^-$ cells do not.

(c) F$^-$ cells are converted to F$^+$ cells by the conjugative transfer of F factors from F$^+$ cells; Hfr cells are formed when F factors in F$^+$ cells become integrated into the chromosomes of these cells.

16.5 **(a)** F' factors are useful for genetic mapping, for complementation analysis (see Chapter 14), and for studies of dominance relationships.

(b) F' factors are formed by errors during excision of F factors from Hfr chromosomes.

(c) Sexduction occurs by the conjugative transfer of an F' factor from a donor to a recipient (F⁻) cell.

16.6 Generalized transduction:

(1) transducing particles often contain only host DNA

(2) transducing particles may carry any segment of the host chromosome. Thus, all host genes are transduced.

Specialized transduction:

(1) transducing particles carry a recombinant chromosome containing both phage and host DNA

(2) only host genes that are adjacent to the prophage integration site are transduced.

16.7 A prophage is a phage chromosome that has become integrated into the host chromosome. The prophage is dormant in the sense that the phage genes involved in lytic development (replication and maturation) are repressed. The prophage is replicated during host chromosome replication just as any other segment of that chromosome.

16.8 By interrupting conjugation at various times after the donor and recipient cells are mixed (using a blender or other form of agitation), one can determine the length of time required to transfer a given genetic marker from an Hfr cell to an F- cell. Since the chromosome is transferred in a linear sequence, the positions of the genetic markers can be ordered relative to each other.

16.9 Cotransduction refers to the simultaneous transduction of two different genetic markers into a single recipient cell. Since bacteriophage particles can package a maximum of 1 to 2 percent of the total bacterial chromosome, only markers that are relatively closely linked can be cotransduced. The frequency of cotransduction of any two markers will be an inverse function of the distance between them on the chromosome. As such, this frequency can be used as an estimate of the linkage distance. Specific cotransduction-linkage relationships must be prepared for each phage-host system studied.

16.10

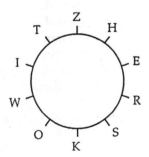

16.11 *lac*Y – *lac*Z – *pro*C.

16.12 *anth* – *A34* – *A223* – *A46*.

16.13 *anth* – *A487* – *A223* – *A58*.

16.14 The donor strain (Hfr) was $a^+b^+c^-d^+$ and the recipient (F⁻) was $a^-b^-c^+d^-$. We conclude this because with $a^+b^+c^-d^+$ as the donor, 2 exchanges are required to produce an $a^-b^+c^+d^-$ recombinant; 4 exchanges are necessary to produce the same recombinant in the reciprocal type of cross, a much rarer event.

16.15 The data indicate that the sequence is a–b–c because the rarest class ($a^+b^-c^+$) is a double crossover where the middle gene has switched position. The distance between a and b is calculated as the number of a^+b^- and a^-b^+ divided by the number of individuals who are either a^+b^-, a^-b^+, or a^+b^+:

$$400 + 2600 + 3600 + 100 = \frac{6700}{19900} = 0.34 \text{ (34 map units).}$$

The distance between b and c is calculated in a similar fashion:

$$700 + 400 + 100 + 1200 = \frac{2400}{18000} = 0.13 \text{ (13 map units).}$$

The distance between a and c is calculated to be:

$$700 + 2600 + 3600 + 1200 + = \frac{8100}{20200} = 0.40 \text{ (40 map units).}$$

The map is thus:

a - - - - 34 - - - - b - - - - 13 - - - - c
← - - - - - - - 40 - - - - - - - - - - - - - →

16.16 (a) The data suggest that the two genes are linked because the double transformation frequency in (B) is $\frac{247}{473} = 0.58$, meaning that the two markers were cotransduced 58 percent of the time, compared to the 2-event kinetics of (A) at $\frac{7}{580} = 0.01$. If the markers were not closely linked, the data from (A) and (B) would be comparable.

(b) The two classes with 130 and 96 represent crossovers between the two genes, so the recombination frequency is $\frac{226}{473} = 0.48$, or 48 map units.

16.17 If the sequence is a–b–c, the reciprocal crosses (A and B) should give the same frequency of + + + recombinants because in both crosses, two exchanges are required:

(A)   Hfr   +   +   c
                            X   X
F⁻ ⟶
                        a   b   +
(B)   Hfr   a   b   +
                                    X   X
F⁻ ⟶
                        +   +   c

16.17 (continued)

If the sequence is *a–c–b*, the reciprocal crosses give different frequencies of + + + recombinants because four exchanges are required in (A) and would be much less frequent than (B), where two are required:

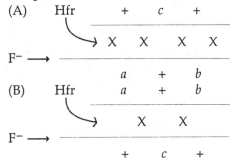

Thus the sequence is: *a–c–b*

CHAPTER 17

17.1   Same orientation: a deletion; opposite orientation: an inversion.

17.2   M cytotype and P transposase.

17.3   The paternally inherited *Bz* allele was inactivated by a transposable element insertion.

17.4   The pair in **(d)** are inverted repeats and could therefore qualify.

17.5   Through crossing over between the LTRs of a *gypsy* element.

17.6   The pair in **(a)** are direct repeats and could therefore qualify.

17.7   Resistance for the second antibiotic was acquired by conjugative gene transfer between the two types of cells.

17.8   The c^n mutation is due to a *Ds* or an *Ac* insertion.

17.9   Cross dysgenic (highly mutable) males carrying a wild-type X chromosome to females homozygous for a balancer X chromosome; then cross the heterozygous F_1 daughters individually to their brothers and screen the F_2 males that lack the balancer chromosome for mutant phenotypes, including failure to survive (lethality). Mutations identified in this screen are probably due to *P* element insertions in X-linked genes.

17.10 *In situ* hybridization to polytene chromosomes using a *TART* probe.

17.11 The transposition rate in humans may be very much less than it is in *Drosophila*.

17.12 Tn3 elements carry a gene that is not essential for transposition.

17.13 In the first strain, the F factor integrated into the chromosome by recombination with the IS element between genes *C* and *D*. In the second strain, it integrated by recombination with the IS element between genes *D* and *E*. The two strains transfer their genes in different orders because the two chromosomal IS elements are in opposite orientations.

17.14 No. IS*1* and IS*2* are mobilized by different transposases.

17.15 Both IS*50* elements should be able to excise from the transposon and insert elsewhere in the chromosome.

17.16 IS*50L* inserted on each side of the cluster of antibiotic resistance genes.

17.17 The *tnpA* mutation: no; the *tnpR* mutation: yes.

17.18 The *Ac* element must be tightly linked to the *Oc* allele.

17.19

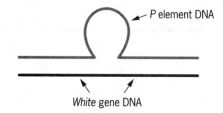

17.20 Factors made by the host's genome are required.

17.21 *TART* and *HeT-A* replenish the ends of *Drosophila* chromosomes.

17.22 No. The intron sequences would be removed by RNA processing prior to reverse transcription into DNA.

CHAPTER 18

18.1 The white phenotype is lethal in whole plants.

18.2 In *Mirabilis*, the chloroplasts are not inherited through the pollen, whereas in *Pelargonium*, they are.

18.3 Color depends on the proportion of wild-type and mutant chloroplasts in the tissue. Female gametes from green sectors should transmit the green color, female gametes from white sectors should transmit the white color, and female gametes from pale green sectors should transmit all three colors. Male gametes should have no effect on the trait.

18.4 Organelle heredity should affect both sexes identically, whereas sex chromosome heredity should affect them differently.

18.5 If *O. hookeri* had yellow plastids, and *O. muricata* had green plastids.

18.6 To prevent self-fertilization in the production of hybrid maize.

18.7 Spectinomycin resistance is encoded by a chloroplast gene.

18.8 Neutral petite.

18.9 Suppressive petite

18.10 Plant mtDNA molecules are much larger than animal mtDNA molecules and sometimes, two or more different mtDNA molecules are present within a single plant cell.

18.11 No. The genetic code for mitochondrial genes is not the same as that for *E. coli* genes.

18.12 Yes. Although the disease is inherited maternally, it can affect either sex.

18.13 Very rarely. Mutations would have to occur in two different cpDNA molecules within the same cell, and then these would have to recombine to produce a molecule that carried both mutations—an extremely unlikely series of events.

18.14 The results show biparental inheritance of chloroplast genes--a phenomenon made possible by irradiating the mt^+ parent prior to mating. Without irradiation, all the progeny would be resistant to spectinomycin and sensitive to streptomycin.

18.15 Half the offspring will be male-sterile. After the intercross, one-fourth of the offspring will be male-sterile.

18.16 All male-sterile.

18.17 Probably not. The polypeptide could be shorter or longer because of differences between the mitochondrial and bacterial genetic codes.

18.18 More than one gene encodes a polypeptide.

18.19 The expression of genes in two different genetic systems must be coordinated.

18.20 Organelles have circular DNA molecules, ribosomes, tRNAs, and special polymerases, some of which are similar to their bacterial counterparts.

CHAPTER 19

19.1 (a) Both introduce new genetic variability into the cell. In both cases, only one gene or a small segment of DNA representing a small fraction of the total genome is changed or added to the genome. The vast majority of the genes of the organism remain the same.

19.1 (b) The introduction of recombinant DNA molecules, if they come from a very different species, is more likely to result in a novel, functional gene product in the cell, if the introduced gene (or genes) is capable of being expressed in the foreign protoplasm. The introduction of recombinant DNA molecules is more analogous to duplication mutations (see Chapter 6) than to other types of mutations.

19.2 Restriction endonucleases recognize and cut specific nucleotide sequences in DNA. Most other endonucleases are not sequence-specific; many cut DNA sequences at random.

19.3 Recombinant DNA and gene-cloning techniques allow geneticists to isolate essentially any gene or DNA sequence of interest and to characterize it structurally and functionally. Large quantities of a given gene can be obtained in pure form, which permits one to determine its nucleotide-pair sequence (to "sequence it" in common lab jargon). From the nucleotide sequence and our knowledge of the genetic code, geneticists can predict the amino acid sequence of any polypeptide encoded by the gene. By using an appropriate subclone of the gene as a hybridization probe in northern blot analyses, geneticists can identify the tissues in which the gene is expressed. Based on the predicted amino acid sequence of a polypeptide encoded by a gene, geneticists can synthesize oligopeptides and use these to raise antibodies that, in turn, can be used to identify the actual product of the gene and localize it within cells or tissues of the organism. Thus recombinant DNA and gene-cloning technologies provide very powerful tools with which to study the genetic control of essentially all biological processes. These tools have played major roles in the explosive progress in the field of biology during the last two decades.

19.4 The nucleotide-pair sequence. Restriction endonucleases recognize a specific nucleotide-pair sequence in DNA regardless of the source of the DNA. In most cases, this is a 4 or 6 nucleotide-pair sequence; in a few cases, the recognition sequence is longer (for example, 8 nucleotide pairs). Most restriction enzymes cleave the two strands of the DNA at a specific position (between the same 2 adjacent nucleotides in each strand) within the recognition sequence. A few restriction enzymes bind at a specific recognition sequence, but cut the DNA at a nearby site outside of the recognition sequence. Some restriction endonucleases cut both strands between the same two nucleotide pairs ("blunt end" cutters), whereas others cut the two strands at different positions and yield complementary single-stranded ends ("sticky or staggered end" cutters). See Table 19.1 for examples.

19.5 Restriction endonucleases are believed to provide a kind of primitive immune system to the microorganisms that produce them—protecting their genetic material from "invasion" by foreign DNAs from viruses or other pathogens or just DNA in the environment that might be taken up by the microorganism. Obviously, these microorganisms do not have a sophisticated immune system like that of higher animals (Chapter 24).

19.6 Microorganisms that produce restriction endonucleases also produce enzymes that modify one or more bases in the recognition sequence for that endonuclease so that it can no longer cleave the DNA at that site. In most cases, the modifying enzyme is a methylase that attaches a methyl group to one or more of the bases in the recognition sequence. For example, *E. coli* strains that produce the restriction endonuclease *Eco*RI, also produce *Eco*RI methylase, an enzyme that transfers a methyl group from S-adenosylmethionine to the 3' adenine residue in each strand of the recognition sequence (5'-GAATCC-3') producing N^6-methyladenines at these positions. *Eco*RI cannot cleave DNA that contains N^6-methyladenine at these positions even if the *Eco*RI recognition sequence is present in this DNA. Thus, if one wishes to digest DNA with *Eco*RI, that DNA must not be isolated from an *E. coli* strain that is producing *Eco*RI methylase.

19.7   A foreign DNA cloned using an enzyme that produces single-stranded complementary ends can always be excised from the cloning vector by cleavage with the same restriction enzyme that was originally used to clone it. For example, if a *Hind*III fragment from the human genome is cloned into *Hind*III-cleaved pUC119, the human *Hind*III fragment can be excised from a plasmid DNA preparation of this clone by cleavage with restriction endonuclease *Hind*III. The human *Hind*III fragment will be flanked in the recombinant plasmid DNA clone by *Hind*III cleavage sites. When terminal transferase is used to add complementary single-stranded ends during cloning, the original restriction endonuclease cleavage sites are destroyed. Thus, the restriction enzyme used to generate the fragment for cloning cannot be used to excise the original fragment from the cloning vector.

19.8   Step 1: Digest genomic DNA isolated from your research organism with *Eco*RI. Step 2: Treat pUC118 DNA with *Eco*RI. Step 3: Mix the *Eco*RI-digested genomic and vector DNAs under annealing conditions and incubate with DNA ligase. Step 4: Transform *amp*^{s} *E. coli* cells carrying the *lacZΔM15* gene with the resulting ligation products. Step 5: Plate the transformed cells on nutrient agar medium containing Xgal and ampicillin. Only transformed cells will produce colonies in the presence of ampicillin. Step 6: Prepare your genomic DNA library by using bacteria from white colonies; these bacteria will contain pUC118 DNA with genomic DNA inserts. Bacteria harboring pUC118 plasmids with no insert will produce blue colonies.

19.9   Most genes of higher plants and animals contain noncoding intron sequences. These intron sequences will be present in genomic clones, but not in cDNA clones since cDNAs are synthesized using mRNA templates and intron sequences are removed during the processing of the primary transcripts to produce mature mRNAs.

19.10   Higher eukaryotes have very large genomes; for example, the genomes of mammals contain approximately 3×10^9 nucleotide pairs. Thus, trying to identify a particular single-copy gene from a clone library is like looking for the proverbial "needle-in-a-haystack." To accomplish this, one needs a nucleic acid hybridization probe specific for the gene or an antibody probe specific for the gene product. Given a specific cell or tissue type producing the mRNA and/or the protein gene product in large amounts, it was relatively easy to obtain pure mRNA or pure protein to use in making a hybridization or antibody probe, respectively, with which to screen a library for the gene or cDNA of interest. These approaches are much more difficult for the majority of the genes that encode products that represent only a small proportion of the total gene products in any given cell type.

19.11   The maize *gln2* gene contains many introns, and one of the introns contains a *Hind*III cleavage site. The intron sequences (and thus the *Hind*III cleavage site) are not present in mRNA sequences and thus are also not present in full-length *gln2* cDNA clones.

19.12   By oligonucleotide-directed site-specific mutagenesis, you can change each of the eight nucleotide pairs in the protein binding site to each of the other three possible nucleotide pairs. You can then examine the binding of the regulatory protein to each of the mutant sequences (24 single nucleotide-pair changes are possible at the eight positions), and you can examine the ability of each of the mutant sequences to mediate induction of transcription of the gene at 45°C *in vivo*. You could study the binding of the protein using synthetic oligonucleotide sequences more easily, but that approach would not let you study induction of transcription of the gene *in vivo*. The site-specific mutagenesis approach will allow you to carry out a saturation mutagenesis analysis of the octameric regulatory protein binding sequence.

19.13 The PCR technique has much greater sensitivity than any other method available for analyzing nucleic acids. Thus PCR procedures permits analysis of nucleic acid structure given extremely minute amounts of starting material. DNA sequences can be amplified and structurally analyzed from very small amounts of tissue like blood or sperm in assault and rape cases. In addition, PCR methods permit investigators to detect the presence of rare gene transcripts (for example, in specific types of cells) that could not be detected by less sensitive procedures such as northern blot analyses or *in situ* hybridization studies.

19.14 Some bacterial viruses package single-stranded DNA rather than double-stranded DNA. These bacteriophages employ a double-stranded DNA intermediate for replication, but then switch to the production of single-stranded DNAs for packaging during phage maturation. When a foreign DNA segment is cloned into the replicative form of the chromosome of one of these phages, only one of the two strands gets packaged during the subsequent maturation processes. In the case of the filamentous single-stranded DNA phages such as M13 and f2, the progeny phage particles are simply extruded through the cell membrane and wall without lysing the host bacterium. Because the phage are so small relative to the size of the bacteria, the phage particles can be separated from the bacteria by a simple low-speed centrifugation step. The bacteria form a pellet at the bottom of the centrifuge tube; the phage remain in the supernatant suspension. The bacterial pellets are discarded, and the phage particles can be collected by high-speed centrifugation or by precipitation with polyethylene glycol and low-speed centrifugation. Pure single-stranded DNA can then be separated from the phage proteins by phenol-chloroform extractions. Many cloning vectors like pUC118 and pUC119 (see Figure 19.14) are phage-plasmid hybrids. They have both plasmid and M13 phage origins of replication as well as M13 phage packaging signals so that they can replicate either as plasmids or as phages and can package single-stranded phage DNA when in the phage mode of replication. The switch from the plasmid mode of replication to the phage mode is accomplished by superinfecting the host bacteria with a mutant "helper" phage.

19.15 All modern cloning vectors contain a "polycloning site"—a cluster of cleavage sites for a number of different restriction endonucleases in a nonessential region of the vector into which foreign DNAs can be inserted. In general, the greater the complexity of the polycloning site—that is, the more restriction endonuclease cleavage sites that are present—the greater the utility of the vector for cloning a wide variety of different restriction fragments. For example, see the polycloning site present in pUC118 and pUC119 shown in Figure 19.14.

19.16 Transcription vectors permit one to synthesize high-specific-activity radioactive RNA transcripts for use as hybridization probes in northern blot and *in situ* hybridization studies. In addition, *in vitro* translation systems can be coupled to the *in vitro* transcription systems and used to synthesize radioactively labeled polypeptide gene products encoded by the cloned gene or cDNA. These polypeptide products can be analyzed by polyacrylamide gel electrophoresis and autoradiography. In some cases, this approach can be used to identify the product of the cloned gene or cDNA.

19.17 Restriction endonuclease *Hpa*I produces fragments with blunt ends. The simplest way to clone the desired *Hpa*I fragment into the *Hin*dIII site of pUC119 would be to use *Hin*dIII linkers. To obtain the desired clone, the *Hpa*I fragment of interest would be isolated by agarose gel electrophoresis, extracted from the agarose in the slice cut out of the gel, ligated to *Hin*dIII linkers, digested with *Hin*dIII restriction enzyme, separated from the remaining linker fragments by agarose gel electrophoresis and reextraction, and ligated into *Hin*dIII-cut pUC119. Amp^s *E. coli* cells would then be transformed with the ligation products and plated on medium containing ampicillin and X-gal (5-bromo-4-chloro-3-indolyl-β-D-galactoside). Only bacteria containing pUC119 plasmids

19.17 (continued)

will be able to grow on the ampicillin-containing medium. Two kinds of colonies will be present: blue colonies that harbor pUC119 plasmids with no foreign DNA inserts and white colonies that harbor plasmids with inserts. A white colony should be used to inoculate rich broth medium containing ampicillin to grow a culture of cells containing the desired clone. Plasmid DNA should then be isolated from the resulting cell culture and used to verify the presence of the *Hpa*I fragment of interest. The entire procedure can be carried out in two or three days.

19.18 The name "western blot" has no literal significance; it is pure laboratory jargon. In 1975, E. M. Southern published a procedure for the transfer of DNA molecules that had been separated by agarose gel electrophoresis to nitrocellulose membranes and their subsequent detection by hybridization to radioactive probes and autoradiography. The resulting autoradiographs showing the positions of the DNA bands that had hybridized to the probes were called Southern blots in reference to E. M. Southern who had developed the technique. When similar procedures were developed for analyzing RNA molecules that had been separated by gel electrophoresis, researchers started calling them northern blots because of their similarity to Southern blots. When the methodology was extended to proteins, the term western blot supposedly was the logical extension of the Southern, northern laboratory jargon. Or, was it? Why wasn't the western blot called an eastern blot?

19.19 Because the nucleotide-pair sequences of both the normal *CF* gene and the *CFΔ508* mutant gene are known, labeled oligonucleotides can be synthesized and used as hybridization probes to detect the presence of each allele (normal and *Δ508*). Under high-stringency hybridization conditions, each probe will hybridize only with the *CF* allele that exhibits perfect complementarity to itself. Since the sequences of the *CF* gene flanking the *Δ508* site are known, oligonucleotide PCR primers can be synthesized and used to amplify this segment of the DNA obtained from small tissue explants of putative CF patients and their relatives by PCR. The amplified DNAs can then be separated by agarose gel electrophoresis, transferred to nylon membranes, and hybridized to the respective labeled oligonucleotide probes, and the presence of each *CF* allele detected by autoradiography. For a demonstration of the utility of this procedure, see B. Kerem et al. "Identification of the cystic fibrosis gene: Genetic analysis." *Science* 245: 1073–1080, 1989. Kerem and coworkers used two synthetic oligonucleotide probes (oligo-N = 3'-CTTTTATAGTAGAAACCAC-5' and oligo-ΔF = 3'-TTCTTTTATAGTA- - -ACCACAA-5'; the dashes indicate the deleted nucleotides in the *CFΔ508* mutant allele) to analyze the DNA of CF patients and their parents. For confirmed CF families, the results of these Southern blot hybridizations with the oligo-N (normal) and oligo-ΔF (*CFΔ508*) labeled probes were often as follows:

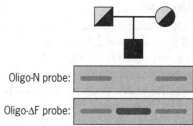

19.19 (continued)

Both parents were heterozygous for the normal *CF* allele and the mutant *CFΔ508* allele as would be expected for a rare recessive trait, and the CF patient was homozygous for the *CFΔ508* allele. In such families, one-fourth of the children would be expected to be homozygous for the *Δ508* mutant allele and exhibit the symptoms of CF, whereas three-fourths would be normal (not have CF). However, two-thirds of these normal children would be expected to be heterozygous and transmit the allele to their children. Only one-fourth of the children of this family would be homozygous for the normal *CF* allele and have no chance of transmitting the mutant *CF* gene to their offspring. Note that the screening procedure described here can be used to determine which of the normal children are carriers of the *CFΔ508* allele: that is, the mutant gene can be detected in heterozygotes as well as homozygotes.

19.20 You could attempt to isolate either a dihydropicolinate synthase (DHPS) cDNA clone or a DHPS genomic clone. Once you have isolated either DHPS clone, it can be used as a hybridization probe to isolate the other (genomic or cDNA) by screening an appropriate library by *in situ* colony hybridization. Four approaches that have proven effective in isolating other eukaryotic coding sequences of interest are the following:

(1) You could obtain a clone of the DHPS gene of a lower eukaryote (a clone of the DHPS gene of *Saccharomyces cerevisiae* is available) or even a prokaryote and use it as a heterologous hybridization probe to screen a maize cDNA library using low stringency conditions. Sometimes this approach is successful; sometimes it is not successful. Whether or not this approach works depends on how similar the coding sequences of the specific gene of interest are in the two species.

(2) You could purify the DHPS enzyme from corn and use the purified protein to produce an antibody to DHPS. This DHPS-specific antibody could then be used to screen a maize cDNA expression library by a protocol analogous to the western blot procedure. (An expression library contains the cDNA coding sequences fused to appropriate transcription and translation signals so that they are expressed in *E. coli* or other host cells in which the cDNA library is prepared.)

(3) You could purify the DHPS enzyme from maize and determine the amino acid sequence of its NH2-terminus by microsequencing techniques. From the amino acid sequence and the known genetic code, you could predict the possible nucleotide sequences encoding this segment of the protein. Because of the degeneracy in the code, there would be a set of nucleotide sequences that would all specify the same amino acid sequence, and you would not know which one was present in the maize DHPS gene. However, the synthesis of oligonucleotides is now routine and quite inexpensive. Thus, you could synthesize a mixture of oligonucleotides containing all possible coding sequences and use this mixture as a set of hybridization probes to screen an appropriate library by *in situ* colony hybridization.

(4) Finally, you might try a simple and very quick genetic approach based on the ability of cDNAs in an expression library to rescue DHPS mutants of *E. coli* or other species that can be transformed at high frequencies. You would obtain a DHPS-deficient mutant of *E. coli* (available from the *E. coli* Genetics Stock Center at Yale University), transform it with your cDNA expression library, and plate the transformed cells on medium lacking diaminopimelic acid (the product of DHPS). DHPS-deficient *E. coli* mutants cannot grow in the absence of diaminopimelic acid; thus, any colonies that grow on your selection plates should be the result of rescue of the DHPS mutant bacteria by corn DHPS encoded by cDNAs in the library. This entire screening procedure can be carried out in three or four days; thus, it is much simpler than the preceding approaches. In fact, David A. Frisch first isolated a maize DHPS cDNA by this simple, but powerful genetic approach. However, note that this approach would only be expected to work in the case of enzymes that are active as monomers or homomultimers; it is not applicable when the active form of the enzyme is a heteromultimer.

19.21 (a) Southern, northern, and western blot procedures all share one common step, namely, the transfer of macromolecules (DNAs, RNAs, and proteins, respectively) that have been separated by gel electrophoresis to a solid support—usually a nitrocellulose or nylon membrane—for further analysis.

(b) The major difference between these techniques is the class of macromolecules that are separated during the electrophoresis step: DNA for Southern blots, RNA for northern blots, and protein for western blots.

19.22 No. The genome of the species in question may contain one, two, or three copies of the gene (or family of closely related genes) encoding this protein. The possibilities are:

(1) one copy of the gene with two *Eco*RI cleavage sites located within intron sequences

(2) two copies of the gene with one *Eco*RI cleavage site located within an intron sequence of one of the copies

(3) three copies of the gene with no *Eco*RI cleavage site in any of the copies, that is, each copy present on a single *Eco*RI restriction fragment.

19.23

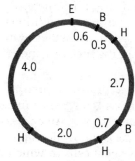

Restriction enzyme cleavage sites for *Bam*HI, *Eco*RI, and *Hin*dIII are denoted by B, E, and H, respectively. The numbers give distances in kilobase pairs.

19.24

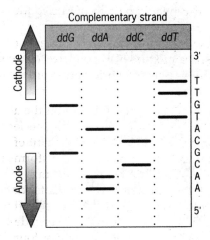

CHAPTER 20

20.1 Genetic map distances are determined by crossover frequencies. Cytogenetic maps are based on chromosome morphology or physical features of chromosomes. Physical maps are based on actual physical distances—the number of nucleotide pairs (0.34 nm per bp)—separating genetic markers. If a gene or other DNA sequence of interest is shown to be located near a mutant gene, a specific knob on a chromosome, or a particular DNA restriction fragment, that genetic or physical marker (mutation, knob, or restriction fragment) can be used to initiate a chromosome walk (see Figure 20.5).

20.2 Chromosome walks are performed by isolating overlapping genomic clones and using them to "walk" along a chromosome or, more accurately, along a chromosomal DNA molecule (see Figure 20.5). Chromosome jumps use large genomic DNA fragments produced by partial digestion of genomic DNA with a restriction endonuclease to "jump" to a new position 100 kb or more away from the original site in a DNA molecule (see Figure 20.6). If there is no molecular marker near a gene of interest, isolating the gene by chromosome walking would take too long. In such cases, chromosome jumps must be used to speed up the process.

20.3 A contig (<u>contig</u>uous clones) is a physical map of a chromosome or part of a chromosome prepared from a set of overlapping genomic DNA clones. An RFLP (<u>r</u>estriction <u>f</u>ragment <u>l</u>ength <u>p</u>olymorphism) is a variation in the length of a specific restriction fragment excised from a chromosome by digestion with one or more restriction endonucleases. A VNTR (<u>v</u>ariable <u>n</u>umber <u>t</u>andem <u>r</u>epeat) is a short DNA sequence that is present in the genome as tandem repeats and in highly variable copy number. An STS (<u>s</u>equence <u>t</u>agged <u>s</u>ite) is a unique DNA sequence that has been mapped to a specific site on a chromosome. An EST (<u>e</u>xpressed <u>s</u>equence <u>t</u>ag) is a cDNA sequence—a genomic sequence that is transcribed. Contig maps permit researchers to obtain clones harboring genes of interest directly from DNA Stock Centers—to "clone by phone." RFLPs are used to construct the high-density genetic maps that are needed for positional cloning. VNTRs are especially valuable RFLPs that are used to identify multiple sites in genomes. STSs and ESTs provide molecular probes that can be used to initiate chromosome walks to nearby genes of interest.

20.4 CpG islands are clusters of cytosines and guanines that are often located just upstream (5') from the coding regions of human genes. Their presence in nucleotide sequences can provide hints as to the location of genes in human chromosomes.

20.5 The gene was named *huntingtin* after the disease that it causes when defective. The gene will probably be renamed after the function of its gene product has been determined.

20.6 The gene encoding dystrophin was identified by map position-based cloning, and the amino acid sequence of dystrophin was predicted from the nucleotide sequences of dystrophin cDNA clones by using the established codon assignments. The CF gene was also identified by map position-based cloning, and, as with dystrophin, the nucleotide sequences of CF cDNAs were used to predict the amino acid sequence of the CF gene product. A computer search of the protein data banks revealed that the CF gene product was similar to several ion channel proteins. This result focused the attention of scientists studying cystic fibrosis on proteins involved in the transport of salts between cells and led to the discovery that the CF gene product was a transmembrane conductance regulator—now called the CFTR protein.

20.7 Once the function of the CF gene product has been established, scientists should be able to develop procedures for introducing wild-type copies of the *CF* gene into the appropriate cells of cystic fibrosis patients to alleviate the devastating effects of the mutant gene. A major obstacle to somatic-cell gene-therapy treatment of cystic fibrosis is the size of the *CF* gene—about 250 kb, which is too large to fit in the standard gene transfer vectors. Perhaps a shortened version of the gene constructed from the *CF* cDNA—about 6.5 kb—can be used in place of the wild-type gene. A second major obstacle is getting the transgene into enough of the target cells of the cystic fibrosis patient to alleviate the symptoms of the disease. A third challenge is to develop an expression vector containing the gene that will result in long-term expression of the introduced gene in transgenic cells. Another concern is how to avoid possible side effects caused by overexpression or inappropriate expression of the transgene in cystic fibrosis patients. Despite these obstacles, many scientists are optimistic that cystic fibrosis will be effectively treated by somatic-cell gene therapy in the future.

20.8 Oligonucleotide primers complementary to DNA sequences on both sides (upstream and downstream) of the CAG repeat region in the *MD* gene can be synthesized and used to amplify the repeat region by PCR. One primer must be complementary to an upstream region of the template strand, and the other primer must be complementary to a downstream region of the nontemplate strand. After amplification, the size(s) of the CAG repeat regions can be determined by gel electrophoresis (see Figure 20.11). Trinucleotide repeat lengths can be measured by including repeat regions of known length on the gel. If less than 30 copies of the trinucleotide repeat are present on each chromosome, the newborn, fetus, or pre-embryo is homozygous for a wild-type *MD* allele or heterozygous for two different wild-type *MD* alleles. If more than 50 copies of the repeat are present on each of the homologous chromosomes, the individual, fetus, or cell is homozygous for a dominant mutant *MD* allele or heterozygous for two different mutant alleles. If one chromosome contains less than 30 copies of the CAG repeat and the homologous chromosome contains more than 50 copies, the newborn, fetus, or pre-embryo is heterozygous, carrying one wild-type *MD* allele and one mutant MD allele.

20.9 Yes. A somatic-cell gene therapy procedure similar to that used for ADA⁻ SCID (see Figure 20.18) might be effective in treating purine nucleoside phosphorylase (PNP) deficiency. White blood cells could be isolated from the patient, transfected with a vector carrying a wild-type *PNP* gene, grown in culture and assayed for the expression of the *PNP* transgene, and then infused back into the patient after the expression of the transgene had been verified.

20.10 The goals of the Human Genome Project are to prepare genetic and physical maps showing the locations of all the genes in the human genome and to determine the nucleotide sequences of all 24 chromosomes in the human genome. These maps and nucleotide sequences of the human chromosomes will help scientists identify mutant genes that result in inherited diseases and, hopefully, will lead to successful gene therapies for at least some of these diseases.

20.11 The transcription initiation and termination and translation initiation signals of eukaryotes differ from those of prokaryotes such as *E. coli*. Therefore, to produce a human protein in *E. coli*, the coding sequence of the human gene must be joined to appropriate *E. coli* regulatory signals—promoter, transcription terminator, and translation initiator sequences. Moreover, if the gene contains introns, they must be removed or the coding sequence of a cDNA must be used, because *E. coli* does not possess the spliceosomes required for the excision of introns from nuclear gene transcripts. In addition, many eukaryotic proteins undergo post-translational processing events that are not carried out in prokaryotic cells. Such proteins are more easily produced in transgenic eukaryotic cells growing in culture.

20.12 You would first construct a chimeric gene containing your synthetic gene fused to a plant promoter such as the 35S promoter of cauliflower mosaic virus and a plant transcription termination and polyadenylation signal such as the one from the *nos* gene of the Ti plasmid. This chimeric gene would then be inserted into the T-DNA of a Ti plasmid carrying a dominant selectable marker gene (for example, 35S/NPTII/*nos*, which confers resistance to kanamycin to host cells) and introduced into *Agrobacterium tumefaciens* cells by transformation. Tissue explants from *Arabidopsis* plants would be co-cultivated with *A. tumefaciens* cells harboring the recombinant Ti plasmid, and plant cells that carry T-DNAs inserted into their chromosomes would be selected by growth on medium containing the appropriate selective agent (for example, kanamycin). Transgenic plants would then be regenerated from the transformed cells and tested for resistance to glyphosate.

20.13 DNA fingerprints are the specific patterns of bands present on Southern blots of genomic DNAs that have been digested with particular restriction enzymes and hybridized to appropriate DNA probes such as VNTR sequences. DNA fingerprints, like epidermal fingerprints, are used as evidence for identity or nonidentity in forensic cases. Geneticists have expressed concerns about the statistical uses of DNA fingerprint data. In particular, they have questioned some of the methods used to calculate the probability that DNA from someone other than the suspect could have produced the observed DNA fingerprints. This concern is based in part on the lack of adequate databases for various human subpopulations and the lack of precise information about the amount of variability in DNA fingerprints for individuals of different ethnic backgrounds. These concerns can be best addressed by the acquisition of data on fingerprint variability in different subpopulations and ethnic groups.

20.14 Neither F1 nor F2 could be the girl's biological father; only individual F3 could be the child's father.

20.15 The T in T-DNA is an abbreviation for "transferred." The T-DNA region of the Ti plasmid is the segment that is transferred from the Ti plasmid of the bacterium to the chromosomes of the plant cells during *Agrobacterium tumefaciens*-mediated transformation.

20.16 Disarmed retroviruses are lacking genes essential for reproduction in host cells, but can still integrate into the DNA of the host cell in the proviral state (see the Technical Sidelight: Proviruses as Vectors for Human Gene Therapy in Chapter 21). The retroviral genomes are small enough to allow them to be manipulated easily *in vitro* and yet will accept foreign DNA inserts of average gene size (see Figure 20.7). The retroviruses contain strong promoters in their long terminal repeats that can be used to drive high levels of transcription of the foreign gene insert.

20.17 Transgenic mice are usually produced by microinjecting the genes of interest into pronuclei of fertilized eggs or by infecting pre-implantation embryos with retroviral vectors containing the genes of interest. Transgenic mice provide invaluable tools for studies of gene expression, mammalian development, and the immune system of mammals. Transgenic mice are of major importance in medicine; they provide the model system most closely related to humans. They have been, and undoubtedly will continue to be, of great value in developing the tools and technology that will be used for human gene therapy in the future.

20.18 An antisense RNA molecule is an RNA that is complementary to the pre-mRNA or mRNA ("sense" sequence) of a gene. In some cases, antisense RNAs can be used to block gene expression and study the effect(s) of the absence of a particular gene product on the growth and development of an organism. In agriculture, antisense RNAs have been used to block the synthesis of specific gene products to achieve desired changes in the phenotype of an organism. The best known example is the FlavrSavr™ tomato, which remains firm longer during the ripening process and allows the tomatos to remain on the vines longer before picking. The FlavrSavr™ tomato was produced by using antisense RNA to decrease the rate of expression of the gene encoding the enzyme polygalacturonase, which breaks down cell walls and causes softening during the ripening process. An organism that synthesizes an antisense RNA is usually produced by inverting the coding sequence of a gene relative to the promoter and transcription termination signal and reintroducing the altered gene into the host organism by transformation. With the coding sequence inverted, the nontemplate strand will be transcribed — rather than the template strand — producing antisense RNAs (see Figure 20.29).

20.19 Post-translationally modified proteins can be produced in transgenic eukaryotic cells growing in culture or in transgenic plants and animals. Indeed, transgenic sheep have been produced that secrete human blood-clotting factor IX and α1-antitrypsin in their milk. These sheep were produced by fusing the coding sequences of the respective genes to a DNA sequence that encodes the signal peptide required for secretion and introducing this chimeric gene into fertilized eggs that were then implanted and allowed to develop into transgenic animals. In principle, this approach could be used to produce any protein of interest.

20.20 The noncoding regions of the ten actin genes can be subcloned and used as gene-specific hybridization probes to measure individual gene transcript levels in various organs and tissues of developing plants by either northern blot or *in situ* hybridization experiments. Alternatively, the gene-specific noncoding regions can be used to design PCR primers, and PCR can be used to measure individual gene transcript levels in organs and tissues. Perhaps the most powerful approach would be

(1) to fuse the promoter regions of the individual genes to the coding region of a reporter gene that encodes a product that can be detected by histochemical assays of plant organs and tissues

(2) to introduce these chimeric reporter genes into *Arabidopsis* plants by *Agrobacterium tumefaciens*-mediated transformations

(3) to analyze the temporal and spatial patterns of expression of the reporter genes in the transgenic plants. Indeed, Meagher and colleagues have already used these approaches to document the striking temporal and tissue-specific patterns of actin gene expression in *Arabidopsis*.

CHAPTER 21

21.1 By studying the synthesis or lack of synthesis of the enzyme in cells grown on chemically defined media. If the enzyme is synthesized only in the presence of a certain metabolite or a particular set of metabolites, it is probably inducible. If it is synthesized in the absence but not in the presence of a particular metabolite or group of metabolites, it is probably repressible.

21.2 Repression occurs at the level of transcription during enzyme synthesis. The end-product, or a derivative of the end-product, of a repressible system acts as an effector molecule that usually, if not always, combines with the product of one or more regulator genes to turn off the *synthesis* of the enzymes in the biosynthetic pathway. Feedback inhibition occurs at the level of enzyme *activity*; it usually involves the first enzyme of the biosynthetic pathway. Feedback inhibition thus brings about an immediate arrest of the biosynthesis of the end-product. Together, feedback inhibition and repression rapidly and efficiently turn off the synthesis of both the enzymes and the end-products that no longer need to be synthesized by the cell.

21.3

	Gene or Regulatory Element	Function
(a)	Regulator gene	Codes for repressor
(b)	Operator	Binding site of repressor
(c)	Promoter	Binding site of RNA polymerase and CAP-cAMP complex
(d)	Structural gene Z	Encodes β-galactosidase
(e)	Structural gene Y	Encodes β–galactoside permease

21.4 (1) Constitutive synthhesis of the *lac* enzymes
(2) Constitutive synthesis of the *lac* enzymes
(3) Uninducibility of the *lac* enzymes
(4) No β-galactosidase activity
(5) No β–galactoside permease activity.

21.5 **(a)** 1, 2, 3*, and 5
(b) 2, 3*, and 5; *3 may be either noninducible or constitutive, depending on whether the specific O^c mutation eliminates binding of the I^s "superrepressor."

21.6 **(a)** Inducible, this is the wild-type genotype and phenotype
(b) constitutive, the O^c mutation produces an operator that is not recognized by the *lac* repressor
(c) constitutive, same as for **(b)**
(d) inducible, I^+ is dominant to I^-
(e) constitutive, no active repressor is synthesized in this bacterium.

21.7 **(a)** $\dfrac{I^+O^cZ^+Y^-}{I^+O^+Z^-Y^+}$

(b) $\dfrac{I^sO^cZ^+Y^-}{I^sO^+Z^-Y^+}$

21.8 **(a)** The O^c mutations map very close to the Z structural gene; I^- mutations map slightly farther from the structural gene (but still very close by; see Figure 21.4).

(b) An $\dfrac{I^+O^cZ^+Y^+}{I^+O^+Z^+Y^+}$ partial diploid would exhibit constitutive synthesis of β-galactosidase and β-galactoside permease an $\dfrac{I^+O^+Z^+Y^+}{I^-O^+Z^+Y^+}$ partial diploid would be inducible for the synthesis of these enzymes.

(c) The O^c mutation is *cis*-dominant; I^- is *trans*-recessive.

21.9 The system could have developed from a series of tandem duplications of a single ancestral gene. Mutational changes that make the system more efficient and, therefore, favored in selection could have brought the system to its present level of efficiency.

21.10 Catabolite repression has apparently evolved to assure the use of glucose as a carbon source when this carbohydrate is available, rather than less efficient energy sources.

21.11 Possibly by directly or indirectly inhibiting the enzyme adenylcyclase, which catalyzes the synthesis of cyclic AMP from ATP.

21.12 Positive regulation; the CAP-cAMP complex has a positive effect on the expression of the *lac* operon. It functions in turning on the transcription of the structural genes in the operon.

21.13 Yes; in the gene coding for CAP. Some mutations in this gene might result in a CAP that binds to the promoter in the absence of cAMP. Also, mutations in the gene (or genes) coding for the protein (or proteins) that regulate the cAMP level as a function of glucose concentration.

21.14 Negative regulatory mechanisms such as that involving the repressor in the lactose operon block the transcription of the structural genes of the operon, whereas positive mechanisms such as the activator in the arabinose operon or the CAP-cAMP complex in the *lac* operon promote the transcription of the structural genes of the operon.

21.15 Uninducible, but with a higher level of baseline synthesis of the arabinose enzymes; or, stated differently, a low level of constitutive synthesis of the enzymes. The product of the arabinose regulator gene is required for induction; in its absence, induction could not occur. However, in the absence of arabinose, the regulator gene product represses the level of *ara* operon transcription. This effect would also be eliminated in such a deletion mutant, resulting in a higher baseline level of synthesis (or a low level of constitutive synthesis).

21.16 0.1; 100; 25.1; 200.

21.17 Constitutive synthesis of the *araB*, *araA*, and *araD* proteins would occur at basal or noninduced levels because the *araC* protein could no longer bind at $araO_2$ and form the loop structure (see Figure 21.15c).

21.18 Repression/derepression of the *trp* operon occurs at the level of transcript initiation, modulating the frequency at which RNA polymerase initiates transcription from the *trp* operon promoters. Attenuation modulates *trp* transcript levels by altering the frequency of termination of transcription within the *trp* operon leader region (*trpL*).

21.19 Deletion of the *trpL* region would result in the levels of the tryptophan biosynthetic enzymes in cells growing in the presence of tryptophan being increased about tenfold because attenuation would no longer occur if this region were absent.

21.20 First, remember that transcription and translation are coupled in prokaryotes. When tryptophan is present in cells, tryptophan-charged $tRNA^{Trp}$ is produced. This allows translation of the *trp* leader sequence through the two UGG Trp codons to the *trp* leader sequence UGA termination codon. This translation of the *trp* leader region prevents base-pairing between the partially complementary mRNA leader sequences 74-85 and 108-119 (see Figure 21.13b), which in turn permits formation of the transcription-termination "hairpin" involving leader sequences 108-119 and 126-134 (see Figure 21.13c).

21.21 No. Attenuation of the *trp* operon would now be controlled by the presence or absence of $Gly-tRNA^{Gly}$.

21.22 A prophage is a chromosome of a bacteriophage after it has become integrated into the chromosome of the host bacterium. The lambda prophage is present as a transcriptionally inactive linear structure in the *E. coli* chromosome. During lytic infections, the lambda chromosome begins its reproductive cycle by replicating as a transcriptionally active circular DNA molecule.

21.23 The autogenously maintained repression of λ lytic genes during lysogeny occurs at the level of transcriptional initiation, whereas the λ lytic cascade is regulated at the level of transcriptional termination. In the first case, a repressor protein binds to the promoter and blocks the transcription of the λ lytic genes. In the second case, two key antiterminator proteins act to prevent the termination of transcription at sites distal to the immediate early and delayed early coding sequences. These antitermination events result in the transcription of delayed early and late λ genes.

21.24 The operator region of the *lac* operon in *E. coli* contains a single repressor binding site, whereas the lambda operators O_L and O_R each contain three repressor binding sites that are differentially occupied under different conditions. The differential affinity of lambda repressor for the three binding sites in these operators plays an important role in the maintenance of the lysogenic state of a lambda prophage in *E. coli*.

21.25 Autogenous means self-generated. The λ repressor regulates its own synthesis; the presence of λ repressor stimulates the synthesis of more repressor. This autoregulatory mechanism enhances the stability of the lysogenic state by virtually assuring that once repressor has been synthesized in an infected bacterium, it will continue to be synthesized in amounts that are sufficient to keep the l lytic genes repressed.

21.26 The precise temporal pattern of gene expression—immediate early genes followed by delayed early genes and then late genes—during the lytic cycle of phage lambda is controlled at the level of transcriptional termination. The two immediate early genes, *cro* and *N*, are adjacent to the origins of transcription at P_L and P_R (see Figure 21.20). The product of gene *N* is a transcriptional antiterminator that causes transcription to continue past the termination signals t_{L1} and t_{R1} and on through the delayed early genes (see Figure 21.22). Similarly, the product of one of the delayed early genes, *Q*, functions as a transcriptional antiterminator at site t_{R3}, resulting in the continuation of transcription through the late genes.

21.27 The N protein, the product of the first gene transcribed from promoter P_L, prevents the termination of transcription at two sites (t_{R1} and t_{L1}), which results in the transcription of the delayed early genes located downstream from these sites. N protein binds to *nut* sites just upstream from the termination sites, and, together with the Nus protein and ribosomal protein S10, modifies the specificity of RNA polymerase so that termination does not occur at sites t_{R1} and t_{L1} (see Figure 21.22).

21.28 This mutation will produce the so-called clear-plaque phenotype. In the absence of functional lambda repressor, lambda phage can only reproduce lytically. Lysogeny requires the lambda repressor to be present to keep the lytic genes of the prophage repressed. Thus all infected cells will be killed and lysed during the lambda lytic cycle, resulting in clear plaques rather than the turbid plaques formed by wild-type lambda

21.29 Whether a λ phage will undergo lytic growth or enter the lysogenic pathway when it infects an *E. coli* cell is determined by a genetic switch controlled by two regulatory proteins—λ repressor and Cro protein—that bind to the λ operators and govern transcription of the λ genome. If Cro protein occupies these sites, lytic development occurs. If λ repressor occupies these sites, lysogeny results. The C_I gene encodes the λ repressor. Thus lytic development occurs in cells infected with λ phage that carry a deletion of the C_I gene. In contrast, lysogeny will occur in bacteria infected with λ phage that harbor a deletion of the *cro* gene.

21.30 No. Since transcription (nucleus) and translation (cytoplasm) are not coupled in eukaryotes, attenuation of the type occurring in prokaryotes would not be possible.

CHAPTER 22

22.1 In multicellular eukaryotes, the environment of an individual cell is relatively stable. There is no need to respond quickly to changes in the external environment. In addition, the development of a multicellular organism involves complex regulatory hierarchies composed of hundreds of different genes. The expression of these genes is regulated spatially and temporally, often through intricate intercellular signaling processes.

22.2 Both types of hormones bind to a receptor protein. The steroid hormone-receptor complex then proceeds to the nucleus where it acts as a transcription factor to regulate gene expression. The peptide hormone-receptor complex remains at the cell's surface, where it interacts with other proteins to convey a signal into the cytoplasm; eventually, this signal induces a transcription factor to enter the nucleus and regulate gene expression.

22.3 By monitoring puffs in response to environmental signals, such as heat shock, or to hormonal signals.

22.4 An enhancer can be located upstream, downstream, or within a gene and it functions independently of its orientation. A promoter is almost always immediately upstream of a gene and it functions only in one direction with respect to the gene.

22.5 By alternate splicing of the transcript.

22.6 The two genes are controlled by different enhancers, one active in roots and the other in vascular tissues.

22.7 Northern blotting of RNA extracted from plants grown with and without light, or PCR amplification of cDNA made by reverse transcribing these same RNA extracts.

22.8 It may make an RNA but it not a functional polypeptide.

22.9 The green fluorescent protein will be made after the flies are heat shocked.

22.10 Probably not unless the promoter of the *gfp* gene is recognized and transcribed by the *Drosophila* RNA polymerase independently of the heat shock response elements.

22.11 This enzyme plays an important role in photosynthesis, a light-dependent process. Thus it makes sense that its production should be triggered by exposure to light.

22.12 A zinc finger is a short peptide loop that forms when two cysteines in one part of a polypeptide and two histidines in another part nearby jointly bind a zinc ion. A leucine zipper is a peptide sequence with a leucine at every seventh position.

22.13 The flies would be phenotypically normal males.

22.14 Both XX and XY animals would develop as intersexes.

22.15 Yes. Enhancers are able to function in different positions in and around a gene.

22.16 Polytene chromosome puffs.

22.17 No

22.18 Yes

22.19 H4 is acetylated

22.20 The *mle* gene is not functional in females.

22.21 They possess introns.

22.22 The products of these genes play important roles in cell activities.

22.23 Mutations in these codons cause amino acids changes that activate the Ras protein.

22.24 A promoter mutation could alter the rate of transcription of the oncogene, causing inappropriate expression; a mutation in the gene's coding region could alter or abolish the function of the polypeptide product; a rearrangement mutation could put the gene under the control of a different enhancer, causing inappropriate expression.

22.25 Ras protein is an activator of cell metabolism; RB protein is a suppressor of cell metabolism.

CHAPTER 23

23.1 Cell division forms the blastula; cell movement forms the gastrula.

23.2 Unequal division of the cytoplasm during the meiotic divisions; transport of substances into the oocyte from surrounding cells such as the nurse cells in *Drosophila*.

23.3 Determination establishes cell fates; differentiation realizes those fates.

23.4 No. The primary germ layers are ectoderm, endoderm and mesoderm.

23.5 Collect mutations with diagnostic phenotypes; map the mutations and test them for allelism with one another; perform epistasis tests with mutations in different genes; clone individual genes and analyze their function at the molecular level.

23.6 Mitotic division is so rapid that there is not enough time for membranes to form between cells.

23.7 Imaginal discs

23.8 1/4

23.9 Mate homozygous *unc* hermaphrodites that carry one of the *dumpy* mutations with males that carry the other *dumpy* mutation and score the non-uncoordinated (*unc*/+) F_1 hermaphrodites for the dumpy phenotype. If these worms are dumpy, the two mutations are allelic; if they are wild-type, the two mutations are not allelic.

23.10 The mutation would kill males but not females.

23.11 Cross heterozygous carriers separately with *tra*/+ and *tra2*/+ flies. If any of the XX progeny of these matings are transformed into sterile males, the new mutation is an allele of the *tra* or *tra2* tester.

23.12 X:A ratio → *Sxl* →*tra* → *dsx* → differentiation as male or female
                       tra2

23.13 Intersex. The *dsx* mutation is epistatic to the *tra* mutation.

23.14 The X:A ratio is 2:3—between the 1:2 ratio needed for normal male development and the 1:1 ratio needed for normal female development. By chance, the autosomal denominator proteins will be in excess in some cells of the triploid, causing the *Sxl* gene to be inactivated; these cells will develop male characteristics. In other cells, the X-linked numerator proteins will be in excess, causing the *Sxl* gene will be activated; these cells will develop female characteristics. Thus, the triploid fly will be a mosaic of male and female cells—that is, an intersex.

23.15 XX, male phenotype; XO, female phenotype.

23.16 The *xol-1* gene is needed in *Caenorhabditis* males but not in hermaphrodites; the *Sxl* gene is needed in *Drosophila* females but not in males.

23.17 The fly is a gynandromorph. It probably arose through the loss of a wild-type X chromosome early in the development of a *y*/+ zygote.

23.18 Somatic crossing over between the *white* gene and the centromere. The *white* gene functions autonomously.

23.19 The lethal is autonomous.

23.20 Between positions 5 and 8.

23.21 (a) Wild-type
    (b) embryonic lethal
    (c) embryonic lethal
    (d) wild-type
    (e) wild-type

23.22 Some structures fail to develop in the posterior portion of the embryo.

23.23 Female sterility.

23.24 Screen for lethal mutations that prevent regions of the embryo from developing normally.

23.25 $ey \rightarrow boss \rightarrow sev \rightarrow$ R7 differentiation

23.26 Extra mouse eyes

23.27 Northern blotting of RNA extracted from the tissues at different times during development. Hybridize the blot with gene-specific probes.

CHAPTER 24

24.1 The genetic information specifying antibody chains is stored in sequences of nucleotide pairs encoding sequences of amino acids, just like the genetic information specifying other polypeptides. However, the information specifying an antibody chain is stored in bits and pieces that are assembled into functional genes encoding antibody chains by genome rearrangements (somatic-cell recombination events) occurring during the development of the B lymphocytes (the antibody-producing cells). See Figures 24.12–24.14.

24.2 4; 2; 1.

24.3 (1) The joining of different V, D, and J gene segments by somatic recombination during B lymphocyte development (see Figures 24.12–24.14)

   (2) variability in the exact location of the joining reaction during V-D-J joining events (see Figure 24.16)

   (3) somatic mutation.

24.4 Antibody class is determined by the the type of heavy chain that is present. Antibodies of classes IgA, IgD, IgE, IgG, and IgM have heavy chains of type α, δ, ϵ, γ, and μ, respectively. The different types of heavy chains are specified by different gene segments encoding the heavy-chain constant regions.

24.5 At the DNA level; class switching occurs by somatic recombination during B lymphocyte differentiation (see Figure 24.14*c–e*).

24.6 (a) B lymphocytes, T lymphocytes, and macrophages.

   (b) B lymphocytes differentiate into the plasma cells that synthesize the antibodies responsible for the humoral immune response; T lymphocytes synthesize the T-cell antigen receptors responsible for the cellular immune response; and macrophages ingest and degrade antigen-antibody complexes.

24.7 The *MHC* locus encodes the transplantation antigens that play a major role in the rejection of foreign tissues after organ and tissue transplant operations.

24.8 (a) The *MHC* genes are said to be highly polymorphic because each is represented by a large number of alleles segregating in human populations.

(b) The polymorphic nature of the *MHC* genes makes it difficult to match the tissue types of potential organ donors and organ recipients so as to avoid rejection of the foreign organ following transplant surgery.

24.9 This response occurs by a process called clonal selection during which antibodies on the surface of the B lymphocyte producing them bind to the antigen. This in turn stimulates the cell to divide and produce a population (clone) of plasma cells all producing the same antibody.

24.10 Yes, the T lymphocyte antigen receptor proteins (see Figure 24.18).

24.11 Both contain polypeptides with variable regions that form the antigen-binding domains and constant regions that facilitate interactions with cell membranes and other components of the immune response.

24.12 After the exposure of virgin B and T cells to an antigen, they develop into activated cells and memory cells. The activated cells differentiate into antibody-producing plasma cells and receptor-producing T cells, respectively, that are responsible for the primary immune response. The long-lived memory cells are also activated and remain in the activated state until a subsequent exposure to the same antigen triggers their rapid differentiation into antibody-producing plasma cells and receptor-producing T cells. Thus a secondary immune response is faster than a primary immune response because the memory B and memory T cells do not have to go through the activation processes required with virgin B cells and T cells.

24.13 Nonspecific immune responses include inflammation of infected tissues resulting in increased blood flow to the tissues and recruitment of phagocytes to ingest and destroy the infecting agent(s). The two major specific immune responses are the production of antigen-specific antibodies and killer T cells.

24.14 Most of the cells of the immune system have short life spans—usually a few days to a week. In contrast, memory cells have long life spans—months to years.

24.15 In both cases—SCID and AIDS—the disease symptoms and eventually death are the result of a nonfunctional immune system. SCID is inherited; AIDS is caused by a virus (HIV).

24.16 (a) Monoclonal antibodies are produced by fusing B lymphocytes from mammals that have been exposed to a specific antigen with cancer cells called myelomas to produce immortal hybrid cells called hybridomas. Each hybridoma clone produces a single antibody. The most difficult step in monoclonal antibody production is screening the resulting population of hybridoma cells for the clone producing the desired antibody. Once identified, the antibody-producing cells can be cultured indefinitely as a result of their cancerous heritage.

(b) Monoclonal antibodies are invaluable for use in identifying individual polypeptides separated by gel electrophoresis (see Figure 19.24) and in determining the location of specific macromolecules in tissues and cells by immunolocalization techniques. They are also used to immunoprecipitate proteins during purification processes and to diagnose human diseases.

24.17 (a) HIV mutates at an extremely high rate. As a result, infected individuals contain a heterogeneous population of viruses, which helps them escape the immune response of their hosts.

(b) Because HIV is spread by sexual contact, blood transfusions, or other forms of blood transfer, the spread of AIDS can best be prevented by practicing safe sexual behavior (abstinence, sexual encounters only with uninfected individuals, and the use of condoms), by not sharing the use of hypodermic needles, and by testing all blood used in transfusions to make certain that it is free of HIV.

24.18 The human immunodeficiency virus, which causes AIDS, infects and kills helper T cells. Because helper T cells stimulate T and B lymphocytes to differentiate into killer T cells and antibody-producing plasma cell, respectively, an effective immune response cannot occur once the concentration of helper T cells falls below a critical level. At this stage, the AIDS patient becomes susceptible to infections by opportunistic microorganisms—microbes that would not cause serious infections in a healthy individual.

24.19 Some forms of SCID can now be treated by bone marrow transplants or by infusions of functional T lymphocytes. In addition, many scientists are optimistic that certain types of SCID will be effectively treated by somatic-cell gene therapy in the near future.

24.20 One. Mammals are diploids. Thus, in the absence of allelic exclusion, each plasma cell could produce two different heavy chains because heavy-chain antibody genes could be assembled from the gene segments on both homologous chromosomes. Allelic exclusion assures that only heavy-chain gene segments on one chromosome — the maternal or the paternal chromosome — are assembled into a functional antibody heavy-chain gene.

24.21 When helper T cells encounter antigens on the surfaces of macrophages, they secrete cytokines, lymphokines, and interleukins that stimulate B and T lymphocytes to differentiate into antibody-producing plasma cells and killer T cells, respectively. Thus helper T cells are involved in both the antibody-mediated and T cell-mediated immune responses.

24.22 One type of phagocyte, the macrophage, plays a key role in initiating the immune response in mammals. Macrophages ingest bacteria, viruses, and other foreign substances and partially degrade them. Antigens produced by the degradation process become anchored on the surfaces of the macrophages, where they are recognized by the helper T cells that mediate both the humoral and cellular immune responses. Other phagocytes ingest and destroy foreign cells, viruses, and antibody-antigen complexes. The complement proteins also lyse foreign cells and degrade antigens that have IgG or IgM antibodies bound to them. Although phagocytes do ingest and destroy viruses, bacteria, and other pathogenic cells in a nonspecific manner, they are most effective in destroying foreign substances that have been complexed with antibodies.

24.23 utoimmune diseases are disorders in which the immune system of an individual attacks specific tissues or cells of her or his own body ("self" cells). The combining term "auto" means self; they are called autoimmune diseases because of the immune system's attack on the patient's own cells.

24.24 1,088,000,000 different antibodies.

24.25 Because helper T cells are required to stimulate the development of both antibody-producing plasma cells and killer T cells, the toxicity of deoxyguanosine—the substrate of purine nucleoside phosphorylase—to T lymphocytes would be expected to interfere with both the humoral and cellular immune reponses in an individual lacking this enzyme.

CHAPTER 25

25.1 The phenotypes are determined by the number of primed alleles in the genotypes. This is a classic dihybrid cross with no dominance. $\frac{1}{16}$ of the F_2 will have 0 primed alleles; $\frac{4}{16}$ will have 1 primed allele; $\frac{6}{16}$ will have 2 primed alleles; $\frac{4}{16}$ will have 3 primed alleles; and $\frac{1}{16}$ will have 4 primed alleles.

25.2 This is a difficult question with no obvious or clear answer. Geneticists continue to wrestle with developing models to explain genetic data. Data are often consistent with more than one model. A reasonable approach to the question is to construct a detailed history of the inheritance pattern and then determine which genetic models could best explain this pattern.

25.3 This is a trihybrid cross because of the $\frac{1}{64}$th total in one of the classes. The parental genotypes were AABBCC 3 A'A'B'B'C'C', and the F_1 was A'AB'BC'C. We expect $\frac{1}{64}$ of the F_2 progeny to be white and to produce all white offspring, but in a sample of 78, white is so low in frequency $\left(\frac{1}{64} \times 78 = 1.2\right)$ that it may not have been produced.

25.4 (a) the mean is 20.45
   (b) the variance is 2.37
   (c) the standard deviation is 1.54.

25.5 Because $\Sigma(x_i - \text{mean}) = 0$.

25.6 For the F_1, $V_g = 0$ because they are all genetically identical and heterozygous; for the F_2, $V_g > 0$ since genetic differences result from the segregation and independent assortment of chromosomes.

25.7 $\frac{3.17}{6.08} = 0.52$.

25.8 To estimate V_g for both traits, $V_g = V_t - V_e$;
For wing span, $H^2 = \frac{200.2}{271.4} = 0.74$; for beak length, $H^2 = \frac{520.5}{627.8} = 0.83$.

25.9 V_e can be estimated by the average of the variances of the inbreds; V_g is obtained as the difference between the variances of the randomly pollinated population and the inbreds; thus the broad-sense heritability is $\frac{(26.4 - 9.4)}{26.4} = 0.64$.

25.10 $t = g + e$ and $t' = g + e'$; therefore,

$$\text{Cov}(t,t') = \sum \frac{(t)(t')}{n-1}$$

$$= \sum \frac{(g = e)(g + e')}{n-1}$$

$$= \sum \frac{g^2 + ge + ge' + ee'}{n-1}$$

$$= \frac{\Sigma g^2 + \Sigma ge + \Sigma ge' + \Sigma ee'}{n-1}$$

$$= V_g + \text{Cov}(g,e) + \text{Cov}(g,e') + \text{Cov}(e,e').$$ If g is uncorrelated with e and e', and if e is uncorrelated with e', then the last three terms are each 0.0, and $\text{Cov}(t,t') = V_g$.

25.11 Broad-sense heritability must be greater than narrow-sense heritibility. $H^2 = Vg/Ve$; $h^2 = Va/Vt$, and $Va \leq Vg$.

25.12 $(125 - 100)(0.4) + 100 = 110$.

25.13 $(15 - 12)(0.3) + 12 = 12.9$

25.14 $(90 - 100)(0.2) + 100 = 98$.

25.15 $\dfrac{12.5 - 10}{15 - 10} = 0.5$; selection for increased growth rate should be effective.

25.16 $(40)(0.3)(10) + 2000 = 2120$

25.17 The heritabilities for body size and antler size in the mule deer are high, which means that both traits will respond strongly to selection. The strong positive correlation means that if selection is for larger antler size, the body size will increase too, and vice versa.

CHAPTER 26

26.1 Yes; in a mating of $Uu\ Rr\ \times\ Uu\ Rr$, $\dfrac{1}{16}$ of the offspring will be $uu\ rr$.

26.2 This approximates a 2:1 phenotypic ratio, suggesting that it is a modified 1:2:1 ratio. The A'A' class is lethal, AA' is zigzag, and AA is normal.

26.3 Certainly among the problems associated with heritability and IQ is the testing procedure used to determine IQ. Other problems include making heritability determinations, quantifying the interaction between the genotype and the environment, estimating additive genetic variation, applying heritability studies on one population at one point in time to other populations at different times, and using heritability studies in educational programs.

26.4 *Sensitive* is probably a recessive mutation. When *sen* was crossed to a wild-type from the same population, the wild-type was probably heterozygous; *sen* × *sen* would be expected to produce all *sen* offspring if it was recessive; and in wild-type × wild-type crosses, we would expect some of the crosses to involve two heterozygotes and thus produce sensitive offspring.

26.5 Highly inbred lines are homozygous for most alleles. The greater the degree of homozygosity, the less successful the selection will be. Thus selection for maze-bright and maze-dull lines in a highly inbred line would not be effective.

26.6 Both are pleiotropic because they affect several systems. PKU, for example affects neurological functions, pigment formation, head diameter, phenylalanine metabolism, and other traits. Lesch-Nyhan syndrome affects behavior, neurological functions, purine metabolism, and viability.

26.7 Examples of genes that affect behavior indirectly are described in 26.6. Both PKU and Lesch-Nyhan syndrome are metabolic disorders that affect the metabolic activity of several tissues and organ systems. The metabolic abnormalities associated with these disorders cause brain cells to malfunction, which in turn produces behavioral abnormalities.

26.8 The *per* gene affects only behavior and has no other phenotypic affect, as far as we know.

26.9 Different *per* alleles alter the periodicity of the *Drosophila* song-burst pattern. When *per* alleles are transferred from one *Drosophila* species to another, the recipient species exhibits the song pattern of the donor species.

26.10 The PER protein must complex with other proteins or with itself before it is able to regulate transcription.

26.11 Down syndrome, Klinefelter syndrome, Turner syndrome, and other aneuploid conditions all have an impact on neurological functions, some more dramatically than others.

26.12 Both PKU and Lesch Nyhan syndrome are recessive metabolic disorders with pleiotropic effects. Their affect on brain function is indirect. PKU can be successfully managed by dietary control. HD is a fatal, dominant disorder with no treatment possible at this time and appears to affect only a specific area of the brain. There are other differences.

26.13 Two important conclusions are:

  (1) genetic factors exert a profound influence on behavioral variability

  (2) the effect of being raised in the same home environment is negligible for most psychological traits.

26.14 Among the implications that should be considered are the following: Intelligence, as measured by standardized IQ tests, is strongly influenced by genetic factors; culture does not greatly constrain the development of individual differences in psychological traits; monozygotic twins raised apart are similar psychologically because they seek out similar environments.

26.15 Studies with mice indicate that preference for alcohol varies among different inbred lines as does susceptibility to the effects of alcohol. Twin and adoption studies in humans also suggest a genetic link. Research has recently suggested that variation in dopamine receptors or dopamine transport are related to alcoholism.

26.16 One problem is the failure to replicate Blum's findings, and contradictory findings in other studies. Other problems involve sampling bias and the classification of alcoholism into type I or type II. Blum's research has not been invalidated, but its conclusions have been challenged.

26.17 Amyloid β protein, which accumulates to abnormal levels in the brains of Alzheimer's patients, is derived from APP through a process of cleavage.

26.18 Down syndrome is caused by an extra copy of chromosome 21. Down syndrome patients invariably develop Alzhiemer's disease. The APP gene maps to chromosome 21, implying a gene dosage effect.

26.19 Twin studies suggest a genetic link, but Hamer's study goes much further, identifying a region of the X chromosome that appears to be related to homosexual behavior.

CHAPTER 27

27.1 $(2)(.2)(.8) = 0.32$ = heterozygote frequency

27.2 The recessive allele frequency is the square root of $0.0004 = 0.02$.

27.3 The recessive mutant allele frequency is the square root of $\dfrac{1}{25,000} = 0.006$; the normal allele frequency is $1 - 0.006 = 0.994$; the carrier frequency is estimated to be $2\,(0.006)(0.994) = 0.012$.

27.4 Central American: $L^M = 0.78$, $L^N = 0.22$; North American: $L^M = 0.39$, $L^N = 0.61$.

27.5 $M = (0.78)^2 = 0.61$, $MN = 2(0.78)(0.22) = 0.34$, $N = (0.22)^2 = 0.05$

27.6 The heterozygote classes are: $A_1A_2 = 2(0.6)(0.3) = 0.36$; $A_1A_3 = 2(0.6)(0.1) = 0.12$; $A_2A_3 = 2(0.3)(0.1) = 0.06$. The expected heterozygote frequency is: $0.36 + 0.12 + 0.06 = 0.54$.

27.7 The frequency of homozygous tasters is $(0.4)^2 = 0.16$, and the frequency of heterozygous tasters is $2(0.4)(0.6) = 0.48$. Thus $\dfrac{0.16}{0.16 + 0.48} = 0.25$ = the probability that a taster is homozygous.

27.8 $t = 0.6$ and $T = 0.4$; thus, $tt = 0.36$, $Tt = 0.48$, and $TT = 0.16$. The average plant height in this population is determined by weighting the frequency of the genotypic class by the height of that class: $(0.36)(50) = 18$; $(0.48)(100) = 48$; $(0.16)(100) = 16$; $16 + 48 + 18 = 82$ cm.

27.9 An expectation of 8 per 1000 females is quite small, so finding 3 per 1000 may simply be the result of sampling error. There is another way to view this problem. Nearly all color blind females will come from matings between carrier mothers and color blind fathers. The probability of such a mating is $2pq \times q = 2pq^2 = 2(0.91)(0.09)^2 = 0.015$. Among the daughters of this type of mating, half will be color blind. Thus, the frequency of color blind individuals among all females in the population will be essentially $(0.5)(0.015) = 0.0075$, a little less than 8 per 1000. Of course, color blind females can be produced by matings between color blind women and color blind men. However, the probability of this type of mating is $q^2 \times q = q^3 = (0.09)^3 = 0.0007$, which is insignificant compared to the probability of a mating between a carrier woman and a color blind man.

27.10 To a certain extent this is true for dominant alleles because selection acts on the heterozygote. But for recessive alleles, selection does not necessarily act on the heterozygote so the recessive allele may persist in the population. If a mutant recessive allele is lethal or deleterious, it may be maintained and actually increase in frequency in the population in a heterozygous state if it confers a selective advantage on the heterozygote (see example of sickle cell anemia and malaria). Moreover, selection operates on phenotypes, not on individual genes.

27.11 The evolution of the eye is complex, but remains consistent with Darwinian principles. Darwin argued that all known organs could have indeed evolved in small steps. The eye is a case in point. At its most primitive level, the eye is a simple spot of pigmented cells. At its most complex level, the eye has a cornea, a lens, a retina, an iris, and other structures. There are all stages of eye evolution in between. Many of the genes involved in eye development control developmental processes that are reversible. In total darkness, the eye becomes useless and may even have a negative adaptive value if it increases the risk of infection or injury. Random mutations in developmental genes causing eye reduction could be selected for under these conditions. In small populations, these mutations could be fixed by a combination of drift and selection.

27.12 The frequency of A = frequency of a = 0.5.

27.13 On the assumption that AA is indistinguishable from Aa:

Genotype:	AA	Aa	aa
Fitness (w):	1	1	0.22
Selection pressure (s):	0	0	0.78

27.14 No. There is selection against Aa as well as aa, because the loss of the lethal allele is greater than expected based on selection against only aa.

27.15 They will not change.

27.16 After three generations, the frequency of a will increase to about 0.997 and the frequency of A will decrease to about 0.003.

27.17 Initially, the e frequency was 0.5 but decreased to 0.11 after six generations. Either ee was effectively lethal or selection acted against ee and Ee.

27.18 In the males, $s = 0.44$ and $s^+ = 0.56$; in females, $s = 0.43$ and $s^+ = 0.57$. The allele frequencies in the entire population are: $s = 0.44$ and $s^+ = 0.56$. Using these allele frequencies:

		ss	ss^+	s^+s^+	sY	s^+Y
females:	Observed	0.26	0.34	0.40	—	—
	Expected	0.19	0.49	0.32	—	—
males:	Observed	—	—	—	0.44	0.56
	Expected	—	—	—	0.44	0.56

The population is not in equilibrium because the observed female genotype frequencies do not match the expected genotype frequencies.

27.19 $A = a = 0.5$

27.20 The lethal phenotype frequency in population 1 is 0.0036; in population 2 it is 0.0009; in the merged population, the allele frequencies are halved, assuming population 1 has no b allele and population 2 has no a allele. Thus the a frequency in the merged population becomes 0.03, and the b frequency becomes 0.015. There are three lethal genotypes: aa, bb, and $aabb$. The frequencies of these classes in the new population are 0.0009, 0.0002, and 0.0000002, respectively. The frequency of the lethal phenotype in the merged population is thus approximately 0.0011 (we can ignore the double lethal).

27.21 $\dfrac{(27/108)}{(582/457)} = 0.197 = $ approximate fitness

27.22 94,075 children = 188,150 alleles; the mutation rate $= \dfrac{8}{188,150} = 0.0000425$.

27.23 The frequency of t is $\sqrt{\dfrac{14}{25}} = 0.75$, and the frequency of T is $1 - 0.75 = 0.25$. Thus, because the frequency of heterozygotes should be $2(0.25)(0.75) = 0.38$, 9 to 10 (on average 9.46) of the 11 tasters are expected to be heterozygous.

27.24 The population is not in equilibrium. A chi square test shows this to be the case because observed frequencies do not match expected frequencies. There is a shortage of AA and DD, and an excess of AD, suggesting that AA and DD are semilethal relative to AD.

27.25 (a) The frequency of $aa = 0.16$; the frequency of $a = $ square root of $0.16 = 0.4$; therefore the frequency of $A = 0.6$.

(b) With no selection, we expect $AA = 0.36$; $Aa = 0.48$; and $aa = 0.16$. If the fitness of aa is 0.95, then the selection coefficient is 0.05 and the effective contribution of aa to the next generation decreases to 0.15 (0.95 3 0.16 = 0.15). The contributions of AA and Aa increase accordingly. Thus the a frequency decreases to about 0.39, and the A frequency increases to about 0.61.

27.26 This is a balanced polymorphism. At equilibrium, $H = h = 0.5$.

27.27 The Kerr and Wright experiment with small populations of *Drosophila* shows how drift and selection operate together on a population.

CHAPTER 28

28.1 At one level, morphological similarities should be evaluated. They should also be tested for reproductive isolation by trying to interbreed them. If they do interbreed, do they produce viable offspring? Are the offspring fertile?

28.2 Allopatric speciation requires geographic separation prior to the acquisition of reproductive isolation. In sympatric speciation, the attainment of reproductive isolation does not require geographic separation. In parapatric speciation, new species form from a population that is contiguous with the ancestral species' geographic range. In phyletic speciation, a single species changes over time to become a new species. This form of speciation does not involve a splitting of the ancestral species into two or more new species, so it does not produce species diversity. Quantum speciation is the budding off of a new and very different daughter species from a semi-isolated peripheral population of the ancestral species in a cross-fertilizing organism.

28.3 It is controversial because (except under conditions of self-fertilization or polyploidy) it is difficult to demonstrate that reproductive isolation develops without geographic separation. In a freely interbreeding population, new genetic combinations and new mutations would get "swamped" out through outcrossing. In spite of the controversial nature of sympatric speciation, there may be examples in nature where it has occurred.

28.4 Race or subspecies is a population concept, involving gene frequency differences in populations or groups of populations that inhabit parts of the distribution area of a polytypic species. A single individual does not represent a gene frequency distribution.

28.5 Race or subspecies can be considered a valid taxonomic category characterized by special genotypic and phenotypic characters and often by localization in a given geographic region; however, because of the large amount of genetic and phenotypic variability in most species, the number of racial distinctions that can be made is often arbitrary. Different races of a species can still interbreed successfully with the remainder of the species.

28.6 They may still be distinct species because in nature they may be reproductively isolated from each other due to geography, courtship behavior, seasonality, and other mechanisms. Furthermore, the F_1 hybrid produced in the laboratory may experience hybrid breakdown.

28.7 If one considers the formation of a polyploid to be a single event, then the answer is yes.

28.8 Another interpretation of this fossil series is that A'' had a wide geographic distribution. Along the margin of A'', a small number of individuals split off. These individuals possessed a unique set of genetic characteristics. This small group expanded into a new niche and underwent rapid speciation, perhaps by chromosome repatterning, that led to species B and perhaps C and D, all of which were phenotypically very different from A''. Species B then could have reinvaded the area occupied by A'' and replaced it. The key speciation events would have occurred at a site or in a region far removed from the fossil site.

28.9 All members of the human species can successfully interbreed.

28.10 A Mendelian population is a geographically localized group of sexually reproducing individuals in a species that share a common gene pool. It is important because the biological species concept is based on reproductive isolation between Mendelian populations that are themselves interbreeding.

28.11 The human and the gorilla share a common ancestor, but each species has undergone extensive evolutionary divergence since that ancestor existed.

28.12 The most reliable way to assess speciation would be to try to interbreed the members of the two populations to see if viable, fertile offspring can be produced. If they interbreed and produce viable, fertile offspring, then they might be considered different races of a single species. If they do not interbreed or if they do but produce inviable or infertile offspring, then they could be considered different species. Further studies of intermating in their natural setting would be necessary to verify information collected in the lab.

28.13 The ancestral sequence is 176598234. This ancestral sequence evolved in one direction as follows (underlined markers are the inverted sequences): 176598234 → 178956234 → 172659834; The ancestral sequence evolved in another direction as follows: 176598234 → 128956734 → 128376594.

28.14 Character displacement is the divergence of a character between two species when their distributions overlap in the same geographical zone, compared to the similarity of the character in the two species when they are geographically isolated from each other. In Darwin's finches, the beak size of finch species that feed primarily on seeds has become significantly displaced. When a small and medium-sized finch species share the same habitat on the larger islands, beak sizes differ widely. On the smaller islands, where only one of the two finch species exists, the beak sizes are intermediate in size.

28.15 Most commonly, the first event in the establishment of reproductive isolation is geographic separation of two populations. This would be followed by one or more of the other prezygotic isolating mechanisms, although postzygotic mechanisms could come into play here as well.

28.16 In animals, mechanisms of sex determination are severely disrupted by polyploidy, so that when polyploidy occurs, survival and reproduction are seriously jeopardized. This is not usually the case with plants.

28.17 If a hybrid between two species is viable, it may not be fertile because the chromosomes do not have homologous partners and so do not go through a normal meiosis. However, if the hybrid undergoes chromosome duplication, it would have homologous chromosome pairs and be reproductively isolated from the parental species because it has a new chromosome number. The evolution of wheat is an example of this phenomenon.